Justice in Young Adult Speculative Fiction

Advance praise for *Justice in Young Adult Speculative Fiction*

"Marek Oziewicz's superb book breaks new ground in the critical field of literature for young and old. Not only does he develop an original theoretical approach based on recent discoveries in cognitive science, but he also demonstrates how young readers (and adults) are scripted to respond to fiction that raises questions about fairness and injustice. This book is essential reading for anyone seeking to understand how YA speculative fiction has become so compelling and urgent in its appeals for social justice."

—*Jack Zipes, University of Minnesota, USA*

"In *Justice in Young Adult Speculative Fiction*, Marek Oziewicz discovers a new way to read—and to value—fantastic literatures. His blend of cognitive science, social critique, and narrative analysis is a bold and convincing reformulation of genre studies."

—*Brian Attebery, Idaho State University, USA*

"Oziewicz draws on cognitive science to illustrate how the brain writes and rehearses scripts that make behavior and even ethics natural to us. Rather than simply applying science to literature, he proposes that justice itself is a script and that stories of the impossible have become a central site for revising it. It's a breakthrough in how and why we talk about children's and YA literature."

—*Joe Sutliff Sanders, Kansas State University, USA*

"This far-reaching and original study places YA literature in the context of world-wide social and moral evolution. Marek Oziewicz argues that speculative fiction offers justice scripts that both explore and influence human behavior. If he's right, the future of our planet depends on whether we learn to embrace scripts based on dialogue and forgiveness, recognition of human rights, sensitivity to the environment and global togetherness."

—*Nancy Farmer, author of* The House of the Scorpion

This book is the first to offer a justice-focused cognitive reading of modern YA speculative fiction in its narrative and filmic forms. It links the expansion of YA speculative fiction in the twentieth century with the emergence of human and civil rights movements, with the communitarian revolution in conceptualizations of justice, and with spectacular advances in cognitive sciences as applied to the examination of narrative fiction. Oziewicz argues that complex ideas such as justice are processed by the human mind as cognitive scripts; that scripts, when narrated, take the form of multiple indexable stories; and that YA speculative fiction is currently the largest conceptual testing ground in the forging of justice consciousness for the twenty-first-century world.

Drawing on recent research in the cognitive and evolutionary sciences, Oziewicz explains how poetic, retributive, restorative, environmental, social, and global types of justice have been represented in narrative fiction, from nineteenth-century folk and fairy tales through twenty-first-century fantasy, dystopia, and science fiction. Suggesting that the appeal of these and other non-mimetic genres is largely predicated on the dream of justice, Oziewicz theorizes new justice scripts as conceptual tools essential to helping humanity survive the qualitative leap toward an environmentally conscious, culturally diversified global world. This book is an important contribution to studies of children's and YA speculative fiction, adding a new perspective to discussions about the educational as well as social potential of non-mimetic genres. It demonstrates that the justice imperative is very much alive in YA speculative fiction, creating new visions of justice relevant to contemporary challenges.

Marek C. Oziewicz is the Marguerite Henry Professor of Children's and Young Adult Literature at The University of Minnesota–Twin Cities. His recent book *One Earth, One People* (2008) was the winner of the 2010 Mythopoeic Scholarship Award in Myth and Fantasy Studies.

Children's Literature and Culture

Jack Zipes, *Founding Series Editor*
Philip Nel, *Current Series Editor*

For a full list of titles in this series, please visit www.routledge.com

1 **The Role of Translators in Children's Literature**
Invisible Storytellers
Gillian Lathey

2 **The Children's Book Business**
Lessons from the Long Eighteenth Century
Lissa Paul

3 **Humor in Contemporary Junior Literature**
Julie Cross

4 **Innocence, Heterosexuality, and the Queerness of Children's Literature**
Tison Pugh

5 **Reading the Adolescent Romance**
Sweet Valley High and the Popular Young Adult Romance Novel
Amy S. Pattee

6 **Irish Children's Literature and Culture**
New Perspectives on Contemporary Writing
Edited by Valerie Coghlan and Keith O'Sullivan

7 **Beyond Pippi Longstocking**
Intermedial and International Perspectives on Astrid Lindgren's Work s
Edited by Bettina Kümmerling-Meibauer and Astrid Surmatz

8 **Contemporary English-Language Indian Children's Literature**
Representations of Nation, Culture, and the New Indian Girl
Michelle Superle

9 **Re-visioning Historical Fiction**
The Past through Modern Eyes
Kim Wilson

10 **The Myth of Persephone in Girls' Fantasy Literature**
Holly Virginia Blackford

11 **Pinocchio, Puppets and Modernity**
The Mechanical Body
Edited by Katia Pizzi

12 **Crossover Picturebooks**
A Genre for All Ages
Sandra L. Beckett

13 **Peter Pan's Shadows in the Literary Imagination**
Kirsten Stirling

14 **Landscape in Children's Literature**
Jane Suzanne Carroll

15 **Colonial India in Children's Literature**
Supriya Goswami

16 **Children's Culture and the Avant-Garde**
Painting in Paris, 1890–1915
Marilynn Olson

17 **Textual Transformations in Children's Literature**
Adaptations, Translations, Reconsiderations
Edited by Benjamin Lefebvre

18 **The Nation in Children's Literature**
Nations of Childhood
Edited by Kit Kelen and Björn Sundmark

19 **Subjectivity in Asian Children's Literature and Film**
Global Theories and Implications
Edited by John Stephens

20 **Children's Literature, Domestication, and Social Foundation**
Narratives of Civilization and Wilderness
Layla AbdelRahim

21 **Charles Dickens and the Victorian Child**
Romanticizing and Socializing the Imperfect Child
Amberyl Malkovich

22 **Second-Generation Memory and Contemporary Children's Literature**
Ghost Images
Anastasia Ulanowicz

23 **Contemporary Dystopian Fiction for Young Adults**
Brave New Teenagers
Edited by Carrie Hintz, Balaka Basu, and Katherine R. Broad

24 **Jews and Jewishness in British Children's Literature**
Madelyn J. Travis

25 **Genocide in Contemporary Children's and Young Adult Literature**
Cambodia to Darfur
Jane M. Gangi

26 **Children and Cultural Memory in Texts of Childhood**
Edited by Heather Snell and Lorna Hutchison

27 **Picturebooks: Representation and Narration**
Edited by Bettina Kümmerling-Meibauer

28 **Children's Literature and New York City**
Edited by Pádraic Whyte and Keith O'Sullivan

29 **Entranced by Story**
Brain, Tale and Teller, from Infancy to Old Age
Hugh Crago

30 **Discourses of Postcolonialism in Contemporary British Children's Literature**
Blanka Grzegorczyk

31 **Fantasy and the Real World in British Children's Literature**
The Power of Story
Caroline Webb

32 **Children's Literature and the Posthuman**
Animal, Environment, Cyborg
Zoe Jaques

33 **Justice in Young Adult Speculative Fiction**
A Cognitive Reading
Marek C. Oziewicz

Justice in Young Adult Speculative Fiction

A Cognitive Reading

Marek C. Oziewicz

LONDON AND NEW YORK

First published 2015 by Routledge

2 Park Square, Milton Park, Abingdon, Oxfordshire OX14 4RN
711 Third Avenue, New York, NY 10017

Routledge is an imprint of the Taylor & Francis Group, an informa business

First issued in paperback 2018

Library of Congress Cataloging-in-Publication Data

Oziewicz, Marek.
Justice in young adult speculative fiction : a cognitive reading / by Marek C. Oziewicz.
pages cm. — (Children's literature and culture ; 103)
Includes bibliographical references and index.
1. Young adult fiction—History and criticism. 2. Speculative fiction—History and criticism. 3. Justice in literature. 4. Fantasy fiction—History and criticism. I. Title.
PN3443.O95 2015
809.3'00835—dc23 2014045549

ISBN: 978-1-138-80943-7 (hbk)
ISBN: 978-1-138-54779-7 (pbk)

Typeset in Sabon
by codeMantra

Contents

Acknowledgments

As is appropriate for a book about our evolving understanding of justice, I have benefited greatly from conversations with friends, colleagues, and books whose authors I may never get to meet in person. I have deep debts of gratitude to institutions and individuals who have helped me bring this project to completion. This book took many unexpected turns in the five years I spent writing and rewriting it; it would not have acquired its present shape without helpful criticism, suggestions, and encouragement from friends and colleagues, including Brian Attebery, Joe Sutliff Sanders, Andrzej Izyk, Anastasia Ulanowicz, Bradley Fern, John Stephens, Jack Zipes, Daniel Hade, Roni Natov, Tanja Nathanael, Karel Rose, Stephanie Branson, Devin Brown, and others. A year of research spent in State College, Pennsylvania, and Pocatello, Idaho, was invaluable. It provided me with resources and intellectual space to work out my ideas and made this book possible in more ways than I can say. I am grateful to the University of Wroclaw for sponsoring my 2011–12 sabbatical and to the Pennsylvania State University and Idaho State University for hosting me as a visiting researcher at that time. I also appreciate the assistance and the intense intellectual stimulation I found in my new home institution, the University of Minnesota–Twin Cities. Our superb staff, helpful librarians, and erudite colleagues not only made my transition from Poland easy but helped me find an ideal space—and almost absurdly beautiful surroundings—for continuing my writing. The fact that the twin cities area is a bustling center for children's literature events, justice-related projects, and intense debates about the meaning of education and literature in our diverse, multicultural world was itself a powerful inducement to reflect on the meaning of justice in the twenty-first century as well as its narrative representations.

In the past two years, especially, I have been very fortunate in institutional support. Yet, behind the work of institutions there are people who make things happen. Many of them I have not met but I want to name a few whose assistance has been particularly important to me: Jean Quam, Lynn Slifer, Deborah Dillon, Lee Galda, David O'Brien, Cynthia Lewis, Thom Swiss, Tim Lansmire, Lori Helman, Nina Asher, Kim Clarke, Lisa von Drasek from the Kerlan, Vicki and Steven Palmquist from the Winding Oak, Bonnie Houck, Rob Reid, and other friends from the Children's Literature

Network—all of them helped me clarify my ideas about justice in ways they may not even suspect. My heartfelt gratitude also goes to the two anonymous reviewers who went over the early draft with meticulous care and to Philip Nel, the Routledge series editor, who brought to the manuscript his characteristic intensity and penetrating intelligence.

I have left for last my acknowledgment of the most fundamental indebtedness to my family. Writing can be a lonely endeavor, but it would not be possible without their support and forbearance. As always, my sons, Jeremy and Teo, whom I harassed with a never-ending stream of books and films, have listened patiently and prodded me in new directions through questions about justice in what we read or watched together. I reflect with gratitude on all I owe to my parents, including their reassurance that books must sometimes take precedence over other pursuits. Most importantly, I want to thank my wife, Basia, who has been essential in every way to my meeting the challenges of this arduous project and whose love has sustained me along the way.

Introduction

During one search of our apartment by the Communist secret police, an agent going through my room picked up a copy of *The Lion, The Witch, and The Wardrobe* lying on my desk. "You read *this*?" he asked with a sneer. As he tossed the book on my bed, he spat out, "Fairy tales!" I was fourteen years old but I remember being offended by his suggestion that the story was escapist fodder for gullible minds. Although I found no words to formulate my reply, even then I knew that *Lion* was not a foolish book and that freedom was not a fool's dream. The agent's contempt masked the fear of this dream, for dreams will inspire people to resist injustice in whatever form it manifests.

As a teenager growing up in Communist Poland, I was drawn to speculative fiction largely for its liberating and transformative potential. To my generation, living on the cusp of liberty yet still in the shadow of soon-to-collapse Communism, The Chronicles of Narnia and *The Lord of the Rings* offered a politically-relevant reading experience. When I first read the Chronicles, two years prior, I linked the White Witch's attempt to freeze Narnia in eternal winter with the introduction of martial law in Poland in December, 1981—a winter that lasted almost two years. Our apartment was frequently raided by the secret police. Between 1982 and 1985 my father, an active member of Solidarity, would be gone for days or weeks at a time, arrested or detained like the political opponents of the Witch. In this context, to read about Mr Tumnus' wrecked and empty house, about the Beavers escaping the raid of Maugrim's wolf police, or about Frodo leaving the Hill threatened by Black Riders, was nothing short of subversive. I never saw the Witch but I knew who the wolves were.

As my love of books became my professional career, I have continued to be struck, time and again, by how profoundly plot structures, themes—and, as I now believe, the very appeal—of young people's literature are predicated on the dream of justice. In literary texts justice issues are given a face and a voice; they are transposed from abstract principles to specific representations, overlapping in some ways with discourses of philosophy and law, but reducible to neither. In fact, storied representations of justice—as in the Bible and other ancient writings—grounded human understanding of this concept long before the emergence of law. Even today the overwhelming majority of people conceive of justice in terms of specific narrative examples rather than through abstract rules that are the domain of law. This, as I suggest throughout this book, is even more true for young people for whom law is an alien domain.

Before I introduce these matters in more detail, I want to start with two definitions. First, my concern is with young people's literature in general and speculative fiction in particular. The distinction between children, adolescents, and young adults is important in some studies, but not in this project, where all examples I discuss are works read or watched by young people aged, say, ten and above. Although terms such as "kiddult" (Falconer 32) or "crossover" fiction (2) could also be used, I prefer a neutral inclusive term "young people's literature" that points at its primary audience without making specific age distinctions. This non-compartmentalizing approach has a long history. When "children's literature" was first used in the 19th century, the term "children" referred to all young people, including adolescents and young adults. Adolescence as a concept was defined only by the end of the 19th century (Trites 8), and young adulthood followed about a decade later. It then took another fifty years, Trites argues, before the young adult novel[1] established itself in the 1960s as a distinct literary genre within the children's literature field. At that time another trend was under way, whereby a defining factor of the YA novel was a reliance on adolescent protagonists as they strive to understand themselves and learn to function in relation to various forms of institutionalized power that define their existence. In their central focus on "foregrounding the relationship between the society and the individual" (20), YA and adolescent novels are practically synonymous[2] and I will treat them as such, especially since the search for justice is an important dimension of the relationship between the individual and a society's power structures. One implication of this perspective is that it sets YA/adolescent novels apart from "preadolescent" children's literature, with its preoccupation with the Self and self-discovery. This specific rift largely corresponds with a difference in format, placing my primary source materials—the novel and the graphic novel—firmly on the adolescent/YA part of the spectrum. Since in this project I adopt a reader's perspective, young people's literature will be represented by novels and graphic novels that are taken into young people's lives and have a potential to transform them. The concern with narrow age classifications, important for the publishers and marketing executives, is not something with which engaged readers tend to bother.

Young people's literature is a broad category not merely due to a range of audience ages, but also due to a range of its formats and modalities. Again, while there are important differences between audio literature, interactive media, film, and the printed text, today's young people—Digital Natives, or Millennials—"routinely expect to apply their informally developed [vernacular literacy] strategies into any literate zone where they seem appropriate or productive" (Mackey, "Case" 100). In an era of media convergence, the meaning of literature as a language-based medium for stories is wider than it used to be. It includes numerous formats, with the printed text occupying an important, but no longer a privileged place among others[3]. Inasmuch as discussing young people's literature within any thematic focus requires using examples—and examples from different genres, modalities, or historical periods often call for different forms of contextualizing—I have narrowed the scope of this book to young people's speculative fiction: the novel, graphic novel, and film. Although each of these employs different ways of telling the story, all of them are processed by the same structures of our cognitive architecture and feed into a continuum of our narrative understanding. For instance, reading William Steig's *Shrek* (1990) and watching the *Shrek* movie (2002) are different experiences but fall into into the same blended mental space.

The other large category I need to define is speculative fiction. Although commonly taken to be a two-star system orbiting around fantasy and science fiction, speculative fiction is in fact a galactic-span term for a great number of nonmimetic genres such as the gothic, dystopia, zombie, vampire, and postapocalyptic fiction, ghost stories, superheroes, alternative histories, steampunk, magic realism, retold or fractured fairy tales and so forth. Most of these genres are either derivatives of fantasy and science fiction or hybrids that elude easy classifications. The relationship between fantasy and science fiction has been one of the most hotly debated issues in literary criticism and there are certain advantages to seeing either as primary to the other[4]. Since it is not my intention to engage in border squabbles, I use speculative fiction as a supercategory and a generic fuzzy set. As described by Brian Attebery, genres can be thought of as categories "defined not by any clear boundary or any defining characteristics but by resemblances to a single core example or [narrative] strategies" (*Stories* 33). In line with Attebery's idea of "degrees of membership" (33), my capacious definition of speculative fiction also comprises traditional genres that featured the fantastic or the supernatural, like myth, legends, folk and fairy tales. These are best defined as speculative fiction's "taproot texts" (Clute 338) because they employ rhetorical devices, themes, characters, plot types, and other elements identified with speculative fiction, but do so within a different mental universe, informed by magical or other beliefs that are not meant to be taken as fictional but factual.

Further broadening the category, speculative fiction includes genres addressed to young people and adults alike. It draws from romance, heroic adventure story, mystery and other supergenre conventions (Hogan, *How* 29), operating not only as printed texts but in other formats as well. In its filmic form, speculative fiction spans anything from *Spongebob* to *Avengers*, *Thor*, the *Ice Age* movies and *The Hunger Games* to a *Southpark* spoof of *The Game of Thrones*. In all its forms, speculative fiction occupies the non-mimetic part of the literary spectrum[5]. Characterized by extended thought-experiments that involve the hypothetical, the supernatural, or the impossible, speculative fiction is more cognitively stimulating than the mimetic genres (Nikolajeva 50). Unlike the mimetic genres, it offers no pretense of being factual or accurate, a denial that endows it with a deeper potential for ethical considerations (195). Where realistic fiction works metonymically, describing representative examples, speculative fiction is "inescapably metaphoric" (Attebery, *Stories* 21). Because, as demonstrated by Lakoff and Johnson, the human mind thinks in metaphors and figurative language—but also because it is seriously fun—speculative fiction, especially fantasy, has always been part of the fabric of literary traditions across the world. In its capacity to describe the alternative past, the alternative present, and the inherently alternative—i.e. not-yet-realized—future, speculative fiction is conceptually larger than mimetic fiction, whose domain is only the actual past and the actual present. In short, speculative fiction is as essential to our humanness as conditionals, past, and future tenses are for our language. They keep our dreaming alive, including the dream of justice.

This book grew out of my investigation of the complex relationship between YA speculative fiction and justice. It argues that complex ideas such as justice are stored and processed by the human mind as scripts—standardized sequences of causally linked events—and that actualized scripts, when described, take the form of multiply indexable stories. It asserts that narrative fiction, including young people's literature, has been an important carrier of justice scripts. It then focuses on the medium of YA speculative fiction, examining the historical evolution and dominant types of the poetic, retributive, restorative, environmental, social, and global justice scripts actualized in these narratives. My argument throughout this book is that YA speculative fiction ought to be recognized as one of the most important forges of justice consciousness for the globalized world of the 21st century, an argument with clear implications for both literary criticism and educational practice.

To make a case for this thesis, I tie together three large strands: the recent revolution in conceptualizations of justice, spectacular advances in cognitive sciences as applied to literary studies, and the snowballing expansion of YA speculative fiction. Each of these three processes has generated an extensive body of research; nevertheless, they have not yet been examined in conjunction. Each began independently in the second part of the 20th century, each accelerated since the 1990s, and each has fed ideas into the same pool of collective social and cultural imagination. Even though each process continues to follow its own trajectory, by 2015 the overlap among the three suggests connections that have not yet been appreciated.

My departure point is what I brazenly propose to call "the big bang of justice": the revolutionary explosion of new conceptualizations of justice that began after WWII and has been accelerating ever since. I find reasons to believe that the big bang of justice of the second part of the 20th century is no less revolutionary than the "big bang of human culture" that occurred some 40,000 years ago (Mithen 171) and emerged as a result of the newly developed capacity for what Fauconnier and Turner call "double-scope blending" or conceptual integration (389–91). Just as the Middle-Upper Paleolithic revolution made available art, ritual and religion—thus opening undreamt-of possibilities for the evolution of the human species (Mithen

171–210)—the big bang of justice has made available a number of new conceptions of justice, thus offering modern humanity versatile and much needed tools for coping with challenges arising from a major transition from local to global humanity.

As a result of this recent big bang, the monologist retributive justice that dominated Western justice-thinking since the antiquity until the early 20th century has gained a powerful alternative in the dialogical forms of restorative justice, which seeks resolutions of conflicts on the basis of principles of repentance, forgiveness, and reconciliation. On another level, the big bang generated new cultural adaptations of justice in their environmental, social, and global varieties. Environmental justice emerged as a reflection of a novel understanding of rights and obligations stemming from human relationship to the environment. Social and global justice, in turn, developed from the many human rights and justice movements of the 1960s, but underwent fine-tuning in the 1990s—processes I contextualize in Chapters Seven and Eight. Social justice seeks the achievement of justice on every level of society's functioning, locally and nationally, targeting social, gender, racial and other forms of discrimination and exclusion. Global justice, in turn, builds on a set of similar imperatives on the international and planetary level, seeking a just response to the ills of profit-driven globalization, global racism, poverty, and other issues. Because the unprecedented transformation currently sweeping through the world is a revolution both in consciousness and technology, it demands tools for dealing with changes in both areas. The new communitarian formulations of justice, as well as the emergence of new blends or types of justice, all belong in the consciousness-change category. They are, I believe, important conceptual tools that will help humanity survive the qualitative leap into the era of environmentally conscious, culturally diversified, digital and global world.

Even more important for my argument is an attempt to theorize justice as a cognitive category. As we now know, human cognition occurs through conceptual structures variously called scripts, schemata, mental or cognitive models, mental spaces, and conceptual frames or worlds. These structures are created through the operation of conceptual blending or integration—"invisible, unconscious activity involved in every aspect of human life" (Fauconnier and Turner 18)—that occurs at lightning speed and without our conscious awareness. These blends or integrated networks, which I will refer to as scripts, frame our understanding, intelligence, memory, and expectations about the world, especially about causally-linked event sequences. A number of cognitive studies demonstrate that the human cognitive architecture is hardwired for a script-based narrative understanding. Drawing on the work of Schank, Turner, Herman, Hogan and others—discussed in detail in Chapter Two—I propose that complex ideas such as justice are processed by our minds as scripts and can best be theorized as scripts. A script is a standardized sequence of events in a particular context, activated by specific cues, proceeding toward expected goals, and consisting of specific slots—including character roles—that have to be filled in with relevant content. This technical definition applies to scripts both as short event sequences, a preferred analytical material for cognitive non-literary scholars, but also to scripts understood as large structuring narrative patterns, or themes, informing entire novels or genres[6]. As demonstrated through Fauconnier and Turner's examination of the creation processes, scripts are subliminal and universally human cognitive mechanisms that structure our ideas and behavioral protocols, making them unreflexively available to our minds in their everyday functioning.

As generalizable, fill-me-in knowledge structures, scripts *per se* are invisible, but can be examined in their actualizations in real life or in the stories we tell. Seen from this angle, narrative fiction is a data bank of actualized scripts that are packaged in bundles called stories. As claimed by affective neuroscientists, the most meaningful source of our learning is episodic memory that positions us in relation to events we remember as personally meaningful and fully formed—with the details of what happened, where, when, how and why (Panksepp and Biven 214). In other words, we are hardwired to detect and remember stories, both those we were part of, and those we have participated in vicariously, such as literary and filmic narratives. This cognitivist explanation accounts for the importance of narrative fiction as an evolutionary adaptation that recalibrates minds by making available to us a set of prompts, cues, and expectations for recognizable types of purposeful action sequences in the real world. It also accounts for fiction as a kind of exercise program for Theory of Mind: the cognitive program built into the human mind, thanks to which we are able to understand others as intentional agents.

Culling these findings, I posit that human understanding of justice is script-based too. In cognitive terms, justice is a domain: a system of circuits clustered around the lexical entry for the term "justice," in addition to related entries such as fairness, hurt, reward and punishment, as well as their relationships in various structured/lexical complexes. Of these complexes, the fundamental ones are scripts and their actualized bundles called stories. Our understanding of justice is shaped by exemplary, even stereotypical, causal chains of events that involve beliefs, motivations, and other mental states of their actors, but these stereotypical, exemplary scripts continually evolve. Because scripts are established through cultural practice, the content of specific justice scripts has changed and will likely change in the future. What has not changed, and are unlikely to change, are the cognitive systems and processes through which the human mind makes available to consciousness such complex concepts and associated thought- and action-protocols as justice.

My claim that specific clusters of ideas about justice, or justice blends, coalesce into scripts implies three other corollaries. First, any justice-seeking activity involves enacting any one or a combination of, justice scripts one is familiar with. Second, scripts are updated whenever we acquire new information about different types of justice, about what constitutes justice and injustice, and about how specific types of injustice can be redressed. Third, narrative fiction, as a carrier of actualized scripts, spreads and reinforces expectations related to these scripts. In our lives we use a great number of justice scripts. There are justice scripts that apply to everyday circumstances and those activated in extraordinary situations; there are justice scripts for kin and friends, and those for strangers and animals; there are justice scripts for dealing with domestic issues as well as those that help handle community and international ones. Among this multiplicity, I have selected six scripts associated with six types of justice—poetic justice, retributive justice, restorative justice, environmental justice, social justice, and global justice—that seem to me statistically dominant in YA speculative fiction. Each of these scripts comes with its own specialized tracks. Justice scripts also interact with and feed into one another, creating practical hybrids that blend elements of two or more scripts. Even though any single novel of film is likely to be structured on one predominant justice script, its subplots may be actualizations of different scripts, sometimes even pitting one script against another in an extended narrative thought-experiment.

While this book is the first ever attempt to theorize justice as a cognitive script, the path it takes has not been entirely untrodden. For example, in *Why Fairy Tales Stick* (2006) and then in *The Irresistible Fairy Tale* (2012), Jack Zipes argues that folk and fairy tales are cultural memes—or cognitive prototypes— relevant to specific aspects of biological and cultural evolution and that they have replicated themselves, as mental representations, through processes described by an epidemiological approach to culture. In a different study, *Understanding Nationalism* (2009), Patrick Colm Hogan approaches nationalism as a cognitive category operating on one of the three narrative structures—heroic, sacrificial, and romantic tragicomedy—and declares that specific nationalisms are instantiated variations of the same larger prototypical pattern that can also be theorized as a script. In *How Authors' Minds Make Stories* (2013), Hogan again reiterates claims about certain story structures as universal genres, each defined by specific goals the protagonists pursue and based on activation of different affective/motivational systems—a description that captures the essence of scripts' operation in narratives. Also, in her recent *Reading for Learning* (2014)—the first study focused on the advantages of cognitive criticism for the study of children's literature—Maria Nikolajeva states that the three patterns of spatio-temporal structuring dominant in children's narratives are best labeled as scripts (33). In yet another study, *Racial Innocence* (2011), Robin Bernstein examines how actual items of material culture script historically-specific normative behaviors in real life. Looking at *Uncle Tom's Cabin* and dolls as two examples, Bernstein identifies them as two repertoire archives of "scriptive things" crucial to the installation of white-raced childhood innocence in American racial projects until the civil rights movement.

While each of these authors' arguments is extremely complex and warrants elaboration that I am unable to offer here, all of them highlight the importance of scripts to our understanding. Zipes' work on the adaptationist and evolutionary history of specific cultural genres, Hogan's examination of the cognitive substrate of a specific cultural concept and of large generic templates, Nikolajeva's identification of prelapsarian, carnivalesque, and postlapsarian scripts as underlying temporality structures in children's literature, and Bernstein's study of material objects and narrations about them as scripting people's actual performance are among recent publications whose thrust overlaps with my argument in this book. Like cultural memes such as folk and fairy tale, justice has evolved from specific human capacities for social interaction, its various forms woven in the fabric of cultures and traditions of all human societies. Like the concepts of nationalism or temporality, justice operates and can be studied as a type of narrative structure, or in other words, a cognitive script. Like racial relations inscribed in the narrative of *Uncle Tom's Cabin* and reinforced through the repertoire of material scriptive things, justice is a cultural performance with its own objects and practices that surround living people as well as with its own historically located scripts/repertoires recoverable in narrative fiction. Scripting is not something we do in addition to living our lives. It is rather our means to engage with the world, which we do by performing multiple scripts at once. While this capacity is universally human and inscribed in our cognitive architecture, the content of scripts is cultural and evolves in response to the changing circumstances of our social life.

Like any typology, a taxonomy of justice scripts carries a risk of compartmentalizing stories into neat categories, implying not just that these categories are more real than the actual novels, but also that the category is all that needs to be known about a given work. However, such a reductive approach is not my aim. For example, to see Le Guin's *Voices* (2006) as an actualization of the restorative justice script *and nothing more* is reductive and offensive to the story's complexity. At the same time, as I argue in Chapter Five, the story *can well be read* as a complex and unique actualization of the restorative justice script: the script our minds recognize through similar motifs, character roles, and action sequence slots in other works. A justice scripts taxonomy—like any other typology when used prudently—can enhance appreciation of the story in the larger context of the cultural moment when it emerged. It can add to the reader's understanding of the novel and perhaps go some way toward explaining her reactions to it. Thus, the ambivalent appeal of narratives that engage with the retributive justice script can be linked to the fact that this script is currently seen as problematic even while it appears unavoidable. The large questions raised by the retributive justice script include what amount of violence is necessary, how violence happens, and under what circumstances it is legitimate. By the same token, stories informed by poetic justice—such as Andersen's "Little Mermaid" and many other fairy tales—are jarring to the contemporary audience precisely because Western culture has largely jettisoned the poetic justice script. Looking at literary and filmic narratives through the lens of justice scripts does not deny any story's uniqueness, but helps us appreciate that authors do not make up these stories completely on their own. Rather, such stories draw on larger narrative and cognitive patterns—recognizable clusters of plot, character, and setting, along with cues about conflict and its resolution. Although no story is a mere actualization of the script and authors are infinitely creative, both authors and stories build on a set of story-specific resources, including the narrative patterns whose foundations involve cultural scripts. This evolution of justice scripts is a ground for hope. As Ursula Le Guin aptly pointed out in her 2014 National Book Award acceptance speech: "We live in capitalism. Its power seems inescapable. So did the divine right of kings." An updated justice script does not guarantee the achievement of justice, but it makes it imaginable and speaks to Le Guin's assertion that "[a]ny human power can be resisted and changed by human beings" (np).

Like quarks that can never be observed directly or in isolation, scripts exist in the text both independently of the reader *and* as the reader's projections. On the one hand, certain structural features of the text, such as setting, characters, protagonist, a large theme or even smaller textual elements such as motifs or tropes are not constructed in the encounter of the text and the reader. They may not be consciously noticed, but once the question is posed, any reader will be able to answer that in *Where the Wild Things Are* the protagonist is Max, not someone else; that the story has two settings, Max's house and Monster Island, but no others. On the other hand, however, literary understanding is subjective. In the Sendak example above, the question of what the story is about, or its theme, opens the door to the other part of the reading experience: interpretation, a place where the text and the reader meet. Thus, to some readers *Where the Wild Things Are* will be a story about being a child, about dealing with uncontrollable emotions, or about imagination and its role in human life. All these interpretations are correct, but each highlights a different aspect of the story: what stands out to *me* as a reader. Readers are never passive consumers of stories. The personal, emotional experience of the story each reader creates is grounded in her individual memories, past experiences—real and vicarious—social and cultural background, age, and other factors. Even for the same reader, a given story will have different meanings and resonances when engaged with at different points of our life.

How do these aspects of literary understanding apply to scripts? The answer is that just as scripts are not consciously created by authors, so too they are not consciously registered by readers. At the same time, both authors and readers draw on scripts in structuring and deciphering causal and motivational links in stories. This automatic processing of strings of events into patterned narratives, whose elements resonate—through contrast or analogy—with the already stored patterns and categories in the author's and reader's mind make scripts a Schrödinger's cat of the literary experience.

If a script comes to exist only when it is examined and recedes into indeterminacy during the literary experience itself, the corollary is that it is possible to study scripts through the lens of the author's cognition, through the reader's cognition, or as a textual matter with an implied author and reader. Each of these avenues has been explored[7], but each also calls for a different approach and tools, the description of which would explode this Introduction. The overwhelming majority of literary cognitive studies have been text-focused[8]. They have taken the cognitive approach as a theoretical rather than empirical field, in part because any attempt to explain what happens *during* the reading experience can only be measured by methods that examine an *after-reading* reaction, and in part because the cognitive/affective processes our minds perform during engagement with the story occur with lightning speeds and without our conscious attention. Thus, although I believe that an empirical study of how flesh-and-blood readers respond to different justice scripts is possible, the goal of this book is more theoretical. My focus is on texts, on the historical evolution of justice scripts, and on young people as implied readers equipped with a literary competence sufficient to read for meaning and respond to the cognitive and affective challenges the narrative offers.

The third large question—why speculative fiction rather than other genres of young people's literature—has no single answer. On one level this choice is a matter of my expertise in speculative fiction. On another it is an attempt to limit the material to a manageable scope. The justice-focused analysis of narratives from the mimetic and non-mimetic part of the spectrum—the examination Michelle Ann Abate performs in *Bloody Murder* (2013)—is perfectly legitimate, but offers little scope to bear on my interest, a cross-pollination between the evolution of justice thinking and the rise of speculative fiction. Additionally, in a cognitive account of how our understanding works, the issue of story's realism or fictionality is almost irrelevant, for our brains' cognitive circuits react to factual and fictional events in the same way. Reading fiction—not just realistic or historical, but speculative as well—affects how we interact with the world[9]. The simulation involved in the literary experience is more intense than quotidian simulations we engage in when considering actions and their possible outcomes, and so narrative simulations that go beyond our experiences are a unique form of learning that connects with our real-life, actable knowledge (Hogan, *How* 5). While mimetic fiction features the same justice scripts as speculative fiction, and in some ways seems better suited for representing specific real-life justice issues (Nikolajeva 183–4), this preference for looking at justice issues in mimetic rather than speculative fiction is grounded in the assumption that realistic fiction offers us efferent or factual knowledge, whereas speculative fiction is limited to aesthetic or "impractical" knowledge. This assumption, however, is built on a distinction between efferent and aesthetic knowledge, which, in cognitive terms, is simply incorrect (Hogan, *How* 5–6). As the following chapters demonstrate, our affective and cognitive circuits are tightly interrelated, with the affective being primary to the cognitive. Learning occurs, and is retained in memory, not merely when facts are remembered, but when a certain degree of empathetic identification and engagement with the story is achieved. Social knowledge that is the domain of fiction is far more complex that "factual" knowledge. Obviously, social knowledge also involves "facts," but its acquisition is infinitely more cognitively stimulating than the acquisition of factual knowledge. To understand *why* a character did something presents a cognitive and interpretative challenge that exceeds simply remembering *what* the character did.

On another level, I believe that there are unique advantages to discussing justice through the lens of speculative fiction, and especially that for the young audience. First, I see a deep connection between justice and young people. "Justice" and "injustice" are not common words in young people's literature. However, the cognate terms of "right" and "wrong" are as fundamental to young people's narratives as "this is not fair" is a fundamental sentiment of childhood. The term "justice" belongs with the adult nomenclature that describes aspects of the socio-political power structure called law. Since it is created and maintained in any society by a relatively narrow group of adult decision makers—often, but not always, acting as representatives of the society at large—law and its approximations of justice are abstract concepts to most young people. At the same time, young people on average have a more acute sense of fairness and a more acute perception of injustice than adults who write and execute law. Whether or not this is because the Euro-American society is structured on what Elisabeth Young-Bruehl has recently called childism—"a prejudice against children on the ground of belief that they are property and can (or even should) be controlled, enslaved, or removed to serve adult needs" (37)—young people experience injustice differently than adults. Youths are not innately innocent, but their growth is subject to the many pressures of socialization through the institutions that define them. Thus, as Trites has demonstrated, young people's literature is defined by "the tension between power and repression" (16). Dickens' observation, voiced through Pip, that "[i]n the little world in which children have their existence ... there is nothing so finely perceived and so finely felt, as injustice" (48), is still valid today.

Even those young people's books that are not explicitly about injustice raise questions about rightness and fairness. The statement, often made by young people, that "this is not fair" is a childish sentiment that many adults in many contexts find irrelevant. Yet, such a sentiment reflects a deeper perception about the nature of justice: that what is right and wrong depends on who defines it and in what context; that some behaviors, although not criminalized, are morally wrong while others, although illegal, may be morally right; that just as morality is a cultural and evolving concept, so too is justice, which needs to be questioned for its underlying assumptions and cultural attitudes. The "this is not fair" adolescent impulse may have a selfish side, but it also has the potential to take the young person beyond the black-and-white notion that a particular act is right simply because it is legal or the other way round. Without the "this is not fair" impulse, the history of humanity would not have been the same.

If, as I suggest in Chapter One, justice issues and justice themes lie at the heart of the novel as a literary form, they are especially important in young people's literature. In fiction for the young audience justice and injustice are evoked by synecdoches such as fairness and unfairness, right and wrong, reward and punishment—concepts that even young children recognize as fundamental to human relationships and actions. In a footnote to his seminal essay "On Fairy-Stories" (1948) J. R. R. Tolkien, the father of four, commented that young readers do not have a special wish to believe that the story is true, but "are more concerned to get the Right side and Wrong side clear" (39). Fairy-stories—Tolkien's term for fantasy, which largely applies to speculative fiction as well—are narratives not about the actual, but about the desirable. They succeed only if they awaken "*desire*" (39). This, I believe, also applies to one of these deep desires of the human heart, which is a dream of justice.

Drawing on this wellspring, the category of young people's literature commonly referred to as YA speculative fiction enjoys a number of special advantages over mimetic fiction that make it a unique medium for thought-experiments on justice issues. First, speculative fiction can test-run scenarios that are theoretically possible but have not yet happened: what would life be on the planet after a nuclear holocaust, as in Jeanne Du Prau's *City of Ember* series? What if we could clone children for replacement organs for powerful drug lords, as in Nancy Farmer's *The House of the Scorpion*? What would you do if, in the face of a global environmental disaster, you found a species thought extinct, but whose feces contains bacteria that eat up toxic waste and can save the world, as in Susan Fletcher's *Ancient, Strange and Lovely*? Or what if genetic engineering enabled us to eliminate perceptions of difference and create a population of happy, peaceful idiots as in Lois Lowry's *The Giver*? Only speculative fiction offers the framework to engage in this sort of investigation, which often is a metadiscourse on justice writ large. If works of fantasy are metadiscursive explorations of religion and belief, works of science fiction are similar explorations of science. All categories of speculative fiction are thus, to a different degree, venues for an essentially philosophical speculation on the human condition that involves issues of justice and moral responsibility.

Second, speculative fiction is the best tool we have to conduct thought-experiments on that which is, and will likely remain impossible, but which expands our cognitive flexibility beyond the limits of the given. To imagine what could happen if we resurrected dinosaurs, if animals could talk, if we could shapeshift and fly, or if Lee had won the battle of Gettysburg are complex extrapolations about the possible consequences of our actions. Unlike mimetic narratives, speculative fiction provides an imaginative distance from which a given issue can be approached. Stories about talking animals, aliens or ghosts, for example, raise questions of identity, agency, and power relations in a way than no realistic fiction can. Stories about alternative dimensions, alternative pasts, futures, or different worlds with non-human inhabitants invite us to rethink our uniqueness as a species, our place in the universe, and the precariousness of causal links that constitute the present.

Third, speculative fiction is the only means to tackle the big questions that we have always asked, but to which science, at least so far, has no answers. Is there a soul—and, if it exists—what happens to it after death? Do ghosts and astral presences exist? Are there other planes of existence or intelligent life forms in the universe? How about parallel universes? Is there destiny and purpose in life? How about free will? What if we could travel through time? Or what if powers of the extended mind like telepathy, precognition, telekinesis, hypnosis and so on, were real? In this capacity speculative fiction reconnects us with archaic mysteries and expands our world beyond what we thought possible, sometimes seriously as in the Narnia series, and sometimes playfully, as in Nickelodeon's *Penguins of Madagascar*. By asking difficult questions and exploring them through specific yet fictional lives, speculative fiction continues the tradition of inquiry that lies at the heart of philosophy and science, yet does so in a popular, nonacademic format accessible to nonprofessional and young audiences. Not limited to real-world cases and to what happened, speculative fiction offers liberating projections that testify to the need to imagine a more just world before one can even begin to bring it about. As Le Guin has said, "[r]esistance and change often begin in art, and very often in the art of words" (np).

Fourth, speculative fiction delivers to its readers a certain economy of unique pleasure. It reflects our dream for extraordinary things to happen in our lives, allowing us "to imagine our way into the realms of wonder and mystery" (Attebery, *Stories* 39). It also translates into how we perceive our own world. C.S. Lewis has once said that reading about enchanted woods does not make us despise real woods; on the contrary, it makes all real woods a little enchanted. Since, unlike adult genres, most works of YA speculative fiction are unapologetically hopeful about the possibility of change, part of this enchantment is a vision of justice. When characters resist oppressive aspects of their societies, challenge and transform them, they envision a more just social order and a better ground for the flourishing of human life. This is rarely explicit, but whenever this happens, readers are asked to take a stand. They are shown why certain choices matter, and how values have consequences.

Last but not least, speculative fiction is global, whereas realist fiction is mostly local. Because it relies on imagined settings, races, characters, and cultures, speculative fiction is more reader-accessible globally than its realist counterparts, which are enmeshed in their local-cultural circumstances and national histories. Even the best realist fiction is read primarily in its country of origin, whereas the best speculative fiction has a planetary reach. Since speculative fiction serves as an alternative to the mainstream canons—including their reliance on the Enlightenment canon of rationality—it records alternative ways of being, seeing, and interacting with the world, where established vested beliefs support the not-infrequently oppressive status quo. Protagonists in speculative fiction are always rebels, never the empire. This subversive potential makes speculative fiction a unique format for raising justice issues.

Studying speculative fiction is also extremely important for anyone who wants to understand today's culture and youth culture in particular. The entire 20th century was marked by the explosion of the fantastic. Since the 1990s, speculative genres have radically overshadowed other categories of fiction, narrative and filmic alike. A glance at the lists of the highest-grossing films available on Wikipedia—including lists by year, decade, or even centuries—reveals that for over twenty years now between 70 and 95 percent of top-grossing movie productions for all age groups have been speculative fiction[10]. Mimetic genres have been doing better in literature, but again, data available on sales figures suggests that since the 1940s or so, speculative fiction has been outselling mimetic works by a large margin[11]. Like the so-called realist literature, YA speculative fiction gives shape to our understanding of justice and arguments for justice. Unlike mimetic fiction, however, it reaches young global audiences counting in millions on each continent. Since it is read and watched by these young people in especially formative years of their lives, speculative fiction is a considerable force in the ongoing public discussions about justice and rights.

Why does the connection between speculative fiction and justice matter? It matters because we are beginning to realize that injustice and inequality are neither unavoidable nor beneficial. It matters because we are heading toward a tightly interconnected world where we either achieve justice or will be doomed to a vicious circle of escalating conflicts. And it matters because the context in which we are asking questions about justice has changed tremendously over the past century. In the late 19th century, at the dawn of speculative fiction, there were only 57 independent states in the world, an all-time low. Colonialism and racism was the norm, women had no voting rights anywhere except for Wyoming (since 1869), and most political systems were monarchies and empires. Even countries with the most democratic systems, such as the UK and the US, were restricted democracies, with large segments of population disfranchised on the basis of gender and race. Today the number of independent states in the world is 195. There are no colonies, racism is widely condemned, women have full voting rights in more than half the countries of the world, and the near quadrupling of states is paralleled by the spread of democracy, which has become a statistical planetary norm—if not in practice, then certainly as a planetary aspiration. Our imperfect world has become more democratic and accepting of difference. As I have claimed in *One Earth, One People* (2008), speculative fiction, while obviously not the source, is definitely part of this ongoing transformation. In this book I suggest that it is one of the best tools we have to meet the challenges of a global and interconnected world of the future.

My argument is developed in two theoretical chapters, followed by examination of specific justice scripts in six analytical chapters. Chapter One offers an overview of the history of justice in the Western civilization as unfolding in three major phases: *Old Justice*, *New Justice*, and *Open Justice*, all of which I contextualize against a timeline of the rise of young people's literature. The trajectory I propose is meant to gesture at a few notable factors about the evolution of the idea of justice. I note that between the rise of the Greek polis and of modern Western democracies there has been an expansion of the scope of justice: from male citizens to women, minorities, and other groups previously excluded from the rubric of rational consideration. On another level, justice expanded from a concept applicable within a single society, to justice applied across societies, and even to justice for animals and the natural world. In the last thirty years or so there has been an increasing stress on a situational rather than absolutist understanding of justice, based on a view of human beings as historically conditioned by specifically realized forms of social life rather than as rational agents or generic abstractions in a social equation. This expansion coextended with a tendency to move away from a uniform scale that reduced all aspects of justice to one form of measurement and to embrace irreducible heterogeneity, in which justice requires attending to distinctive capabilities or freedoms, each of which must be secured and protected in distinctive ways. There has been, finally, a growing awareness of the fact that all justice proposals are provisional, which contrasts sharply with totalizing aspirations of *Old* and *New Justice* theories. Although elements of *Old Justice* continue in their residual forms within the dominant *New Justice* framework, the future, as the remaining chapters of this book suggest, seems to belong to *Open Justice*[12].

Chapter Two moves the focus from political philosophy to cognitive science. It explores the complex relationship between justice and literature—or storytelling in general—but does so by considering the ways in which the human cognitive architecture is hardwired for narrative understanding and thus operates on the basis of scripts and their bundles called stories. By evoking the cognitivist account, I build the case for narrative fiction as an evolutionary adaptation that reflects and gives shape to justice scripts described and defined in the latter part of this chapter.

Chapters Three through Eight zoom in on the six justice scripts I believe are dominant in YA speculative fiction. Chapter Three is a close look at two tracks of a script that is largely extinct today but used to be a normative thought- and action-protocol in the West until the 19th century. The poetic justice script, I claim, supported a cognitive programing about justice as unachievable and endorsed the perception of justice as hierarchical, absolutist, and gradational. It imagined justice to be had, if at all, only by the "deserving." Situating poetic, unachievable justice in the social, material, and cultural conditions of pre-modern Europe, I argue that the distribution of the feudal and the transcendentalist tracks of this script largely overlaps with the division between folk and fairy tales. As I suggest on examples drawn from Perrault, the Grimm brothers, Andersen, and Oscar Wilde, both tracks deny the possibility of achieving justice. The feudal one erases it in the name of "might makes right" whereas the transcendentalist track defers justice through a promise of righting the present injustice at some point in the future.

The examination of the retributive justice script is the goal of Chapter Four. The retributive justice script, the protocol of response to harm that involves force and punishment, informs contemporary criminal law. As evolutionary studies of revenge suggest, retributive justice is a cultural adaptation of a mammalian reaction to harm hardwired into human cognitive architecture (McCullough, *Beyond* 10–12). By describing the retributive justice script, especially in its ties with the phenomena of violence, revenge and punishment, I suggest that it has evolved away from a license for unrestrained violence to violence in its unavoidable minimum. In modern YA speculative fiction, I argue, the retributive justice script functions in three main tracks distinguished by their enactor agents and the scope of violence projected as necessary to achieve justice. These tracks are discussed on examples that include Nancy Farmer's *The Islands of the Blessed* (2009), Brian Jacques' *Redwall* (1986), and Philip Pullman's *The Golden Compass* (1995).

In Chapter Five my focus moves to the most powerful alternative to the retributive justice script that has emerged to date: the restorative justice script. The first of the modern justice scripts to emerge in the second part of the 20th century, restorative justice is based on the idea of forgiveness and reconciliation. It enables people to heal relationships torn by harm, resume cooperation after conflict, and avoid costly and unproductive cycles of revenge. My proposal is that restorative justice can be theorized as a species-universal cultural manifestation of an evolutionary adaptation called the "forgiveness system." Taking a close look at Ursula K. Le Guin's *Voices* (2006), Terry Pratchett's *Nation* (2008), and Jeanne DuPrau's *The People of Sparks* (2004), I argue that in modern YA speculative fiction, this script comes in three tracks determined by the primary narrative focus: on the victim, on the offender, and on the community.

Chapters Six through Eight focus on justice scripts that employ deep structures of retributive and/or restorative justice but do so to address "new" areas of recognizable injustice: particularly those that concern the natural environment, the social environment, and the global environment. Each of these new scripts seeks justice in a specific relationship: humans in relation to their natural environment, humans within a single society, and humans as they interact across various societies.

Chapter Six traces the development and actualizations of the environmental justice script, seeing it as a modern application of the principles of distributive justice for the environmentally conscious world of the 21st century. In this chapter I define environmental justice as a two-layer concept. In its anthropocentric focus environmental justice recognizes the rights of humans to environmentally sound living conditions and challenges environmental racism. In its biospheric form, however, environmental justice encompasses concerns for the rights of non-human animals and the natural environment. On this level, it is used to challenge environmental speciesism, especially its assumptions about the absolute divide between culture and nature and about human superiority over other life forms. This distinctly modern justice script, I argue, recognizes that humans are animals too and extends the idea of environmental rights to non-human animals and the natural world. I suggest that this script functions in YA speculative fiction in three main tracks which I then explore as informing Susan Fletcher's *Ancient, Strange, and Lovely* (2010), Isabel Allende's *City of the Beasts* (2002), and Terry Pratchett's *The Amazing Maurice and His Educated Rodents* (2001).

In Chapter Seven I focus on a type of justice that has enjoyed the longest tradition in Western critical reflection since the Enlightenment, even though as a script it crystalized only in the second part of the 20th century. Social justice is a complex cultural construct that emerged in the early 19th century from the equality imperative central to the *New Justice* paradigm. It includes considerations of distributive and retributive justice—especially allocation of burdens and benefits as well as appropriate responses to harm—but locates these concerns in the context of political economy and social institutions that enhance or hamper the achievement of social justice. If the aspiration of social justice is equality in human relations across the lines of gender, race, and class, what exactly equality means has been historically and culturally varied. The modern social justice script, I suggest, thus emerged only in the last decades of the 20th century and represents a blend of *New Justice* and *Open Justice* concepts. A thought- and action-protocol for effecting positive change in one's own society through eliminating aspects of inequality or unfairness embedded in its structures or modes of functioning, this script functions in two main tracks whose actualizations I examine in Ursula K. LeGuin's *Powers* (2007), Margaret Paterson Haddix's The Shadow Children sequence (1998–2006), and Nancy Farmer's *The House of the Scorpion* (2002).

Finally, Chapter Eight deals with the global justice script. In part the product of the global justice movement that expanded the scope of social justice to relations between and across different national communities, and in part the consequence of a search for global peace, the global justice script is an *Open Justice* concept. Global justice advocates justice for a world of diversity and envisions securing the basic threshold of human rights to everyone across national borders as compatible with establishing fair terms of cooperation among different societies. Through this planetary vision, global justice reflects the search for the Nash Equilibrium of social relations in the global world—a solution that enables self-interested societies or cultures to cooperate fairly. The two tracks of this script—examined in Pete Docter's *Monsters, Inc.* (2001), Carol Emshwiller's *The Mount* (2002), Gene Luen Yang's *Avatar, the Last Airbender: the Promise* (2013), and Jonathan Stroud's *Ptolemy's Gate* (2006)—reflect strategies of resistance to the exploitative policies of global capitalism, especially corporations, and defense of universal human rights.

Although theorizing about justice falls under the subfield of moral philosophy called political philosophy, justice has always been one of those ideas too large to contain within a single discipline, area, or profession. Justice is more than a philosophical concern and more than any law. It is a culturally constructed cluster of values at the center of the social space; an idea that has moved people in the past and is likely to move them in the future. For 21st-century humanity, new justice scripts are becoming necessary to deal with the challenges brought by the increasing global interdependence of peoples all over the world. Life in any society is rife with disagreements about right and wrong, justice and injustice, but in democratic societies and in an increasingly democratic world these debates are absolutely central. Elections, culture wars, educational, social, environmental and economic policies are all fought over these disagreements.

In the ongoing public debate about justice and rights, the mediating function of literature is no less important than any act of legislation. Thus, as I contend throughout this study, fiction is one of the main channels for the dissemination and evolution of specific justice scripts. In particular, works of YA speculative fiction function as a phenomenal tool for evoking new concepts of justice in rising generations and making new justice scripts available to millions of young people all over the world. Through a strategy called extrapolation, speculative fiction offers thought experiments about situations that have not yet happened. Through a strategy called analogy, speculative fiction offers thought experiments about situations so familiar yet estranged that they enable fresh insights. If justice is a desirable social goal—and if cognitive scientists are right about the role of scripts in human understanding—justice can be attained only after it has first been imagined as attainable. Although the justice scripts I describe are only one ray of hope for a better tomorrow, it is largely through these scripts that YA speculative fiction empowers readers to question institutions of injustice and respond to the challenges of the contemporary world in a holistic and ethical way.

NOTES

1. Admittedly, the distinction between books for children and young adult readers was first proposed by Sarah Trimmer as early as 1802. However, Trimmer's argument was largely overlooked until adolescence was "rediscovered" in the process described by Trites. For more information see Lee Talley's "Young Adult" in *Keywords for Children's Literature* (2011).
2. The only meaningful distinction between them is the protagonist's level of maturity reached at the end. As Trites stresses, both adolescent and YA novels are novels of development (*Entwicklungsroman*), but only YA novels are coming-of-age novels (*Bildungsroman*), where the protagonist reaches adulthood and becomes fully enfranchised within their culture (Trites 18–19).
3. For the description of Digital Natives and their reading habits see Prensky's "Digital Natives, Digital Immigrants" and Margaret Mackey's empirical studies in *Mapping Literacies* and "The Case of Flat Rectangles." The concepts of media convergence and participatory culture have been best examined by Henry Jenkins in his Convergence Culture, with some practical implications explored in Winograd and Hais' *Millenial Momentum*.

4. Fantasy, admittedly, has been consistently more difficult to define. Its beginnings are murky—traceable to myths, epics, fairy tales, legends, and other pre-modern protofantasy genres (see Mathews, Attebery *Strategies* and *Stories*), and it has long functioned as a literary mode opposed to mimesis (Hume), which makes it a thicker and more sedimented category than science fiction. For example, fantasy is such an expansive category that it usually needs a modifier—animal fantasy, mythopoeic fantasy, time-travel fantasy, and so on. This has not been the case with science fiction, which is perceived as more monolithic and younger too. Science fiction has been traced back to Mary Shelley's *Frankenstein* (1818) and generally the 19th century, with its fascination with progress and technology—even though it also has taproot texts, the earliest of which is perhaps Francis Godwin's *The Man on the Moone* (1638). Although both fantasy and science fiction emerged as reactions to the Enlightenment project of modernizing the world through rationality and technology, they were different and staggered responses. The identifiably modern science fiction emerged first, at a time when the functions of what would later crystallize into fantasy were performed by the fairy tale and the weird tale. The first modern fantasy novels followed only in the 1880s, but the genre—as different from weird tales or horror—came into its own only in the 1950s (see Oziewicz), which is about a century after science fiction. The term "science fiction" was established earlier, through pulp magazines of the 1920s (Attebery *Stories* 33). The term "fantasy"—although used by the editor John W. Campbell as a defining feature of his magazine *Unknown* (1939–43)—emerged as a generic category in common usage only in the early 1960s (Oziewicz 63).
5. The concept of generic spectrum derives from Frye's modes described in *Anatomy of Criticism*. Perhaps the first attempt to apply this spectrum to the study of fantasy was Kathryn Hume's *Fantasy and Mimesis* which argued that fantasy and mimesis are both responses to reality and a literary impulses that spell out our assumptions "about the nature of literature" and about the relationship of literature to life (8). A more recent and successful elaboration can be found in *Strategies of Fantasy*, in which Brian Attebery identified fantasy as a formula, a genre, and a mode.
6. This take on scripts informs, for example, Nikolajeva's *Reading*, in which she theorizes scripts as large narrative patterns "tied to the ways society [its norms, structures, and hierarchies] is reflected in children's literature" (33).
7. A good example of the examination of the author's cognition is Hogan's *How Authors' Minds Make Stories* (2013). Examples of empirical examination of the reader's cognition include Holly Blackford's *Out of this World* (2004) and Margaret Mackey's *Mapping Recreational Literacies* (2007).
8. These include, among others, Mark Turner's *The Literary Mind* (1996), Lisa Zunshine's *Why We Read Fiction* (2006), Suzanne Keen's *Empathy and the Novel* (2007), Brian Boyd's Boyd's *On the Origin of Stories* (2009), Blakey Vermeule's *Why Do We Care about Literary Characters?* (2010), Hogan's *Cognitive Science, Literature, and the Arts* (2003) and *What Literature Teaches Us about Emotions* (2011), and Maria Nikolajeva's *Reading for Learning* (2014).
9. The extended argument about fiction as fine-tuning our Theory of Mind can be found in Zunshine's *Why We Read Fiction* and empirical studies such as Kidd and Castano's "Reading Literary Fiction Improves Theory of Mind."
10. See especially Wikipedia's "List of Highest-grossing films." http://en.wikipedia.org/wiki/List_of_highest-grossing_films. May 6, 2014.

11. See, again, Wikipedia's "List of Best-selling Books." http://en.wikipedia.org/wiki/List_of_best-selling_books. May 6, 2014.
12. The relationship between the philosophical ascendency of justice as a concept and its institutionalization through political and legal practice is complex. In general, it seems that law lags behind ideas advanced by philosophy, and so too was the case with *Old*, *New*, and *Open Justice*. The philosophical advent of *New Justice* in the late 18th century did produce the age of the revolutions, but except in America they were all quelled and the hierarchical *ancient regime* grew even stronger. A trickle of reforms continued in Europe—the abolition of slavery actually happened earlier than in the US—but *Old Justice* not merely survived until WWI, but actually became more entrenched. In political and social terms *Old Justice* as a dominant culture collapsed only under the rubble of WWI: over 130 years after its foundational assumptions were denounced by *New Justice* political philosophers. Nor does it mean that *New Justice* replaced it immediately: after some fumbling in the 1920s and 1930s—whose key achievements were women voting rights and national self-determination, albeit limited to Europe—*New Justice* was almost swept away by the tide of totalitarianism that crested in WWII. In the post war period, it was carried on the banners of decolonization, which took decades to dismantle the *Old Justice* political structures in Africa and Asia, and is still challenged by some of them even today. Despite these upheavals, if any specific date is to be chosen, *New Justice* became the dominant culture with the creation of the Human Rights regime in 1948. It has only strengthened its position since then—from the Civil Rights Movement in the US and all over the world since the 1950s to such large current programs as Obamacare or the search for ways to bridge the Achievement Gap. For all its triumphs, *New Justice* was philosophically challenged in the 1980s. With the death of John Rawls in 2002, it has no champions any more—certainly not as an all-encompassing construct Rawls took it for. How long it will take for *Open Justice* to fully replace *New Justice* in the political and legal domain is a matter of conjecture, but the process is well under way and, I believe, unstoppable.

WORKS CITED

Attebery, Brian. *Stories about Stories: Fantasy and the Remaking of Myth*. New York: Oxford UP, 2014.

Attebery, Brian. *Strategies of Fantasy*. Bloomington, IN: Indiana UP, 1992.

Bernstein, Robin. *Racial Innocence: Performing American Childhood from Slavery to Civil Rights*. New York: New York University Press, 2011.

Boyd, Brian. *On the Origin of Stories: Evolution, Cognition, and Fiction*. Cambridge, MA, Belknap Press: 2009.

Clute, John. "Taproot Texts." *The Encyclopedia of Fantasy*. Ed. John Clute and John Grant. New York: St Martin's Griffin, 1999. 921–922.

Dickens, Charles. *Great Expectations*. Mineola, NY: Dover Publications, 2001.

Falconer, Rachel. *The Crossover Novel: Contemporary Children's Fiction and Its Adult Readership*. New York: Routledge, 2008.

Fauconnier, Gilles and Mark Turner. *The Way We Think: Conceptual Blending and the Mind's Hidden Complexities*. New York: Basic Books, 2002.

Hogan, Patrick Colm. *How Author's Minds Make Up Stories*. New York: Cambridge UP, 2013.

Hogan, Patrick Colm. *Cognitive Science, Literature, and the Arts: A Guide for Humanists*. New York: Routledge, 2003.

Hogan, Patrick Colm. *Understanding Nationalism: On Narrative, Cognitive Science, and Identity*. Columbus, OH: Ohio State UP, 2009.

Hume, Kathryn. *Fantasy and Mimesis: Responses to Reality in Western Literature*. New York and London: Methuen, 1984.

Jenkins, Henry. *Convergence Culture: Where Old and New Media Collide*. New York: New York UP, 2006.

Kidd, David Comer, and Emanuele Castano. "Reading Literary Fiction Improves Theory of Mind." *Science, New Series* 342.6156 (2013): 377–380.

Lakoff, George and Mark Johnson. *Metaphors We Live By*. Chicago: U of Chicago P., 1980.

Le Guin, Ursula K. "National Book Awards Acceptance Remarks." Nov. 20, 2014. Video file. http://www.npr.org/blogs/thetwo-way/2014/11/20/365434149/book-news-ursula-k-le-guin-steals-the-show-at-the-national-book-awards . Nov. 21, 2014.

Mackey, Margaret. *Mapping Recreational Literacies: Contemporary Adults at Play*. New York: Peter Lang, 2007.

Mackey, Margaret. "The Case of Flat Rectangles: Children's Literature on Page and Screen." *International Research in Children's Literature* 4.1 (2011): 99–114.

Mathews, Richard. *Fantasy: The Liberation of Imagination*. New York and London: Routledge, 2002.

Mithen, Steven. *The Prehistory of the Mind: A Search for the Origins of Art, Science, and Religion*. London: Thanes and Hudson, 1996.

Nikolajeva, Maria. *Reading for Learning: Cognitive Approaches to Children's Literature*. Amsterdam and Philadelphia: John Benjamin's Publishing Company, 2014.

Oziewicz, Marek. *One Earth, One People: The Mythopoeic Fantasy Series of Ursula K. Le Guin, Lloyd Alexander, Madeleine L'Engle and Orson Scott Card*. Jefferson, NC: McFarland, 2008.

Panksepp, Jaak and Lucy Biven. *The Archeology of Mind: Neuroevolutionary Origins of Human Emotions*. New York: W.W.Norton, 2012.

Prensky, Marc. "Digital Natives, Digital Immigrants." Part 1. *On the Horizon* 9.5 (September/October 2001): 2–6.

Schank, Roger C. *Tell Me a Story: Narrative and Intelligence*. Evanston, IL: Northwestern UP, 1995.

Tally Lee A. "Young Adult." *Keywords for Children's Literature*. Ed. Philip Nel and Lissa Paul. New York: New York UP, 2011: 228–231

Tolkien, J.R.R. "On Fairy Stories." *Tree and Leaf*. London: HarperCollins, 2001. 1–81.

Trites, Roberta Seelinger. *Disturbing the Universe: Power and Repression in Adolescent Literature*. Iowa City, IA: U. of Iowa P., 2000.

Turner, Mark. *The Literary Mind*. New York: Oxford UP, 1996.

Winograd, Mortley and Michael D. Hais. *Millenial Momentum: How a New Generation Is Remaking America*. Piscataway, NJ: Rutgers UP, 2011.

Young-Bruehl, Elisabeth. *Childism: Confronting Prejudice Against Children*. New Haven, CT: Yale UP, 2012.

Zipes, Jack. *The Irresistible Fairy Tale: The Cultural and Social History of a Genre*. Princeton, NJ: Princeton UP, 2012.

Zipes, Jack. *Why Fairy Tales Stick: The Evolution and Relevance of a Genre*. New York: Routledge, 2006.

Zunshine, Lisa. *Why We Read Fiction: Theory of Mind and the Novel*. Columbus, OH: The Ohio State UP, 2006.

1 Toward Modern Justice Consciousness

An Idiosyncratic History of Justice with a Note on the Rise of Young People's Literature

Until the late eighteenth century an indispensable part of carnival in all European countries was the elaborate torture of animals, especially cats. Cats were used in charivaris; as their fur was being torn away, their claws pulled out, or their paws grilled, the resulting howls gave the name to the practice of mock serenades people performed for fun across civilized Europe. Parisians incinerated cats in sacks or grate boxes; the citizens of Saint Chamond liked to chase flaming cats through the streets; in Burgundy and Loraine crowds danced around burning maypoles with shrieking cats tied to the tops. In other places throughout France—the European pinnacle of civilization in the 1700s—cats were roasted over bonfires, tossed in the air and clubbed on the ground, or were shaved and dressed up to look like whomever the crowd loathed. No matter how a particular celebration started, all captured cats were subsequently executed by hanging, drowning, lacerating, shredding, quartering, or other forms of elaborate mauling (Darnton 83ff). Far from being a sadistic fantasy on the part of a few individuals, the torture of animals was "a deep current of popular culture" and "a popular amusement throughout early modern Europe" (90). These ceremonies involved entire communities, with the educated elite—bishops, mayors, nobility, and dignitaries—shrieking with glee alongside commoners. Although the first steps to abolish public ceremonies of animal torture were taken in some regions of France as early as 1765 (85), popular habits of torture as amusement survived well into the twentieth century.

The French were not uniquely predisposed to this kind of violence. An account of English sports brought to Virginia in the seventeenth century tells of a hierarchy of gory entertainments that a modern scholar has aptly called "a great chain of slaughter" (Fischer 360). These bloodsports flourished throughout the nineteenth century and were access-graded so that virtually every male in the English-speaking world could be ranked according to the size of animals he was allowed to kill for pleasure. One favorite entertainment of farmers in Virginia and in the south of England was gander pulling. After an old male goose was suspended upside-down from a tree, its neck lathered with grease, the contestants galloped past the goose, trying to tear off its head. The game was dangerous. Many riders would be thrown off their horses or lose fingers to the angry beak. However, one player would eventually succeed in ripping off the gander's head and claim the body as his prize. Other sports included killing hares, songbirds, maiming frogs, torturing lizards, and pulling the wings off insects and butterflies (363–4). When Tolkien introduced Bilbo's stone-throwing skills—commenting in an aside that "[a]s a boy he used to practice throwing stones at things, until rabbits and squirrels, and even birds, got out of his way as quick as lightning if they saw him stoop" (Tolkien, *Hobbit* 201)—the remark was meant as praise and alluded to a world where stoning small animals for fun was a common and legitimate form of play.

What do the above examples have to do with the history of justice? Wouldn't a summary of historical court rulings or public debates have better exemplified justice lacking? My answer is a qualified no. Unlike Richard Posner and scholars in the law and literature movement, I am not interested in literary representations of law and its procedures. My concern is with justice as a wider cognitive universe that includes law but draws its creative sap from a youthful sentiment about what makes things fair and right. The accounts of popular, inhumane entertainment that I have listed are as good an entry point as any other. Even more than legal precedent, they speak of the widely accepted cultural standards and sensibility of the early modern period. And if these sports sound like reports from a galaxy far, far away, it is only because they open a window on a world in which ideas of popular entertainment, or stewardship or child rearing, were radically different from our own. Just like gander pulling—legal or otherwise—would today be seen as sadism, so too the many orphan-saving schemes from the past would appear to modern child advocates as child abuse or enslavement. Be it the eighteenth-century practice of placing orphans in workhouses together with and at the mercy of embittered, often depraved or insane adults[1], the nineteenth-century "improved" schemes of shipping orphans to the British colonies (Parr 11ff), or the practice of "plac[ing] them out" in the American rural West (see Holt)—all of these well-meant initiatives strike the modern reader as no less callous than the claims of certain nineteenth-century commentators that poor people do not mind being hungry and homeless because they are well-adapted to these circumstances (Dimock 165). More examples can be provided, of course, but my point is that what counted as doing the right thing in the eighteenth and nineteenth centuries is very removed from ideas about justice and the right we hold today. This chapter will address how the concept of justice has evolved and how this evolution helped create space for the emergence of young people's literature.

THE EXPANDING SCOPE OF JUSTICE

Over the past 25 centuries, an understanding of the idea of justice in the Western tradition has continually expanded. Some developments were built by accretion on preexisting formulations. Others emerged through rejection or reformulation of those earlier notions. In general, the trajectory has been that of expanding scope and depth: from justice as applicable to Greek men, to noblemen, to Christian men, to rational men, to human beings of both genders and all ethnicities, and recently, to all sentient beings and the natural world. This complex and polyvocal process was marked by two large arcs that unevenly informed its various phases. The first arc led from pre-modern communalism through early modern individualism and then on to contemporary communitarianism. The key question in this trajectory was whose needs justice is supposed to serve. The other arc spanned the evolution of the idea of justice and its operative premises from hierarchical absolutist through non-hierarchical absolutist and on to, recently, non-hierarchical situational. The question in this second arc was who defines justice and how is it applied. The two arcs developed at different speeds, their elements morphing constantly into different configurations at any given time. In this overview, however, I suggest the process may be best theorized as forming three major phases: *Old Justice*, *New Justice*, and *Open Justice*. Now, to talk about arcs, phases, and tripartite classifications smacks of neat taxonomies, but any totalizing ambition is the furthest thing from my purpose. Categories do not exist until they are imagined, and I do not believe any history of justice can uncover an ontological given. My proposal is conditional, a situated construct, whose porousness I want to acknowledge upfront. At the same time—while *Old*, *New*, and *Open Justice* represent just one possible grid imposed on a continuous process—they are not arbitrary divisions. At each phase the range of choices and possible perspectives for theorizing justice was significantly broader. Also, each new accretion to thinking about justice can be identified by a pronounced preference for a dominant framework within which justice was conceptualized. As the understanding of justice evolved, certain claims were universally discarded and new ones were adduced to the picture: the legitimacy of slavery being an example of the former, while the equality imperative illustrating the latter.

Two further caveats are necessary. One is that there is, obviously, a difference between philosophical reflection about justice and the legal execution of justice. What philosophers say informs and challenges current ideas of law. Yet, legal practice never fully reflects philosophical postulation and there is a large gap between the two—usually with law lagging behind philosophical argument by decades. Given the distinction between justice and law, mine is not a discussion about the evolution of law or legal systems, but the evolution of *theorizing about* justice. My limited purpose is to demonstrate how the scope of justice in the Euro-American tradition grew and to suggest what direction this expansion took. The second caveat is that the history of justice discussed below is not meant to suggest a sequence where one phase replaces the one before it in such absolute terms as a train departs one station and arrives at another. Rather, my operative metaphor is that of an organic growth by accretion, much like a shrub with stems of various thicknesses and lengths, not all of them blooming at the same time or with the same intensity, but all drawing sap from the same root clump. Thus, *New Justice* added to *Old Justice*, and *Open Justice* added to *New* and *Old Justice*, but all three continued to grow, albeit with uneven speeds. At each stage a number of developments contributed to changing the conceptual domain of justice, often reconfiguring what is central versus what is peripheral. For example, retribution, which was central for *Old Justice*, was displaced by deterrence in *New Justice* and deterrence, in turn, had to surrender before distribution in *Open Justice*.

The cluster of ideas comprised by the shorthand term *Old Justice* began with the ancient Greek philosophers, flourished through Church fathers, and dominated justice thinking throughout the pre-Enlightenment period. In this cognitive universe, justice was conceptualized as hierarchical and absolute. It was hierarchical because it was defined from the perspective and for the purposes of a small elite at the top of a social hierarchy—an elite who believed in a hierarchical natural order and identified order with hierarchy. Justice was also legitimized absolutely by nature or God: an intrinsic, exhaustive, and all-encompassing quality, at once reflective of the world as its indwelling truth and synecdochic for its creator as singular, categoric, and good. Additionally, justice was communal, seen as a value focalized through the needs of a hierarchically ordered community—Greek *polis* or Christian *civitas Dei*—rather than those of an individual.

The stems of *New Justice* appeared in late Renaissance, came to a full bloom during the Enlightenment, and dominated justice thinking throughout the nineteenth and much of the twentieth centuries. In this framework justice continued to be seen as all-encompassing and absolute—predicated on Hobbesian pure necessity, Lockean tacit consent, Kantian pure reason, or hypothetical agreement as later for Mill and Rawls—but was increasingly seen as a non-hierarchical concept. In philosophical reflection, if not in judicial practice, justice was now perceived as derivative of mediation of human relations rather than of a God-mandated virtue, which equated the morally transgressive with criminally sanctionable. Justice now applied equally to all people, irrespective of their social position, gender, or ethnicity. In practice, of course, it took another 140 years for this promise to even begin to be fulfilled. *New Justice* was predicated on rational usefulness, thus a shift from retribution to deterrence, and it was theorized through the lens of individualism. Unlike in the *Old Justice* paradigm, justice was now seen as essential to safeguarding the rights of an individual—albeit theorized as a categorical "rational person," not an actual self—relative to the needs of the community.

Only since the philosophical advent of *Open Justice* in the 1980s has justice finally been uncoupled from its absolutist claims. This phase began roughly after WWII and accelerated in the 1980s as a reaction to the narrowly individualistic contractarian and utilitarian *New Justice* thinking. In this most recent *Open Justice* phase, justice was reformulated as situational and provisional, balanced between the claims of community and those of the individual—with the individual now seen as an actual, gendered, situated person with particular traits. The ideal of non-hierarchical scope of justice was brought more in line with the realities of the world after the human and civil rights revolutions. By the close of the twentieth century *Open Justice* had become the most progressive mode in philosophical, political, and scholarly reflection as well as the ideological fuel for all reform movements that are now transforming the globe. Western legal systems, however, are still dominated by the paradigm of *New Justice*.

One useful way to appreciate the dynamics of internal relations among these three phases is to see them as reflecting what Raymond Williams has called residual, dominant, and emergent features[2]. According to Williams any cultural practice or idea at any historical moment is challenged simultaneously by its past, not-yet-gone understanding and by a host of new, emerging reformulations, one of which will eventually become dominant (Williams 121–4). *Old Justice* was thus a dominant framework until well into the seventeenth century, and only then was it challenged by the emergent culture of *New Justice*. The two coexisted throughout the eighteenth century, until by the end of the nineteenth century *New Justice* had become the dominant culture, while *Old Justice* continued in its many residual forms: "effectively formed in the past but ... still active in the cultural process" (122). No longer seriously challenged by *Old Justice*, *New Justice* triumphed with the enfranchisement of women in the 1920s, in the Universal Declaration of Human Rights in the 1940s, and in the Civil and Minority Rights Movements of the 1950s through the 1980s. Roughly at that time, however, its hegemony was slowly being toppled by the emerging culture of *Open Justice*, which conveyed the realization that the world is not just one and "universal"—the keyword of *New Justice*—but one and "diverse," needing situated, different, and flexible responses rather than a uniform grid posited by *New Justice*. In Williams' terms, *Old Justice* is a residual culture today, *New Justice* is the dominant one—at least in the legal domain—whereas *Open Justice* is the emergent culture, one that creates "new meanings and values, new practices, new relationships and kinds of relationship" (123). Its eventual dominance seems ensured by the fact that it has already taken over the fields where visions of the future are being forged: literature, ethics, political philosophy, cultural studies, and increasingly customizable technology.

Each of the three phases can be identified chronologically but also through its flag bearers. The *Old Justice* period began with ancient Greek reflections about the rights of citizens as well as the obligations that bound them as members of the *polis*. For Plato, Socrates, and Aristotle—concerned with cultivating civic virtue in the context of highly restricted forms of communalism, such as those available to adult male citizens of Greek city-states—justice was necessarily hierarchical and absolutist. It was an excellence embodied within a just person, meaning a male citizen of the *polis* placed at the top of the power structure by the accidents of gender and birth. The holding of justice excluded non-citizens, women, and slaves.

These hierarchical ideas found their best expression in the work of Aristotle, who saw justice as proportionate equality reflecting the principle that equals must be treated equally while unequals should be treated unequally (Pakaluk 203). For Aristotle and his followers justice was intimately bound with questions of virtue and moral desert. If virtues are universal givens that define the most desirable way of life, justice is both teleological and honorific: justice means giving people what they deserve and determines what a person is due by their merit or relevant excellence (Sandel 186–7). Appealing as this sounds to the modern ear, Aristotle's theory of justice was neither democratic not meritocratic. It legitimized social and economic inequality and was concerned solely with the political equality among the male elite. For these relatively few lucky men, political association was a means to cultivate the virtues expected of good citizens and to develop good character. Others—including non-citizens, slaves, and women—were seen as unfit by nature to engage in debates about virtues, nor were they believed to be inherently able to develop virtue. For both groups justice was a matter of fit. Rights were allocated in accordance with the purpose of social institutions, and persons were fit to their "appropriate" roles.

Aristotle's theory of justice was also absolutist in that it was projected as universal—issuing from what he saw as the natural order of things[3]. In this way it resembled theories of justice proposed by patristic writers such as St. Augustine (fourth c. AD) and St. Thomas Aquinas (13th c), both of whom derived justice from God-given natural law. St. Augustine's reflection blended the Judaist Deoteronomic inheritance with the requirements of justice in the Greco-Roman world. Justice for Augustine was action "produced by a love of the just," meaning the love of God. That is why he argued that non-Christians cannot know justice. Even when pagans' actions outwardly conform to what justice requires, Augustine believed that they were motivated by pride, self-interest, or other vices. For Augustine, then, justice was "absent ... from the history of pagan Rome" (MacIntyre, *Whose* 155). In asserting that justice exists only in that republic, which is the city of God, Augustine embraced an idea of justice that was as hierarchical and absolutist as Aristotle's in yet another way: absolutist as a non-negotiable reflection of God's natural law, and hierarchical as achievable solely for Christians.

Aquinas' account of justice shared these assumptions. Considered the first normative theory of justice in the Western tradition, his *Summa Theologiae* described justice as an ethical virtue stemming from natural law, which in itself was a reflection of God's light within the human heart. Aquinas saw justice as an absolute virtue rather than as a virtue midway between two vices as Aristotle did, and he saw injustice as a distinctive vice, rather than a sign of pride as Augustine did. Unlike his predecessors, however, he did not believe in the power of moral education to fine-tune human capacity for justice. For Aquinas, the defining experience of the human existence was disobedience of the divine law. In order to act justly, he concluded, humans needed divine grace; thus, justice is inconceivable without religion. Those who do not know God—i.e., are not Christian—can never achieve perfect obedience to natural law and so cannot achieve justice (MacIntyre, *Whose* 164–208).

Although his theory of justice is different from those of Aristotle or Augustine, in claiming that there is a timeless standard of justice grounded on a theological understanding, Aquinas advanced a conception of justice just as hierarchical and absolutist as those of his predecessors. *Old Justice* was nothing if not exclusivist. Its advocates insisted on only one true understanding of justice—their own—and denied justice to certain groups. Women, children, slaves, non-Greeks, non-Christians, and non-whites were seen as incapable of justice. This hierarchical and absolutist *Old Justice* reflected the realities of the hierarchical social structures of the pre-modern world. It underwrote economic feudalism and political absolutism, with their attendant ills of oppression, wars, crusades, and early colonial conquests. Discredited in Europe since the era of the revolutions, *Old Justice* was then gradually removed to carry out its operations outside of the cultural center. It largely informed American westward expansion as well as European colonialism throughout the nineteenth century.

By then it was over 120 years since the stalk of *New Justice* first emerged in the writings of the Enlightened Age. The two foundational names for *New Justice* are Beccaria and Kant. Both men envisioned justice as non-hierarchical and established a tradition that reached its most complete expression to date in the work of John Rawls. Conceived in the age of revolutions against the unreason of feudalism and absolutism, *New Justice* theories rejected stifling demands of the hierarchical collective to vigorously defend rights of an individual. From Beccaria through Rawls, all of these theories have been contractarian, inasmuch as they deduct justice from the variously conceived idea of the social contract[4]. Also, unlike those of *Old Justice*, all *New Justice* theories have argued about justice by assuming the equal rational capacities of every human being[5].

Just how deeply popular sentiments about what justice ought to be can revolutionize the civil world may be gleaned from the most successful homework ever, Cesare Beccaria's *An Essay on Crimes and Punishments* (1764). Written by a quiet man who worked on a self-improvement assignment and had no first-hand knowledge of what he wrote about[6], Beccaria's essay was so enthusiastically received because it captured the widely shared sentiment of the time: a zeitgeist that had come to see *Old Justice* as unbearably oppressive. Beccaria's set of postulates spelled out the founding principles for the non-hierarchical *New Justice* paradigm and in no time became enshrined as one of the supreme credos of the Enlightenment. The essay was the main intellectual force behind the penal reform that swept across the Western world in the nineteenth century and remains foundational to the modern criminal justice system. Even though, compared to Kant's style, Beccaria's essay reads like the work of a third-grader, many of Kant's conclusions about justice were preceded by those of Beccaria's. Where they had disagreed—such as on the issue of capital punishment, which the Milanese marquis rejected and Kant, echoing the voice of *Old Justice*, defended—it was the Italian's views that were embraced by the Enlightened Europe.

Beccaria advanced two key claims about justice. First, he argued that hierarchical conceptions of law and justice are faulty and must be discarded. "If we look into history," he said, "we shall find that laws, which are, or ought to be, conventions between men in a state of freedom, have been, for the most part, the work of the passions of a few, or the consequences of a fortuitous or temporary necessity" (Beccaria "Introduction" np); such laws, he added, have only produced countless "authorized examples of the most unfeeling barbarity" (np). Second—and leaving no doubt about his non-hierarchical understanding of justice—Beccaria claimed that all people should be equal before the law. Writing about an especially vexing issue of the punishment of nobles, Beccaria states: "I assert that the punishment of a nobleman should in no wise differ from that of the lowest member of society" (Chapter 21, np). Unlike Kant, Beccaria did not explain *why* justice should be non-hierarchical and *why* people should be equal before the law. He simply assumed it. His conclusions are nevertheless the same as Kant's.

In addition to these two points, Beccaria's argument contained other elements especially attractive to his enlightened audience. In bringing to public attention "the groans of the weak, sacrificed to the cruel ignorance and indolence of the powerful," and "the barbarous torments lavished, and multiplied with useless severity, for crimes either not proved, or in their nature impossible" ("Introduction" np), he advanced an emerging idea of humanitarianism[7]. In postulating that punishment should be "proportioned to the crime, and determined by the laws" ("Conclusion" np)—and that laws should be constructed with only one goal in view, "the greatest happiness of the greatest number" ("Introduction" np)—Beccaria was also a forerunner of the utilitarian spirit, that offshoot of *New Justice* crowned as the supreme principle by Jeremy Bentham and central to the penal and legal reforms of the nineteenth century[8].

The most influential *New Justice* theory to emerge in the eighteenth century, however, was that of Immanuel Kant. Kant's account of morality and its implications for justice was a reaction against two bodies of thought: one dominant and one emergent. The dominant culture Kant opposed was *Old Justice*: a cluster of absolutist hierarchical ideas about justice that sought their legitimacy in religion and sacrificed individual rights to rights of the state—which in practice placed Beccaria's "passions of a few" at the top of the social hierarchy[9]. The other body of thought Kant denounced was Bentham's utilitarianism. While Bentham was just as keen as Kant on doing away with *Old Justice*, his approach was, in Kant's view, based on faulty assumptions. Bentham's rejection of the concepts of natural law, natural rights, and social contract reduced justice to what law does: an evasive definition, as Kant saw it[10], and one largely unacceptable, I think, to the twenty-first-century mind as well.

When it appeared, Kant's *Groundwork for the Metaphysics of Morals* (1785) was thus a spirited defense of universal human rights against arbitrary hierarchies of absolutism. It was also a devastating critique of utilitarianism. Speaking in one voice with late-eighteenth-century revolutionaries, Kant affirmed inalienable human rights as the basis of any just political arrangement. In contrast to Bentham, he argued that morality was about respecting persons as ends in themselves and that might does not make right. Kant's theory of justice rests on his conception of moral law that cannot be extracted from empirical considerations or from an individual's particular interests. Instead, moral law must be deducted from pure practical reason, which legislates a priori, "independently of all experience" (Kant, *Groundwork* 23). Since the capacity for pure reason is, according to Kant, a human universal, anyone who exercises pure reason must arrive at the same conclusion: a universal categorical imperative. In his insistence that humanity is an end in itself and that respect for humanity has nothing to do with who people are in particular, Kant's philosophy gave birth to a tradition that conceives justice as non-hierarchical yet absolutist: one justice for all, yet "all" conceived as a community of freely choosing selves.

Postulated by Beccaria and Kant, *New Justice* was a world-altering concept. Throughout the nineteenth century its tenets informed the development of free market capitalism and colonial imperialism as well as movements for universal suffrage, abolitionism and women's rights; in the twentieth century, its various strands inspired human rights movements, capitalism *and* welfare state, decolonization *and* democratization[11]. The most influential twentieth-century exponent of *New Justice* was, of course, John Rawls. First formulated in *Theory of Justice* (1971), revised in 1999, and achieving its final form in *Justice as Fairness: A Restatement* (2001), Rawls' justice as fairness aspires to be a theory for a free society, in which people may have differing views about moral values to live by. In its ambition to eliminate arbitrariness altogether and to offer a universal version of equality, Rawls' proposal is a completion of Kant's reasoning about an imaginary social contract, which the Königsberg philosopher assumed as necessary but failed to describe.

Like Kant, Rawls holds that actual contracts do not justify the terms they produce. If mutual agreement to a deal does not make it fair, justice should be sought through a form of contract that is "both hypothetical and non-historical," made behind the veil of ignorance, and made by the parties free and equal, abstracted from any contingencies of political power, wealth, even native endowments (Rawls, *Justice* 16). This original position, as Rawls calls it, constitutes the lynchpin of his theory, for an agreement made by parties in the above position leads to the emergence of two principles of justice. Principle one, "[e]ach person has the same indefeasible claim to a fully adequate scheme of basic liberties, which scheme is compatible with the same scheme of liberties for all" (42), guarantees basic liberties and takes priority over considerations of social utility and the general welfare. Principle two—"[s]ocial and economic inequalities are ... to be attached to offices and positions open to all under conditions of fair equality and opportunity; and ... they are to be to the greatest benefit of the least-advantaged members of society" (42–3)—guarantees social and economic equality. Using these principles, Rawls unfurls a vision of justice that is free from three main sources of moral arbitrariness. It remedies a contingency of social class and origin, a contingency of one's inborn endowments, and a contingency of "good or ill fortune, or good or bad luck, over the course of life" (55). In seeking to eliminate *all* arbitrariness, Rawls' justice as fairness is the most egalitarian conception of justice to date. It represents the unfurling of *New Justice* that—as the advocates of *Open Justice* would argue—has reached its limits[12].

The works of Beccaria, Kant, and Rawls are the key voices of *New Justice*: theories that see justice as non-hierarchical, binding equally for all people, and stemming from the social contract that also respects one's freedom of choice. These theories are animated by totalizing ambitions and conceive of justice as a categorical imperative. Each posits some absolute principle, by which justice can be universally measured: either, in laissez-faire theories, the actual choices people make in the free market or, in liberal egalitarian theories, the hypothetical choice people would make if, like Kant, they argued from pure practical reason or, like Rawls, they argued in a hypothetical agreement behind the veil of ignorance. Unlike Beccaria's and Kant's early visions of *New Justice*, its modern theories as articulated by Posner or Rawls are concerned with distributive rather than criminal or social justice[13].

Empowering as *New Justice* may be, it is also frustrating. Given that no known society has achieved a total equality of all citizens, irrespective of their birth, social class, talents, careers and vicissitudes in life, the overreaching ambition of Rawls' project is also its major weakness. As a theory, an ideal construct, *New Justice* leaves nothing wanting. In practice, however, not everything is categorical and translatable, subject to the canon of objective adequation. The absolute claims and scope of *New Justice* make its application either inhuman—because its subjects must be purely theoretical, rational persons rather than contingent situated persons—or impossible-though-thinkable: a conceptual equivalent of the Rube Goldberg machine. The reference to a machine is not accidental. The ideal of the rule of law as envisioned by *New Justice*—and as practiced today through "due process of law"—demands robotic objectivity and neutrality. As summed up by a Temple University professor of law, "[j]ustice in this vision is most just when it is most mechanical and leaves least to individual discretion" (Baron 16). Thus, even before Rawls' fine-tuned *Restatement*, a number of critics had taken issue with *New Justice* for its premise about the total reach of the adjudicative power of reason and rational deliberation. Against Rawls' claims that rational choice is noncontingent and universally available, they pointed out that human choices are always contingent, that they are often unchosen choices, and even more often—choices that do not meet the Kantian or Rawlsian standards of rationality. *New Justice*, despite its totalizing claims, leaves out much of the actual human experience, inescapably messy as it is[14].

The recognition of the contingency of justice, of its nonabsolute and plural character, is the signature mark of *Open Justice*: the most recent phase of justice-thinking. Emerging in the 1980s, this non-hierarchical and situational idea of justice has been outlined in a host of approaches referred to as communitarian[15]. Unlike *New Justice* theories, all of which are contractarian, *Open Justice* approaches reject the classical theory of the social contract on the premise that in practice not all parties in a social contract are equal. Instead, they embrace a community-focused, non-absolute, and plural ideal of justice. All of them question whether, as Rawls would have it, one can be neutral toward the conception of virtue and whether individual freedom of choice is an adequate basis for establishing obligations the citizens of a just society owe to one another. Besides criticizing the rationalist *New Justice* tradition for its tendency to reduce the language of justice to the lowest common denominator of absolute, universal, and formalizable rules, communitarian approaches reject the claims of laissez-faire libertarianism, which Rawls, too, attacked. Instead, they conceive justice as incomplete and situational, one that is able to accommodate irregularities and human imperfections, and about which no single conclusion can be drawn.

As befits a polyvocal tradition, *Open Justice* has no one spokesperson but several. It does have an identifiable origin, however, in what seems to me the first work of the communitarian tradition: Alasdair MacIntyre's *After Virtue* (1981). In it, MacIntyre countered the voluntarist liberal conception of a person with a narrative one. Arguing that human beings are story-telling animals for whom the only way to make sense of their lives is to see them as narrative quests, MacIntyre demonstrated that the narrative concept of selfhood involves reflection within and about the larger life stories of which one's life is a part. The narrative concept of the self, in turn, implies that the self is not detachable from its social and historical roles. An attempt to arrive at justice—as Kant and Rawls did—by disregarding moral situatedness of every human being and moral ties that have neither been chosen nor can be traced to the social contract is, as MacIntyre sees it, "utterly doomed to failure" (MacIntyre, *After* 208). In advancing these claims, *After Virtue* was a pioneering argument, showing that *New Justice* cannot provide a rational criterion for deciding between claims based on legitimate entitlement and claims based on need. It also drove home the realization that *New Justice* fails to appreciate that there is no justice and morality for rational beings as such; there is only justice and morality for human beings, as practiced at some time, in some social setting, "in the context of a community whose primary bond is a shared understanding [of values]" (250)[16].

A similar case against a vision of justice as one absolute whole—and an argument for justice as a morally engaged and situated value—can be found in the works of other communitarian philosophers. Michael Sandel, in his *Liberalism and the Limits of Justice* (1982) and then *Justice: What's the Right Thing to Do?* (2009), stresses the importance of particular obligations of membership—obligations that are not universal or voluntarily chosen—for any consideration of justice. Will Kymlicka's theory of group-differentiated rights—developed in *Liberalism, Community and Culture* (1989) and then in *Multicultural Citizenship: A Liberal Theory of Minority Rights* (1995)—focuses on ethnic minorities and on how a just society benefits from recognition of minority rights. Feminist scholars, in turn, have demonstrated why women rights cannot be reduced to universal human rights and have sought to theorize justice from a feminist angle. Carol Gilligan's *In a Different Voice: Psychological Theory and Women's Development* (1982) develops an argument for a feminist ethic of care as an alternative to the masculine ethic of responsibility, describing how each ethic produces a different and separate foundation for thinking about justice. This assumption of separateness is questioned in a re-examination of Gilligan's proposal by Grace Clement, whose *Care, Autonomy, and Justice: Feminism and the Ethic of Care* (1996) uncouples both ethics from their gender anchors and makes a strong case for their complementarity. Finally, perhaps the most recent attempt at describing a viable, communitarian vision of justice has been made by Amartya Sen and Martha Nussbaum in their much-discussed Capabilities Approach.

The essence of the Capabilities Approach is that the quality of life in a just social arrangement should be measured on the basis of notions of human functioning and human capability. The former means ways of doing or being—"the *actual living* people manage to achieve"—but also as "the *freedom* to achieve actual livings that one can have reason to value" (Sen, *Development* 73, italics in the original). The notion of capability, in turn, refers to "the alternative combinations of functionings that are feasible for [a person] to achieve" (75). Starting with the collection of essays they co-edited, *The Quality of Life* (1993), Sen and Nussbaum then developed the Capabilities Approach in their own publications: Sen in *Development as Freedom* and recently *The Idea of Justice* (2009); Nussbaum in her *Poetic Justice* (1995), *Sex and Social Justice* (1998), *Women and Human Development* (2000), *Frontiers of Justice* (2006), and then in *Creating Capabilities: The Human Development Approach* (2011). While their versions of the Capabilities Approach are different—with Nussbaum's view seeking to provide a means for determining whether a society's conditions are unjust and Sen's view focused more on an evaluative framework concerning the quality of life—there is a significant overlap. Both Sen and Nussbaum view justice as practical, focused on identification and redress of "remediable injustices" (Sen, *Justice* vii), rather than as an exercise in idealist speculation. They both argue that justice is provisional and situational rather than absolute and comprehensive. Lastly, they both identify tools and criteria for comparisons of justice and injustice: an individual's advantage is evaluated "by a person's capability to do things he or she has reasons to value" (231). It is no coincidence either that Sen identifies the Capabilities Approach as representing an alternative to the contractarian mode of thinking from Hobbes and Locke to Kant and Rawls and sees it as an opposite of what he dubs "transcendental institutionalism" (5): theories of justice that focused on "transcendental identification of the ideal institutions" (6). Sen's transcendental institutionalism is, of course, another name for *New Justice*. Although still dominant in the legal domain, it is encumbered in problems of feasibility. These problems, as communitarian scholars have repeatedly pointed out, do not plague the more modest proposal of *Open Justice*[17].

Despite differences among the various communitarian approaches, all of them are constructed largely in response to what their authors perceive as deficiencies in the dominant, non-hierarchical, and absolutist *New Justice*. Communitarian approaches theorize *Open Justice* as provisional and contingent, recognizing there can exist several distinct reasons of justice, each legitimate for particular situations[18]. All of them express one version or another of inescapable human situatedness, seeing people as bearers of specific traditions bound by what MacIntyre calls narrative selfhood, Sandel—obligations of solidarity, Kymlicka—ethnic or minority status, Gilligan and Clement—gender ethics, and Sen and Nussbaum—specific available capabilities. If *New Justice* sees people as "thin," interchangeable, theoretical units, *Open Justice* operates in "thick" descriptions, mindful of people as situated individuals with particular histories, allegiances, and beliefs. As a result, *Open Justice* is not neutral on the questions of value. Each of its approaches argues for certain ethical principles that must be considered to give moral validity to any reflection about justice. Each also espouses what Sen calls "a *realization-focused* understanding of justice" rather than "an *arrangement-focused* view" characteristic of *New Justice* theories (Sen, *Justice* 10, italics in the original).

HISTORY OF JUSTICE AS HISTORY OF LITERATURE

How does this trajectory fit in with the evolution of Western literature in general and the history of young people's fiction in particular? Although the farther back one looks, the thinner the causal links seem to be, ideas about justice and social order—as well as the institutions and cultural practices that sustain them—are part of the same historical world that also generates art and literature. While the two domains evolved independently, it was a parallel process not without cross-pollination. It is not the primary function of literature to provide evidence of historical facts. However, when examined retrospectively, literary fiction can be a mother lode of the soft facts of life lived in a historical past: its beliefs, values, and understanding.

Seen from this angle, the *Old Justice* mindset was the dominant framework in heroic, epic, and pre-modern narratives—including fairy tales that will be examined in Chapter 3. From *The Epic of Gilgamesh*, through *The*

Odyssey, *Aeneid*, *Beowulf*, *The Divine Comedy*, *The Faery Queene*, and up to at least *Paradise Lost*, pre-modern works of literature assumed authority flowing from god(s) to king to nobility and only then to common people and the world—a world, in which one's social standing determined one's claim to justice. When Milton's epic poem, aimed at justifying the ways of God to men, ends with Archangel Michael's advice to Adam that "to obey is best and love and fear the only God" (Milton np) the lesson of obedience and acceptance was meant to apply not merely to one's religious sentiments but encompassed other absolutist hierarchies that framed life in pre-modern Europe: hierarchies of gender, social class, religion, and also age.

The latter divide—that between adults and children—was absolute in an especially profound way. Because of high child mortality, it was customary that parents did not invest emotionally in their children until they were likely to make it, which is around the age of seven or so. By then, most children were expected to work alongside adults and in this sense at least they were not seen as children any more. Girls were considered marriageable by twelve and boys ready to take up adult responsibilities by fourteen. The pre-modern world did not recognize children as a "consumer" class. It did not make distinctions between childhood and adolescence[19]—in fact, did not even have the concept of adolescence[20]—and did not see a need for literature for young people. Communities kept their own oral storytelling traditions, some of it likely considered as stories for young ears, but the concept of a book for children—groundbreaking as it might have been—was limited. Comenius' *Orbis Sensualium Pictus* (1658), the *New England Primer* (around 1690), and Newbery's *A Little Pretty Pocket Book* (1744) were all heavily didactic primers. What literate young people read for pleasure were works for adults, like Bunyan's *The Pilgrim's Progress* (1678)—which so impressed eight-year-old Ben Franklin that it remained his favorite book until his death (Fleming 8)—or works like Defoe's *Robinson Crusoe* (1719), which were so widely read by adolescents and provided so many patterns for subsequent children's adventure stories that they feature prominently in history of young people's literature even though they were originally meant for adults. Also, although pre-modern literature was characterized by blending the realistic and the fantastic, none of these works can be considered speculative fiction inasmuch as they were created by authors and read by audiences who did not conceive of them as imaginary. Rather, these narratives are best described as proto-fantasies or "taproot texts" (Clute 921). The mental watershed that enabled the emergence of speculative genres—a self-awareness of engaging with a literary and thus fictional narrative whose claim on reality is at best tentative—came about only in the second part of the eighteenth century[21]. Clute's list of taproot texts ends with Swift's *Gulliver's Travels* (1726), and the earliest example of fantasy he identifies is Goethe's *Faust* (1808). Other scholars have, of course, suggested other titles and dates, but most chronologies point to a breakthrough in the latter part of the eighteenth century. It was then that a transition was made from pre-modern literatures to a map of genres spread between mimetic to nonmimetic poles of the spectrum: a spectrum, which could accommodate realist as well as speculative fiction[22].

It is not a coincidence that the emergence of speculative fiction and young people's literature coincided with the rise of *New Justice* thinking. Both represented non-mainstream voices and both soon adapted to the novel form, one of whose foundational markers—especially pronounced in the eighteenth century—was a reflection on justice and injustice in human relations[23]. The "first avowed novels in English" (Day 229), Henry Fielding's *Joseph Andrews* (1742) and *Tom Jones* (1749), were written by a lawyer and legal reformer. They would be a meaningless babble unless read in the context of the many justice issues they introduce, both as the source of conflict in the plot and as motivations that guide characters in the story. This trend only deepened. Within a century, the realist novel had become the closest "contextual associate" of *New Justice*, building arguments for legal reforms, humanitarianism, abolition of slavery, and other utilitarian projects (Dimock 141)[24]. At the same time, the novel form was also being discovered as a testimony of where *New Justice* failed. It recorded what Dimock has called "residues of justice": the non-corresponding, non-commensurate, and unrationalizable aspects of human life, which the exercise of legal justice forces out of the picture (6–8).

In other ways, however, the non-hierarchical understanding of justice, especially the idea of universal human rights, has been an explosive inspiration for politics and literature alike. Affirmed in Tom Paine's *Common Sense* (1776) and in the American Revolution, individual rights were explored by the Romantics who, following Rousseau and Blake, embraced the equation of childhood with innocence. By so defining the child, they spelled out an aspiration to nourish the "childish" attributes of curiosity, wonder, and innocence throughout one's entire life and created an ideological foundation for the legitimacy of literature written especially for children. Wordsworth's 1802 claim that "the child is father of the man"—together with "Ode: Intimations of Immortality," where it was used as an epigraph—set the stage for the nineteenth-century obsession with idealized childhood as the sole refuge from the corrupt, usually urban, adult existence. This refuge, of course, was sought in the pastoral, the fantastic, and the dreamlike. Not yet fully children's literature, but pronouncedly speculative fiction, the dreamy *Kunstmärchen* by ETA Hoffman and Novalis paved the way for the Grimm brothers' collection of earthly folk tales, which soon mutated into an altogether different project: literature for the young, aimed to socialize German and then European children into bourgeois values. When Andersen, Ruskin, MacDonald, Kingsley, Carroll, Lang, and others entered the scene, speculative fiction became a dominant mode of children's literature. The appeal of these and other nineteenth-century texts for the young audience, perhaps even more in its realist mode as exemplified by Dickens, Montgomery, or Alcott, was centrally predicated on their documenting and evoking the reader's reaction to the existing injustices. The emergence of children's literature in the nineteenth century—and of adolescent and YA literature in the twentieth century—was thus grounded in *New Justice*'s concern for equality in a non-hierarchical social arrangement but derived its impetus from the Romantic poets' translation of these ideas, rather than directly from the Enlightenment philosophers[25].

The focus on the child as a victim of unfair circumstances or an oppressive social system remained a strong undercurrent in young people's literature, realist and speculative alike, throughout the twentieth century. As equality postulates of the *New Justice* paradigm were being extended to women and minorities in the political and legal domain, young people's literature expanded in its range of topics, jettisoned its crude didacticism, became more diversified, stylistically complex, and artistically savvy. This evolution, of course, was not a straightforward process, but the Romantic ideal based on *New Justice* tenets has stood the test of time. Because themes central to young people's fiction concern the formation of individual identity, this literature has remained "dominated by pre-modern conceptions of the individual," grounded in romantic and liberal humanist notions of the unique self and essentialist, shared humanness (McCallum 3). In stressing human and minority rights, free choice, and egalitarian social imperatives, contemporary young people's literature operates largely on the premises of *New Justice*.

Important as it is to recognize this indebtedness, it would be incorrect to assume that the novel, including its subgenres for the young audience, is nothing but the legacy of *New Justice*. On the contrary, literature has always been the carrier of subversive ideas that question the possibility of totalizing, universal justice in a categorically rational world posited by *New Justice*. As Wei Chi Dimock has demonstrated, ideas about the situatedness, plurality, and embarrassing incalculability of justice—the cornerstones of what would later become *Open Justice*—have haunted the novel, in various degrees, since its inception. What justice is achievable for Emily Brontë's Heathcliff, Mary Wollstonecraft's Frankenstein monster, Dickens' Pip, Twain's Huck Finn, Hardy's Tess, and other characters, whose complex stories resist easy conclusions or rationalizable objective adequation that is the hallmark of *New Justice*? This specter of the unquantifiability of justice—of things lost in translation of harm into recompense or of guilt into appreciation—had been the hidden presence there because the novel has always been concerned with the thick characters' actual behavior rather than, like Kantian or Rawlsian political philosophy, with characters as ideal rational agents. And if there is one thing the novel suggests, it is that people's behavior will never entirely comply with the demands of properly functioning, perfectly just institutions; even if such could ever be created.

All of these arguments make it tempting to see the novel form as an evil twin of *New Justice* political philosophy. Although conceived at the same time, the former was drawn to gaps and imperfections, where the latter would look for wholes and measurable equivalences. When *New Justice* would describe ideal social arrangements, the novel would explore actual ones, imperfect even if well-meant. Rather than spell out the conditions that would eliminate injustice altogether—as has been the ambition of *New Justice* from Kant through Rawls—the novel would instead identify and condemn specific cases of redressable injustice, even if the redress is non-generalizable, particular, and incomplete. Last but not least, stories are not theories: even the most ambitious novels have never aspired to be exhaustive arguments in a way that *New Justice* theories did. Novels are open to ever-new interpretations, appearing as evolving clusters of accumulating resonances that unfold in response to whatever perceptual horizon the reader brings to bear upon them. *The Adventures of Huckleberry Finn*, for example, was not meant to be a novel about child abuse, although when read now it certainly is. Unlike political theories, the novel is inherently heteroglottic and polyphonic in the Bakhtinian sense: drawing upon extraliterary socio-ideological discourses, and many-voiced through its characters and narrative voices (McCallum 12). In all these ways, elements that would later become central to communitarian approaches have played important roles in literary narratives long before they were spelled out as philosophical postulates by *Open Justice* advocates.

This thematic and chronological complexity accounts for the fact that *Open Justice*—although foreshadowed in many literary narratives—does not have a definitive arrival moment in young people's literature or YA speculative fiction. The exact date is impossible to pinpoint, but when one looks at Golding's *The Lord of the Flies* (1954) or Lewis' *The Last Battle* (1955)—the former a story *about* adolescents, the latter *for* adolescents, but both being studies of social control theory, which advocates objective normative controls and one standard of justice—and then Babbitt's *Tuck Everlasting* (1975) or Taylor's *Roll of Thunder, Hear My Cry* (1976), where justice is plural, blurred and, if achieved, incomplete and provisional, it becomes clear that an important change occurred some time between the 1950s and the 1970s. I find it interesting, although not a coincidence, that the same period also saw the emergence of adolescent literature and YA novel—Trites uses these terms interchangeably—as marketplace phenomena (Trites 9). What was happening was that a more nuanced and down-to-earth thinking about justice emerged in books for young people, and a conflict between the claims of universality and particularity was increasingly noticeable. A young protagonist was no longer just any youth, their membership in a collective category transferable to any other member of that group. It was now a gendered, ethnically, culturally, and socially situated adolescent, whose particular story was not necessarily representative of young people as a generalizable class but unique.

The growing awareness of this particular embeddedness, part of a larger sense of living in a multicultural and culturally diverse world, brought about two effects, both of them coextensive with the philosophical formulations of *Open Justice* in the 1980s and 1990s. One was the realization of how oppressive and limiting a world envisioned by *New Justice* would be: a realization that spurred the rise of dystopian fiction for the young audience. Exemplified in Lois Lowry's *The Giver* (1993), such YA postapocalyptic dystopias expose the inherent inhumanity of perfectly just social arrangements, where everyone always observes the categorical and generalizing imperatives, where language is purged of emotion and attachment, and where choices are non-contingent. This ideal "sameness," however, is a curse rather than a blessing. It makes human life ethically and emotionally impoverished[26]. As asserted in other dystopias, notably Du Prau's *The City of Ember* (2003), Collins' *The Hunger Games* (2008), or Fisher's *Incarceron* (2010), no happiness is possible without vulnerability, and no appreciation of goodness can happen without access to the messy world of thick relationships with actual, rather than ideal, people, where some choices can turn out to be mistakes or produce unforeseen consequences.

The other corollary of the *Open Justice* turn in young people's literature has been its increasing polyvocalization: a relinquishing of the authorial control in favor of many subjective perspectives that, among other things, enable many angles on what justice and rightness may mean. Walter Dean Myers' *Monster* (1999) is narrated as a film script in the making, with actual courtroom dialogues copied by the teen narrator, who is also on trial. R. J. Palaccio's *Wonder* (2012) uses as many as five different narrators to recount the first year of school experience for the facially deformed protagonist Auggie. First-person narration, often by an outsider protagonist, is becoming more of a norm in fiction, whereas historical nonfiction has expanded its scope in representing large social changes as community- rather than individual-driven[27]. All of these changes have been pronounced trends in young people's literature since the 1990s, reflecting—as I believe—the cultural ascendance of the *Open Justice* mindset. The polyvocalization of adolescent fiction, discussed especially insightfully by Trites in relation to power and McCallum in relation to subjectivity, will thus be seen as *Open Justice* marker in the chapters that follow.

MODERN JUSTICE CONSCIOUSNESS AND THE LAW OF ACCELERATING RETURNS

For all these resonances with literature, perhaps the most spectacular achievement of *Open Justice* is that in it has democratized and diversified the concept of justice. What I mean by this is that the discourse of *Open Justice* freed justice from being the sole domain of philosophy and law and helped it become a less technical and more quotidian value that has been increasingly applied to new areas. When in his *Economics of Justice* (1981) Richard Posner launched the law and economics movement—one of the most powerful trends in modern legal thought—he was following the *Open*

Justice logic, even though his proposal is thoroughly steeped in the *New Justice* mindset[28]. Until the advent of *Open Justice* thinking, justice was an elitist concept beyond the reach of an ordinary person. The egalitarian theories of justice and the idea of universal human rights introduced by *New Justice* were major milestones in broadening the scope of justice, but it was only in the second part of the twentieth century that a true revolution in thinking about justice occurred. One helpful way to theorize this revolution, of which *Open Justice* is merely one manifestation, is to see it as an instance of what Ray Kurtzweil has called "the law of accelerating returns" (35). Although Kurtzweil has coined the term to reflect the soon-to-happen confluence of technological and biological evolution, his argument that the development of technology is exponential rather than linear and that technological change accelerates the transformation of human consciousness, especially during paradigm shifts—"major changes in methods and processes to [conceptualize and] accomplish tasks" (36)—offers a good framework to interpret the ongoing, snowballing revolution in our perceptions of justice. Within this framework, the developments that have occurred under the *Old*, *New*, and *Open Justice* compensations illustrate the law of accelerating returns. If the arrival of *New Justice* was a revolution, the cluster of developments that happened in the second part of the twentieth century may just as well be seen as "the big bang of justice." I am encouraged to apply this concept—not in its astronomical sense as the absolute beginning, but in its cultural-historical sense, as a marker of a major mental breakthrough—by the cognitive archeologist Steven Mithen, who used "the big bang of human culture" to refer to the most important turning point in prehistory: the cultural explosion that enabled *Homo sapiens* to transition from being one among the many to the dominant species[29]. The big bang of justice is a process of different proportions but one whose consequences for our current transition from local to global humanity—perhaps also from humanity to posthumanity—may turn out to be as fundamental as those of the big bang of culture were for Middle-Upper Paleolithic humanity.

As I see it, the big bang of justice began in the wake of WWII, with the creation of a global standard of universal human rights. It accelerated through the 1950s and 1960s, fuelling processes of decolonization, civil rights, and environmentalist movements. It reached a new global level with the collapse of apartheid and the end of the cold war in the 1990s and since then has made justice issues a matter of common public awareness across nations, cultures, and religions. Like the big bang of culture, so too the big bang of justice has involved consciousness and technology. Since the 1950s the spread of new ideas about justice was made easier through the medium of television, and since the 1990s—through the Internet. These and other media helped diffuse new concepts of justice, human rights, and civil rights across the globe. The big bang reverberated through politics, economy, education, law, and culture, redefining everything from family relations to our attitudes to the natural world. In moral and political philosophy it stimulated the fine-tuning of *New Justice* theories, as in the work of Rawls or Posner, but also created conditions for the emergence of *Open Justice* communitarian approaches since the 1980s. In politics, the big bang of justice generated its own body of internationally recognized texts and declarations, starting with the Universal Declaration of Human Rights in 1948. In business and economy, it stimulated the emergence of new organizations, like the US-based Global Justice NGO founded in 2001 at Harvard University or The Human Development and Capability Association co-founded by Nussbaum and Sen in 2004. It also gave rise to new concepts of social responsibility, such as the Corporate Social Responsibility Index adopted by World Bank in 2005, and it gave rise to new norms in managerial education, such as the first non-technical ISO norm released in 2010: ISO 26000, "guidance on social responsibility." In all of these ways, the big bang is not one thing but many interconnected processes, this book and my own thinking about it being one of them. As a Kurtzweilan phenomenon, the big bang is a function of the great cognitive acceleration, which by now has affected all areas of life and whose fractal consequences in social movements and initiatives continue to proliferate at astounding speed.

It is precisely because we are living through the big bang and because its effects shoot through every aspect of our life that it eludes description. It can only be pointed at, rather than proven, through its multiple, seemingly unrelated, manifestations. I believe there are connections between the globalization of ethics[30], the decline of violence[31], Ursula Le Guin's *The Other Wind*, and the *Ice Age* movies; among the spectacular outcropping of communitarian approaches to justice in moral philosophy, the rise of peace and conflict resolution studies[32], Gene Luen Yang's graphic novel *Avatar: The Last Airbender*, and the film *Princess Mononoke*; among the rise of the reparations and apologies movement[33], the pardoning of the National Thanksgiving Turkey, Nancy Farmer's *The House of the Scorpion*, and the movie *Free Birds*. All of these works and phenomena are correlated within a broad cultural resonance field, often with causal-intertextual relations. All of them participate in the big bang in different domains and different bandwidths. The argument for this connection will be developed in the following chapters. At this point, however, I want to close this overview with what appears to me as the most glaring signature marker of the big bang: the recent emergence of new types of justice, each with its own specialized tools, area of application, and organizations devoted to promoting its awareness.

The past thirty years have seen the rise of four foci and a radical reformulation of two foci within the justice field. The four new foci are environmental justice, global justice, transitional justice, and restorative justice. The two revisioned ones are social justice and retributive justice. None of these types of justice is new in all of its aspects. Nevertheless, they are new as emergent phenomena: realization-focused, communitarian visions of justice, each a conceptual cluster assembled to deal with identifiable injustices, for clarifying and furthering a specific type of justice. Some named by area of application, others designated by their preferred method of applying justice, they usually operate as hybrids, such as restorative environmental justice or retributive transitional justice. None is exclusive or marked by absolutist claims: each relates to a specific subfield and defines justice through the goals and methods relevant to these particular applications. Linking them to the big bang is not just their underlying communitarian logic but also their recent provenance. Environmental justice emerged in the late 1980s but became topical only in 1991 after the First National People of Color Environmental Leadership Summit issued its seminal "The Principles of Environmental Justice." Roughly at the same time, an interest in restorative justice was rekindled in the context of the dismantling of apartheid in South Africa. When Howard Zehr published *Changing Lenses* in 1990, restorative justice was a concept in the forming but had become globally recognizable by 1999, which year saw the publication of Desmond Mpilo Tutu's *No Future Without Forgiveness*. The philosophical ascendancy of restorative justice, in turn, challenged the hitherto dominant concept of retributive justice. Starting in the mid 1990s a discussion developed on the deficiencies of the narrow, legalistic definition of crime and on better responses to harm than those offered by the criminal justice's application of retributive justice. Showcased in debates about incarceration and the death penalty, this investigation yielded a number of proposals that seek to minimize retribution to only what is necessary (Capeheart and Milovanovich 45–60). The same decade also saw the emergence of a reformulated concept of social justice. Born out of labors of critical criminology, this new social justice sought to expand understanding of justice to essentially communitarian considerations of socio-economic, historical, and other contexts. It was also in the late 1990s, roughly since the establishment of the International Criminal Court in 1998, that the term global justice gained wide recognition (Glasius 413)[34]. Finally, the term transitional justice—concerned with justice relating to how, after the fall of autocratic regimes, societies on the road to democracy respond to wrongdoings of their relative regimes and the suffering they caused—gained circulation only after 1999 (Elster xii). The 1990s were a marvel decade for justice-thinking, but the last word has not yet been said. Justice has never been more important—or more complicated.

My focus in this book will be on five of the six new types of justice mentioned above, inasmuch as these feature prominently in young people's texts and films. I will also discuss poetic justice, an "old" idea about justice that was the cultural product of the pre-modern world. Although it is almost non-existent in young people's fiction today, it was the dominant justice scaffold for much of the nineteenth-century classics, which helps explain why most of these classics have had to be reworked to appeal to the modern audience. For reasons that will be explained in the next chapter, I will discuss these categories of justice not as types but as scripts.

Although the metaphor of the big bang may seem presentist and somewhat grandiose, it is a metaphor that is supported by facts. Even without other manifestations of the big bang, the emergence of these new types of justice alone deserves recognition as unprecedented. In a truly *Open Justice* way, their many voices introduced specialized subdivisions to, across, and beyond the territory called justice—a field that had thus far been either-or, mapped in terms of distributive versus rectificatory justice, and of a division that roughly overlaps with that between civil and criminal law.

The implications of the big bang will proliferate, as the process is still unfolding. One thing seems clear, though. Early in the 1980s a number of scholars in various disciplines declared that a major transformation in humanity's worldview is taking place. As one of them saw it, the transformation was being informed by the shift from the mechanistic to the holistic conception of reality and was going to bring about "a fundamental change in our thoughts, perceptions, and values ... a transformation of unprecedented dimensions, a turning point for the planet as a whole" (Capra 16). The big bang of justice lies at the heart of this paradigm change. Part of the ongoing transformation from local to global humanity, the big bang offers us conceptual tools to better handle this transition.

NOTES

1. According to Cheryl L. Nixon in her *The Orphan in Eighteenth-Century Law and Literature*, "orphaning was a phenomenon experienced by approximately half of all children" in eighteenth-century England (53). Quoting an appalling 7-percent survival rate of infants taken into workhouses and other eighteenth-century statistics (66), Nixon paints a picture of institutionalized orphanhood that was just one step removed from slavery.
2. This framework has been insightfully applied by Brian Attebery to his discussion of fantasy's historical development and I owe this idea to him. See Attebery, *Stories about Stories* 41–2.
3. This is despite the distinction Aristotle makes between specific cases of justice and injustice on the one hand, and justice and injustice as general terms on the other. The comprehensive definition of justice seems not to have been his goal, but the taxonomy he created, the examples and rationales he provided, all bespeak of a hierarchical and absolutist. See Pakaluk 181ff.
4. A contractarian theory does not need to subscribe to a non-hierarchical idea of justice. In fact, the first contractarian theory—that proposed by Thomas Hobbes in his *Leviathan* (1651)—continued to see justice as a hierarchical concept. Assuming that the natural condition of humanity is war of everyone against everyone, Hobbes reasoned that the only way to keep this tendency in check was to establish, through social contract, an authoritarian government. The sovereign is thus the source of justice and has full authority to dictate rights and judge all claims.
5. In this way, of course, those theories built on earlier formulations, especially on John Locke's *Two Treatises of Government* (1689), in which Locke assumed that people are by nature rational beings who recognize the rights of others to life and liberty and that social contract serves the purpose of resolving disagreements

between people over individual rights. Unlike Hobbes, though, Locke claimed that "authority is held accountable by the people to respect the rights of individuals" (Capeheart and Milovanovic 16). Yet, he did not extend this horizontal idea of justice beyond white, European Christians. For example, in his *Second Treatise* Locke defended racial slavery and the rights of conquerors.

6. Like most of his aristocratic peers, Beccaria was a man with no professional training whatsoever. The essay for which he became notorious was written on an assigned topic and on the basis of information provided by Beccaria's friends as part of their mutual help, self-development work in a group called the "academy of fists." For details, see Beccaria's biographical note preceding the electronic version of "On Crimes and Punishments" available online.
7. In a chapter on Torture, for example, Beccaria claimed that "[t]he torture of a criminal during the course of his trial is a cruelty consecrated by custom in most nations," and called it "ancient and savage legislation" (Beccaria, Chapter 16 np). These and similar observations, while attacking Beccaria in *The Philosophy of Law*, Kant would dismiss as "compassionate sentimentality of the human feeling" which he felt have no place in law (Kant, *Philosophy* 201).
8. See especialy Bentham's *Fragment on Government* (1776) and *Introduction to Principles of Morals and Legislation* (1789), both of which contained proposals for legal reform based on the principle of utility. Against what he dismissed as traditionalisms—and which were essentially the pillars of *Old Justice*—Bentham advocated a philosophy based on three elements: the greatest happiness principle, universal egoism, and the artificial identification of one's interests with those of others. In absolutizing utility as the supreme principle, Bentham was a voice of *New Justice*. He does not offer a sustained analysis of justice but sees it as a consequence of utility.
9. *Old Justice*—the traditional, essentially feudal, aristocratic, and absolutist conception of justice and social organization—was, of course, denounced by all Enlightenment thinkers. Yet, many theories, such as Locke's in *Two Treatises of Government* (1689), still invoked belief in God as the supreme source of moral obligation; other proposals, such as Rousseau's in *The Social Contract* (1762), allowed the collective common good to override individual rights. It was thus only in Kant's arguments that the lasting foundation for rejecting hierarchical and arbitrary understanding of justice was created.
10. Bentham rejected the notion—embraced by most other thinkers before him, except Hobbes, and most thinkers after him, except Nietzsche—that law should be rooted in reason or natural law. For Bentham, law was a command expressing the will of the sovereign and law remained law no matter how morally evil or arbitrary it was. Second, and stemming from his legal positivist understanding of law, Bentham renounced the concepts of natural rights and social contract. Rights, for him, were created by the law rather than natural. Bentham also rebutted the idea of the social contract. This was not just because he claimed that law does not have to be based on consent. Rather, it was because he reasoned that society is a fictitious entity and the nature of the human person can be adequately described without mention of social relationships. This, Kant believed, was a deeply limited and flawed perspective.
11. To avoid creating an impression about its uniformity, it must be noted that already in the nineteenth century this tradition diversified over questions of distributive justice, creating two internally diversified schools, each with a different

understanding of what it means to respect individual rights. The free-market libertarians, or the laissez-faire school, have focused on defending the voluntary choices made by consenting adults, even if these choices are potentially harmful. This line of thought is represented by the work of minimal state, free-market advocates: in John Stuart Mill's defense of humanized utilitarianism in *On Liberty* (1859) and *Utilitarianism* (1861), in Friedrich A. Hayek's *The Constitution of Liberty* (1960), Milton Friedman's *Capitalism and Freedom* (1962) and *Free to Choose* (1980), and Robert Nozick's answer to Rawls' theory in *Anarchy, State and Utopia* (1974). Theorists of a more egalitarian bend, who have argued that unfettered markets are neither just nor free, are often referred to as the fairness school. Their ideas are best represented in the work of John Rawls.

12. Although Rawls' theory has been central to much of Anglophone political philosophy since 1971, it is still debatable whether or not there is an independent argument for fair equality of opportunity. Nor is it clear to what extent Rawls' theory is practical. Aspiring to correct the unequal distribution of talents and endowments, for example, Rawls eliminates the notion of moral desert and replaces it with "legitimate expectations and entitlements" specified by the social contract (Rawls, *Justice* 72). Yet, to completely uncouple the notion moral desert from distributive justice raises questions about accommodating incentives and effort. It can lead, as his critics have noted, to the egalitarian nightmare in which fair equality of opportunity would entail handicapping the talented or—if Rawls would be fully consistent with his proposal—to eliminating the family. If this sounds utopian, it perhaps is.
13. The champions of *New Justice*, Beccaria and Kant were also products of *Old Justice*. For example, both identified justice with law—an *Old Justice* habit of mind that had not yet learned to make the distinction. Neither Beccaria in his 47-chapter-long *Essay* nor Kant in his three-section, 72-page-long *Groundwork* ever used the word "justice." Both used the term "law": the former writing about crimes and punishment, the latter—about morality and ethics. Justice, of course, is implicit in what Beccaria and Kant said, but it took a *New Justice*-bred theorist such as Rawls to make this focus explicit. This shift of focus illustrates the evolution of *New Justice* from its inception to its modern and dominant form.
14. To appreciate what Rawls' justice as total fairness might amount to, see Barbara Goodwin's remarkable mockumentary *Justice by Lottery*, in which she postulates the Total Social Lottery—"the social contract as ideally envisaged" (Goodwin 30)—as the ultimate distributive principle that takes Rawls' postulates to their logical conclusions by abolishing literally all human distinctions.
15. Communitarian scholars avoid the term "theory" and prefer "approach" to highlight that their ideas are just one, and limited, way of approaching the subject. The word theory, especially in conjunction with "justice," has a totalizing *New Justice* twang to it and, as used by Rawls or Posner, suggests an absolute scope.
16. Expanding on this acknowledged moral dependence, MacIntyre's later studies—*Whose Justice? Which Rationality?* (1988), *Three Moral Versions of Moral Enquiry* (1990), and *Dependent Rational Animals* (1999)—championed and fine-tuned this communitarian and situational understanding of justice as a non-singular concept.
17. The outline of the Capabilities Approach offered by Nussbaum in her *Creating Capabilities* is more specific than Sen's but also less philosophically situated

and thus directly critical of transcendental institutionalist *New Justice*. The lynchpin of Nussbaum's proposal is that what Sen speaks of as capabilities can be broken into specific categories, such as hierarchically embedded combined, internal, and basic capabilities. This tripartite division Nussbaum then supplements with a host of intrinsic values, which in themselves cannot be deduced from anything else but hold a primitive place in her theory: freedom (Nussbaum, *Capabilities* 25), human dignity (26), active striving (30), and equality before the law (31). Building on those, Nussbaum then proposes ten Central Capabilities, which "a decent political order must secure to all citizens at least [on] a threshold level" (33). Her version of the Capabilities Approach is thus structured around basic protection of "areas of freedom so central that their removal makes a life not worthy of human dignity" (31). By her own admittance, the Capabilities Approach constitutes "a partial theory of social justice" (40)—in that it specifies an ample social minimum but does not address inequalities in society after the threshold of Central Capabilities has been met—but this caveat is exactly what distinguishes *Open Justice* proposals from the totalizing ones possible within the *New Justice* universe.

18. For exploration of these see, for example, such studies as Wei Chi Dimock's impressive and nuanced *Residues of Justice: Literature, Law, Philosophy*, Theodore Ziolkowski's erudite, if at times tedious *The Mirror of Justice: Literary Reflections of Legal Crises*, and Georgia Warnke's engaged *Legitimate Differences: Interpretation in the Abortion Controversy and Other Public Debates*.
19. The modern concept of childhood emerged in the mid-1800s, and only by the end of the nineteenth century "the period defined as childhood had become extended beyond the age of five or six, recognizing the adolescent as a member of the child, not adult, world" (Holt 15).
20. According to Roberta Selinger Trites, the word "adolescent" emerged in the US in the 1870s, but adolescence as a social concept was defined only with the publication of G. Stanley Hall's *Adolescence* (1905). The slow institutionalization of adolescence and its recognition as a psychological phenomenon took another four decades (Trites 9).
21. For a discussion of some exceptions, see Attebery's *Stories about Stories*, 22–26.
22. Although none of such border and genre disputes are conclusive, the basic division between mimetic and nonmimetic genres remains useful. I am not certain what the best candidate for the first work of fantasy is—perhaps Goethe's *Faustus* mentioned by Clute or perhaps Novalis' 1800 novel *Heinrich von Ofterdingen* (Attebery, *Stories* 26)—but I feel pretty confident in identifying Mary Shelley's *Frankenstein* (1818) as the first work of science fiction.
23. The novel was, of course, conceived as a genre for adults and functioned so until the second part of the nineteenth century. At the same time, since most of its early examples followed a *Bildungsroman* or *Entwicklungsroman* models—in which, respectively, an adolescent protagonist grows, or grows and matures to adulthood—the novel was naturally malleable to serve the purposes of the young audience. For a distinction between *Bildungsroman* or *Entwicklungsroman*, see Trites, *Disturbing* 10–19.
24. The rise of the novel and the arrival of drama and theatrics in murder and other serious trials happened at the same time. The *New Justice* procedures—including "eloquent opening statements and closing arguments, clever cross-examination of witnesses, and a sustained battle of wits between the opposing

sides" (Abate 26)—transformed the courtroom into a dramatic stage, with conflict built into the proceedings, with roles people play without knowing what the outcome will be, and with life-and-death issues distilled into personal stories. Throughout the nineteenth century, trials became highly entertaining reality shows and eloquent lawyers were transformed into celebrity orators whose performances drew large crowds. Both were extensively covered in daily press, which further spread the excitement of a courtroom drama and are at the heart of the Judge Judy effect even today.

25. If, like Kant, one identifies rationality as grounds for moral personhood, what then is the status of small children, or people with severe mental disabilities, or anyone who fails to meet Kant's rational standard? There is a debate on whether some Enlightenment philosophers would hesitate to consider children as fully human inasmuch as their rational nature—the capacity to set goals according to reason and act on these considerations—is only a potentiality. Kant, who never married and did not have children, spoke about family as a contract in which obligations must be discharged only out of duty, not out of sentiment. In a larger sense, the Enlightenment mind does not understand human emotions, always reducing them to something else. Love and friendship, for example, are instances duty (Kant, *Groundwork* 66), rational "reciprocity" (Rawls, *Theory* 494), economic exchange or efficiency (Posner, *Economics* 6). Such arguments have little appeal to "thick" people engaged in real relationships with others. For a critique of this aspect of *New Justice* argument, see Dimock 110–113, and 140ff.
26. Madeleine L'Engle's *A Wrinkle in Time* (1962), although not a dystopia, features an episode on a dystopian planet Comazotz, which includes all of the oppressive elements of sameness that have later been denounced in later YA dystopias.
27. This communitarian shift can be seen, for instance, by comparing Russell Freedman's groundbreaking *Lincoln: A Photobiography* (1989) with his recent *Abraham Lincoln and Frederic Douglas: The Story Behind an American Friendship* (2012). While both books deal with processes that led Lincoln to understand the abolition of slavery as an imperative, *Photobiography* highlights Lincoln's central role in the process, inadvertently ignoring African American efforts toward the liberation cause. *The Story Behind an American Friendship*, by contrast, amends the picture by showcasing how Lincoln's thinking was influenced by his interaction with Douglas. The latter book affirms that large-scale change does not trickle down from top to bottom but is a communitarian result of multiple factors and many voices triggering change. The same has been the case with the many books about Martin Luther King. Whereas earlier ones expanded on the role of Dr. King alone, recent texts, such as Andrea Davies Pinkeney's *Martin and Mahalia: His Words, Her Song* (2013), bring from the shadow many other characters whose role in the Civil Rights Movement was just as crucial, even if on a different scale. Pinkney's *Sit-In: How Four Friends Stood Up by Sitting Down* (2010) is another good example of how *Open Justice* mindset encourages seeing large social processes through many lenses.
28. Posner's totalizing reductionism also informs his ambitious *Law and Literature* (1988), where he advocates the study of "the great 'legal' works of literature" (547)—that is, works that describe trial scenes and legal procedures—while dismissing its other aspects and what he calls popular fiction. These he downplays as "rarely a fruitful subject" for scholarly reflection, inasmuch as they are

not rewarding readings and their treatment of law is "unlikely to be insightful" (549). The exemplary list of titles Posner then provides includes only one, but by now a classic, work for the young audience—*Alice's Adventure's in Wonderland* and only for its trial scene. The two benefits of studying literature by lawyers Posner identifies include the fact that fiction offers "valuable background knowledge concerning subjects of legal regulation" and "fascinating hypothetical situations for testing legal principles" (548). For criticism of Posner's position, see Dimock 140–181.

29. Mithen describes the big bang as a cultural revolution that occurred between 60 and 30 thousand years ago and resulted in fundamental changes in human lifestyles. Because it occurred without any apparent changes in anatomy or brain size, it was a change in the nature of the mind, one that made humans "modern," enabled free flows among different cognitive domains, thus laying the foundation for art, religion, abstract, philosophical, and hypothetical thinking such as evident in anthropomorphism and totemism. See Mithen 151–216.
30. The globalization of ethics is a process whereby moral norms deriving from various ethical traditions have entered into a conversation at a global level. The idea of global ethics is, obviously, a *New Justice* concept but one that has recently been reformulated, through the *Open Justice* lens, into a two-level phenomenon, framed, on the one hand, by international discourse of human rights and, on the other, by diverse cultural and ethical traditions. At this second, culturally specific level, articulation and propagation of different ethical views is seen as a space for dialogue and mutual learning across cultures, where "interchange can take place, without presupposing or imposing a single ethical perspective, and without limiting ethical debate to the thin and minimal discourse of international human rights" (Kymlicka and Sullivan 4). For discussions about the implications and application of global ethics, see, for example, Will Kymlicka and William M. Sullivan. Eds. *The Globalization of Ethics: Religious and Secular Perspectives*, J. Michael Adams and Angelo Carafagna's *Coming of Age in a Globalized World: The Next Generation*, David Korten's *The Great Turning: From Empire to Earth Community*, and Raimond Gaita's *A Common Humanity: Thinking about Love and Truth and Justice.*
31. The decline of violence is a global trend, which, following Steven Pinker, I see as "the most significant and least appreciated development in the history of our species" (Pinker 692). Contrary to the impression created by modern media preoccupied with violence and danger; contrary to human moral psychology which thrives on a sense of danger; and contrary to social theories from Durkheim to Foucault that have emphasized the criminogenic nature of modern society, an increasing number of studies demonstrate that violence has been diminishing over long stretches of human history. What Norbert Elias has called the civilizing process has brought radical changes in human toleration of violence, as a result of which we live in the least violent times ever. Having almost been tarred and feathered at two conferences when I spoke about the decline of violence—and by literary scholars who are otherwise kind and gentle people—I realize how controversial this claim is and I leave it to the reader to assess the scholarly discussion on this topic so far. However, I believe this trend is notable in young people's literature and will discuss its examples in chapters on restorative, environmental, social, and global justice. For discussion of this trend and its manifestations, see, for example, Norbert Elias' *The Civilizing Process: Sociogenetic*

and Psychogenetic Investigations, Stephen Moore and Julian Simon's *It's Getting Better All the Time: 100 Greatest Trends of the Last 100 Years*, Julius Ruff's *Violence in Early Modern Europe 1500–1800*, William Ury's *Must We Fight? From the Battlefield to the Schoolyard: A New Perspective on Violent Conflict and Its Resolution*, Gregg Easterbrook's *The Progress Paradox: How Life Gets Better While People Feel Worse*, Matt Ridley's *The Rational Optimist: How Prosperity Evolves*, Steven Pinker's *The Better Angels of Our Nature: Why Violence Has Declined*, Peter Diamandis and Steven Kotler's *Abundance: The Future Is Better than You Think*, and recently Robert Bryce's *Smaller Faster Lighter Denser Cheaper: How Innovation Keeps Proving the Catastrophists Wrong*. Although the scope of each book is different, each of them adds unique evidence that supports the global tendency of the decline of violence.

32. The emergence of new subfields related to peace studies and non-violent conflict resolution strategies began in the 1950s in disciplines such as sociology, history, anthropology, and political sciences to crystallize into a separate field roughly by 1980 (Harris xxii). By 2007 there were over 450 programs based at some 390 institutions representing 40 countries on six continents, 257 of them in 38 US states (Meyer and Shuster v). This exponential growth of academic programs was reflected in the mushrooming of peace-oriented organizations, journals, and initiatives. The US Institute of Peace was created in 1986. The year 1987 saw the founding of the Peace Studies Association, renamed, in 2001, as Peace and Justice Studies Association. Programs for nonviolence and the culture of peace are now run locally, nationally and globally—the United Nations and UNESCO declared the Year 2000 and the Decade 2001–2010 the Year and Decade for a Culture of Peace and Nonviolence for the Children of the World. All of these initiatives promote the culture of peace and gesture toward an understanding of peace not merely as an absence of war but as a dynamic state of a just social functioning. The widespread interest in peace studies reflects a new awareness of justice and new imperatives for justice, including promoting sustainable peace and diminishing various forms of colonial, cultural, domestic, structural, and civil violence.
33. The reparations and apologies movement reflects the change from retribution toward reconciliation and dialogue-based framework, and as such, it is also part of the decline of violence. The trend is informed by the *Open Justice* communitarian logic that acknowledges a nation's and community's history as morally burdened with historic injustices. Theorized as a form of historical justice, the movement is based on recognition that certain actions in the past, even if fully legitimate at that time, were legally committed wrongs. In the US the apology trend began in 1988, when President Ronald Reagan issued an official apology to Japanese Americans for their confinement in internment camps during WWII. In 1993 Congress apologized for an overthrow of the Kingdom of Hawaii a century before, but it was only in 2008 that the House passed a resolution "apologizing to African Americans for slavery and for the Jim Crow era of racial segregation" (Sandel 210). By that time, starting in 2007, state legislatures in Virginia and other former slave states had already issued similar apologies. Also in 2008 Prime Minister of Australia Kevin Rudd apologized to the aboriginal people for decades of kidnappings-and-institutionalization of mixed blood children, the so-called "stolen generation" (209). A year later the government of Brazil offered an official apology to the Brazilian nation and especially to the victims for military persecution in the period 1964–1985. In May 2011, during

the first visit of a British monarch to the Republic of Ireland in a century, Queen Elizabeth II apologized to all Irish victims of English domination from the early seventeenth century onwards. Such examples are legion: for details see the Political Apologies and Reparations Website, which lists a total of 1160 apologies, most of them offered in the past thirty years. What they all suggest is that the apologies and reparations movement is a recent phenomenon grounded in the *Open Justice* logic. The recognition of a duty of contemporary people to bear the moral burden of historic injustices projects justice as a forward-oriented process, in which foundations for a better future require acknowledging and apologizing for the wrongdoings of the past. Although not specifically an apology, the political custom of pardoning the National Thanksgiving Turkey by the US President—formalized since 1989—is part of the same trend.

34. Global justice refers to operations of international law, especially to punishments of human rights violators, but also to issues of distributive justice on a planetary level. In this sense it goes back to Rawls' 1971 *Justice as Fairness* and conceivably to Kant, even though both of them at best pointed at global justice as a logical consequence of the equality imperative but did not attempt to define it.

WORKS CITED

Abate, Michelle. *Bloody Murder: The Homicide Tradition in Children's Literature*. Baltimore: The John Hopkins UP, 2013.

Adams, Michael J. and Angelo Carafagna. *Coming of Age in a Globalized World: The Next Generation*. Bloomfield, CT: Kumarian Press, 2006.

Attebery, Brian. *Stories about Stories: Fantasy and the Remaking of Myth*. New York: Oxford UP, 2014.

Baron, Jane B. "Storytelling and Legal Legitimacy." *Un-Disciplining Literature: Literature, Law, and Culture*. Eds. Kostas Myrciades and Linda Myrciades. New York: Peter Lang, 1999: 13–27.

Beccaria, Cesare. *An Essay on Crimes and Punishments*. 1764. http://www.constitution.org/cb/crim_pun.htm. June 4, 2014.

Bryce, Robert. *Smaller Faster Lighter Denser Cheaper: How Innovation Keeps Proving the Catastrophists Wrong*. New York: Public Affairs, 2014.

Capeheart, Loretta and Dragan Milovanovic. *Social Justice: Theories, Issues, and Movements*. Piscataway, NJ: Rutgers UP, 2007.

Capra, Fritjof. *The Turning Point: Science, Society, and the Rising Culture*. New York: Simon and Schuster, 1982.

Clute, John. "Taproot Texts." *The Encyclopedia of Fantasy*. Eds. John Clute and John Grant. New York: St Martin's Griffin, 1999: 921–922.

Darnton, Robert. *The Great Cat Massacre and Other Episodes in French Cultural History*. New York: Basic Books, 1984.

Day, Martin S. *History of English Literature 1660–1837*. Garden City, NY: Doubleday and Company, 1963.

Diamandis, Peter and Steven Kotler. *Abundance: The Future Is Better than You Think*. New York: Free Press, 2012.

Dimock, Wei Chi. *Residues of Justice: Literature, Law, Philosophy*. Berkeley and Los Angeles: U. of California P., 1996.

Easterbrook, Gregg. *The Progress Paradox: How Life Gets Better While People Feel Worse*. New York: Random House, 2003.

Elias, Norbert. *The Civilizing Process: Sociogenetic and Psychogenetic Investigations*. Transl. Edmund Jephcott. Malden, MA: Blackwell Publishing, 2000.

Elster, Jon. *Closing the Books: Transitional Justice in Historical Perspective*. New York: Cambridge UP, 2004.

Fleming, Candace. *Ben Franklin's Almanac: Being a True Account of a Good Gentleman's Life*. New York, Atheneum Books, 2003.

Fischer, David Hackett. *Albion's Seed: Four British Folkways in America*. New York: Oxford UP, 1989.

Gaita, Raimond. *A Common Humanity: Thinking about Love and Truth and Justice*. New York: Routledge, 2000.

Glasius, Marlie. "Global Justice Meets Local Civil Society: The International Criminal Court's Investigation in the Central African Republic." *Alternatives* 33.4 (2008): 413–33.

Goodwin, Barbara. *Justice by Lottery*. Chicago: the U. of Chicago P., 1992.

Harris, Ian M. "Introduction." *Global Directory of Peace Studies and Conflict Resolution Programs*. 7th ed. Prescott, AZ: The Peace and Justice Studies Association, 2007: xi–xv. http://www.peacejusticestudies.org/globaldirectory. June 5, 2014.

Howard-Hassmann, Rhoda E. Ed. *Political Apologies and Reparations Website*. Waterloo, Ontario: Wilfrid Laurier University. https://political-apologies.wlu.ca/index.php. Jan 15, 2015.

Holt, Marilyn Irvin. *The Orphan Trains: Placing Out in America*. Lincoln, NE: U. of Nebraska P., 1992.

Kant, Immanuel. *The Philosophy of Law: An Exposition of the Fundamental Principles of Jurisprudence as the Science of Right*. 1796. Transl. W. Hastie. Edinburgh: T. and T. Clark, 1887. http://www.heinonline.org.ezp3.lib.umn.edu/HOL/Page?collection=beal&handle=hein.beal/tphol0001&type=Text&id=3. May 27, 2014.

Kant, Immanuel. *Groundwork for the Metaphysics of Morals*. 1785. Transl. and ed. by Mary J. Gregor and Jens Timmermann. Cambridge: Cambridge UP, 2012.

Korten, David. *The Great Turning: From Empire to Earth Community*. San Francisco, CA: Berrett-Koehler Publishers, 2006.

Kurtzweil, Ray. *The Singularity Is Near: When Humans Transcend Biology*. New York: Viking Press, 2005.

Kymlicka, Will and William M. Sullivan. Eds. *The Globalization of Ethics: Religious and Secular Perspectives*. New York: Cambridge UP, 2007.

MacIntyre, Alasdair. *Whose Justice? Which Rationality?* Notre Dame, IN: U. of Notre Dame P, 1988.

Meyer, Matt and Amy L. Shuster. "Preface." *Global Directory of Peace Studies and Conflict Resolution Programs*. 7th ed. Prescott, AZ: The Peace and Justice Studies Association, 2007: v–vii. http://www.peacejusticestudies.org/globaldirectory. June 5, 2014.

McCallum, Robyn. *Ideologies of Identity in Adolescent Fiction: The Dialogic Construction of Subjectivity*. New York: Routledge, 1999.

Milton, John. *Paradise Lost*. 1666. April 7, 2007. http://www.gutenberg.org/cache/epub/26/pg26.html. June 6, 2014.

Mithen, Steven. *The Prehistory of the Mind: The Cognitive Origins of Art, Religion, and Science*. New York: Thames and Hudson, 1996.

Moore, Stephen and Julian Simon. *It's Getting Better All the Time: 100 Greatest Trends of the Last 100 Years*. Washington, DC: Cato Institute, 2000.

Nixon, Cheryl. *The Orphan in Eighteenth-Century Law and Literature: Estate, Blood, and Body*, Farnham, UK: Ashgate, 2011.

Nussbaum, Martha. *Creating Capabilities: The Human Development Approach*. Cambridge, Mass.: Harvard UP, 2011.

Pakaluk, Michael. *Aristotle's Nicomachean Ethics: An Introduction*. New York, Cambridge UP, 2005.

Parr, Joy. *Labouring Children: British Immigrant Apprentices to Canada, 1869–1924*. Montreal: McGill-Queen's UP, 1980.

Pinker, Steven. *The Better Angels of Our Nature: Why Violence Has Declined*. New York: Viking, 2011.

Posner, Richard A. *The Economics of Justice*. Cambridge, MA: Harvard UP, 1981.

Posner, Richard A. *Law and Literature*. 1988. Cambridge, MA: Harvard UP, 2009.

Rawls, John. *Justice as Fairness: A Restatement*. Cambridge, MA: Harvard UP, 2001.

Rawls, John. *A Theory of Justice*. Cambridge, MA: Harvard UP, 1971.

Ridley, Matt. *The Rational Optimist: How Prosperity Evolves*. New York: HarperCollins, 2010.

Roland, Jon and the Constitution Society. "Cesare Beccaria." Oct 31, 2013. http://www.constitution.org/cb/beccaria_bio.htm. Jan 15, 2015.

Ruff, Julius. *Violence in Early Modern Europe 1500–1800*. Cambridge: Cambridge UP, 2001.

Sandel, Michael. *Justice: What's the Right Thing to Do?* New York: Farrar, Strauss and Giroux, 2009.

Sen, Amartya. *Development as Freedom*. New York: Alfred A. Knopf, 1999.

Sen, Amartya. *The Idea of Justice*. Cambridge, MA: Harvard UP, 2009.

Tolkien, J. R. R. *The Hobbit: Or There and Back Again*. 1937. *The Annotated Hobbit*. Rev. Ed. Annotated by Douglas A. Anderson. London: HarperCollins, 2003.

Trites, Roberta Seelinger. *Disturbing the Universe: Power and Repression in Adolescent Literature*. Iowa City, IA: U. of Iowa P., 2000.

Ury, William. Ed. *Must We Fight? From the Battlefield to the Schoolyard: A New Perspective on Violent Conflict and Its Resolution*. New York: Josey-Bass, 2002.

Warnke, Georgia. *Legitimate Differences: Interpretation in the Abortion Controversy and Other Public Debates*. Berkeley and Los Angeles: U. of California P., 1999.

Williams, Raymond. *Marxism and Literature*. Oxford: Oxford UP, 1977.

Ziolkowski, Theodore. *The Mirror of Justice: Literary Reflections of Legal Crises*. Princeton, NJ: Princeton UP, 1997.

2 How We Know What We Know

Justice Scripts in Literary and Filmic Narratives

Anyone who has seen a two-year-old "hiding" by covering her eyes can imagine how peculiar social interaction would be if we did not have a fully developed Theory of Mind. Without understanding the relative subjectivity of our sensory perceptions, we would not be able to anticipate what others know and how they might behave. We would not be able to connect causes and effects or decipher connections between intentionality, beliefs, and actions. We would live in a world in which the simplest everyday tasks would constantly require learning anew.

Theory of Mind is one of the many cognitive programs executed by the human mind. Like memory, perception, language, and other evolved capacities, it is biologically programmed to develop in healthy human beings at a specific developmental stage. While a detailed map of human cognitive architecture is still some way off, research into aspects of cognition has opened up new avenues that shed light on our cognitive operations in a variety of contexts. The previous chapter suggested an evolution in the understanding of justice that helped open, albeit indirectly, possibilities for the arrival and development of young people's literature. In this chapter, I examine the connection between the vehicle of the story and the human cognitive architecture, which in the remainder of this book I will then apply to representations of justice in YA fiction. The goal of this chapter is threefold. First, to demonstrate that on the most fundamental level—the level of basic structures and processes that organize the mind in terms of extraction and storage of information—human understanding is a script-based narrative understanding. Second, to argue that the two fundamental types of knowledge structures humans bring to bear on understanding task are scripts as well as script-based, content-rich information clusters commonly referred to as stories. Third, to suggest that lexical complexes such as justice are processed by our cognitive-affective apparatus as scripts and can best be theorized as scripts. By suggesting that our understanding of justice is script-based, I am referring to the general cognitive-affective mechanism that structures ideas of justice in our mind and makes them available to us in our everyday functioning. This unreflective knowledge is different from self-conscious theorizing, philosophizing and rationalizing about justice peculiar to scholarly, reflexive thought, even though it may also underlie those more elaborate forms of reasoning about justice.

Accepting that the human mind is hardwired for narrative understanding and operates on the basis of scripts and stories implies a reevaluation of narrative fiction. When seen from a cognitive angle, fiction emerges as an evolutionary adaptation that recalibrates the mind, sharpens social cognition, and offers multiple benefits that fine-tune our other capacities. An important function of fiction is to spread and reinforce expectations related to specific scripts, including scripts about justice.

STORIES, LITERATURE, AND THE HUMAN COGNITIVE ARCHITECTURE

When Alasdair MacIntyre attacked "the fiction of shared, ... universal standards of rationality" (*Whose* 400) and instead advocated multiple rationalities, he identified two major types of resources where people can look for clues about how to understand justice: modern academic philosophy and communities of shared belief. Acknowledging that philosophy rarely goes beyond articulating disagreements, MacIntyre then turns to communities of shared belief—intellectual traditions that have informed the evolution of Western culture—claiming that only embedding in these traditions offers people different concepts of practical rationality, each with its own legitimate understanding of justice[1]. As to how people from different traditions can decide among the sometimes-incompatible accounts of justice, one possible avenue MacIntyre suggests is through "acts of empathetic conceptual imagination" (395).

Although MacIntyre does not develop this concept, the mention of acts of empathetic conceptual imagination points to yet another resource for thinking about justice—a resource he and other *Open Justice* advocates use profusely: namely, literature. The complex relationship between justice and literature goes back at least to Homer and has been instrumental, as I suggested in the previous chapter, in the emergence of the novel form. Informed by progressive *New Justice* ideals, the overwhelming majority of eighteenth- and nineteenth-century novels were seen by their authors and readers alike as part of public discourse or "public imagination" (Nussbaum, *Poetic* 3) about the common good. From Fielding's *Tom Jones* (1749), through Stowe's *Uncle Tom's Cabin* (1852), Dickens' *Great Expectations* (1861), Hardy's *Tess of the d'Urbervilles* (1891), and at least up to Dreiser's *An American Tragedy* (1925), the novel was seen as a socially engaged form that showcased the limits of law, the complexities of justice, and the need for change in the social and legal domains. In twentieth-century literature this imperative seems just as strong—just to mention the novels of Toni Morrison, Milan Kundera, Zadie Smith, Arundhati Roy, and others—but in many of these works justice-seeking shares the same space with a number of alternative objectives.

One marker of the big bang of justice in the second part of the twentieth century was the rise of literary critical studies of justice in literature. In recent decades, perhaps following Posner's example, scholars have tended to avoid the fuzzy word justice and focused instead on the more measurable concept of law: thus the law and literature movement. This field has been dominated by two approaches: law *in* literature and law *as* literature. The focus of the former has been on how law, lawyers, and the legal process of justice-seeking have been represented in works of literature. Martha Nussbaum's *Poetic Justice* (1995), Stephen Greenblatt's *Hamlet in Purgatory* (2001), or Michelle Ann Abate's *Bloody Murder* (2013) are some of the many examples that can be seen as reflecting this diversified tradition. The law *as* literature approach, by contrast, has focused on legal rather than literary texts, examining how literary devices such as metaphor or focalization function in legal texts[2]. The use of narrative study as a model for legal study has not been widely embraced. Literary scholars, especially, criticized it as adding little or even impeding the reader's understanding of law (Hogan, "Fictive" 271). I, for one, do not believe that making law understandable is or should be an aspiration of fiction. Writing in 1999, Hogan was certainly right when he pointed out that the overblown promise of law and literature had blinded scholars to areas far more promising for this pursuit, "such as cognitive science" (273).

The confluence of cognitive science and literature—although relatively recent if taken to begin with Mark Turner's *Reading Minds* (1991)—already has a rich history. This richness is due primarily to the interdisciplinarity of the cognitive angle, where the representational and computational capacities of the human mind are studied in many related fields: psychology, artificial intelligence, linguistics, behavioral science, rhetorics, and, of course, literary and cultural studies. Born in the 1950s but gaining wider recognition only in the 1970s, cognitive science has a history, which I want to briefly outline as proceeding in two phases. Each phase was characterized by a different relationship between the cognitive paradigm and literary studies.

The first phase extended roughly until the 1990s and comprised efforts focused through the lens of Artificial Intelligence. Built on the information-processing paradigm that perceives the working of the human mind akin to that of a computer, AI and its sister discipline cognitive psychology[3] explored how intelligent systems employ complex knowledge structures that allow them to understand, interpret, and remember events so that they can learn from them. This phase may be adequately illustrated by the work of Roger Schank, whose goal has been to understand the extraordinary workings of "ordinary" human intelligence. Starting in 1972 with a theory of Conceptual Dependency, which sought to explain meaning representation for events on the level of a sentence, in 1977 Schank had moved to higher-level knowledge structures in *Scripts, Plans, Goals, and Understanding* (1977). Co-authored with social psychologist Robert Abelson, this book proposed four nested levels in the hierarchy of human knowledge structures. It saw scripts, plans, goals, and themes as basic structures for "the intentional and contextual connections between events, especially as they occur in human purposive action sequences" (Schank and Abelson 4), but awarded primacy to scripts as the basic blocks of human cognition[4]. By 1983 Schank had realized that the picture was more complex. In *Dynamic Memory* (1983), he introduced the concept of dynamic memory, one that changes in response to new data and accounts for the processes of script modification. He also offered new evidence for how understanding new events is possible only in the context of previously understood ones. This, in turn, led Schank to the largest and most complex knowledge structure processed by the human mind: stories. In *Tell Me a Story: Narrative and Intelligence* (1990) Schank theorized stories not only as a kind of memory mechanism—detail-packed and multiple indexable bundles of scripts—but also as reflecting the basic nature of intelligence[5]. What makes humans intelligent, according to Schank, is the ability to "correlate the story we are hearing with one that we already know" (Schank, *Tell* 21). His *Tell Me a Story* presents intelligence as a capacity for storytelling, dependent upon a more fundamental capacity for registering events as stories, which in turn depends on script-creation and recognition mechanisms. Arguing that "the bulk of what passes for intelligence is no more than a massive indexing and retrieval scheme" (84–5), Schank thus links narrative, intelligence, and understanding.

In both content and direction, the trajectory of Schank's research exemplifies the achievements and limitations of the AI angle. The limitations stemmed primarily from taking the "human mind *as* computer" analogy too literally, which tended to reduce understanding to a multifactor yet somewhat mechanical process[6]. Within this paradigm, there are no tools to account for the emotive response to literature—a response that differs from reader to reader—and no tools to explain why readers are moved by events and characters they know to be fictional. I find it symptomatic, though not surprising, that after rediscovering narratology by a route starting from an exploration of how memory works and how it makes understanding possible, Schank saw the topic exhausted, changed gears, and never returned to what seems to be the conclusive argument of *Tell Me a Story*. In this sense, the AI perspective shared limitations with inescapably reductionist evolutionary psychology, whose self-confirming accounts—as offered by Tooby, Cosmides, or Pinker—have tended to elevate broad generalizations into "laws" about the working of human thought. This is not to suggest that the AI phase did not yield significant achievements. These attainments include the recognition of the role of scripts and schemata in human understanding[7], the development of Theory of Mind, and an articulation of a broad consensus that has since informed cognitive science. Namely, that the task of understanding requires activation of multilevel knowledge structures, that specific knowledge structures are mutually interrelated and operate in nested hierarchies, that they constantly evolve, but their content lends itself to change more than their form, and that knowledge structures are developed and shaped largely through verbal mediation: narrative, discourse, storytelling[8].

The second phase of cognitive science research began in the mid-1990s. Suspicious of the AI-based mechanical metaphor of the mind as a computer, this ongoing phase has been marked by the search for a middle ground between culture and biology, questing for cross-disciplinary "consilience" (Wilson 8) or vertical integration: "an integrated, 'embodied' approach to the study of human culture" (Slingerland 9). The organic metaphor of the embodied mind reflects the logic that also informs *Open Justice* and inspires the search for what Donna Haraway has referred to as "situated knowledge," where the goal is to hold "*simultaneously* an account of radical historical contingency for all knowledge claims and knowing subjects, a critical practice for recognizing our own 'semiotic technologies' for making meanings, *and* a no-nonsense commitment to faithful accounts of a 'real world' ..." (Haraway 187). Spearheaded by Turner's second and still heavily linguistic book *The Literary Mind* (1996)—but also drawing from groundbreaking ToM studies such as Simon Baron-Cohen's *Mindblindness* (1995) and neurobiological studies such as Antonio Damasio's *The Feeling of What Happens* (1999)—this application of the cognitive perspective to literary and film studies has been so fruitful that evoking cognitivism today requires an explanation about what cognitive components are being referred to and how they will be examined. What, then, are my premises in this study?

First, as may be obvious from the above, I subscribe to the schema-theoretic view of cognition. Because language is linear and the human mind is geared to parallel processing carried out at great speeds and multilevel complexity, an accurate descriptive account of the cognitive process may be unachievable. However, when minuscule aspects of our cognition are frozen for analytical purposes—or when people with cognitive deficiencies are examined for the hows and whys of their conditions—a larger picture of the human cognitive architecture reveals itself. This complex, perhaps best discussed by Hogan, is theorized as a three-level cognitive apparatus. It consists of structures that can be thought of as departments of the mind, then contents—representations or "internal lexical entries" (Hogan, *Cognitive* 30)—and finally processes, in which certain structures activate representations and move them within or across domains. Thus, various types of memory are examples of structures. These structures define what content is stored, how, and in what relation to other content. Processes, in turn, are operations executed on a specific content within one or more structures, for example, activation of stored content to bear on a new experience. This tripartite architecture, Hogan is careful to note, has been studied from four distinct angles, depending on where one's interests lie. A neurobiological or/and connectionist study such as these done by Baron-Cohen, Damasio, or Panksepp can demonstrate which centers in the mind fire up during specific activities and which human capacities are inhibited after damage to a particular part of the brain. Given the complexity of the brain's hardware, however, neurological terms are not sufficient to explain its software. The mind, with its intentions, beliefs, emotions, and interpretations, eludes the neurological nomenclature. Thus, while some neurological research sheds light on aspects of literary understanding, the most traveled cognitive route for literary scholars is what Hogan calls representationalism: an approach that examines representationalist cognitive architecture—especially structures and operations of memory—and their connections to the meaning-making processes (30–35).

My understanding of representationalism is distilled from the accounts offered by Schank, Turner, Herman, and Hogan, but I dispense with their extremely detailed terminological taxonomies[9] in favor of three hierarchically nested types of Schankian knowledge structures—what Hogan identifies as both structures and processes of memory[10]. The first two, schemas and scripts, are layers in the lexical feature complex—a subsystem of long-term memory—essential to bear on any understanding tasks (Hogan, *Cognitive* 42). The third level, the story, is a thick cluster of actualized schemas and scripts, which Schank and socio-narratologists have identified as the most complex and versatile knowledge structure that human beings construct. Each of these three has a long history in cognitive narratology, and each describes an identifiable level of information processing within a larger, distributed, and organic cognitive process.

Schemas, also called simple image schemas (Turner, *Literary* 16), static repertoires (Herman, *Story* 89), or representational schemas (Hogan, *Cognitive* 44), are the "genes" of understanding. They are feature lists and default hierarchies of features that help us identify objects, events, and agents: swamp or castle, visit or bath, Shrek or Fiona. Scripts, also called complex image schemas (Turner, *Literary* 16), dynamic repertoires (Herman, *Story* 89), or procedural schemas (Hogan, *Cognitive* 45), are what cells are to genes: higher-level units built from schemas but having emergent properties beyond that of a schema—thus the qualifiers such as "complex," "dynamic," or "procedural." Scripts are stereotypical sequences of action in a particular context made of slots and requirements about what can fill these slots. Scripts perform many functions at once. They are *memory structures* that help us store knowledge about standard situations (Schank, *Tell* 8). They are *procedural protocols*, "nonconscious instructions to execute particular physical or mental actions" (Hogan, *Cognitive* 45). They are also *experiential repertoires* that "reduce the complexity and duration of processing tasks" (Herman, *Story* 89). As economy devices, they obviate the need to think and enable "automatic" understanding of sequences of events, especially those connected in causal chain by the directed inference process.

One overlap between schemas and scripts—and a source of confusion too—is that both scripts and schemas are scalable. Schank, for example, ignores schemas, treating them either as "nominal concepts" (Schank and Abelson, 10) or as a synonym for basic scripts. Herman likewise gives short shrift to schemas, or frames in his terminology. Instead, his focus is on scripts as "molecular narratives" (*Story* 85) and their relationship to fleshed narratives, or stories, which both employ and deviate from the stereotyped sequences that scripts provide. Hogan, in turn, on the one hand subsumes scripts under the larger label of *representational schemas*. On the other, he distinguishes the type of script where "all the defaults are in place" (*Cognitive* 46) as a separate category of the *prototype*—something that Schank calls a "story skeleton" (*Tell* 158)—and then scales it up even further to the category of "models" (*Cognitive* 40) or "*exemplars*" (46, italics in the original): a name he assigns to fully fleshed, specific actualizations of scripts as embodied in exemplary stories, i.e., stories clearly dominated by one script.

This proliferation of terms is not always helpful, and I find it more useful to stick to the schema and script division. However, I also recognize their inherent scalability, which can best be visualized along what Herman calls the "narrativity" line (Herman, *Story* 91). In *Shrek 2*, a simple schema such as "visit"—when scaled up into a narrative sequence of the eight screen minutes of Shrek, Fiona, and Donkey's invitation, preparations, travel, and arrival in *Far, Far Away*—grows in narrativity to the level of an actualized script of "traveling." By the same token, a script of traveling to a foreign place—when scaled down or fast-forwarded to the character arriving at the destination, as is the case with the King's envoy in the same film who shows up at Shrek and Fiona's door—may reach a minimal narrativity of a non-narrative sequence that characterizes schemas. Thus, although schemas and scripts are distinguishable layers of the cognitive process, sometimes which is which depends on how it is positioned in the story.

This brings me to the highest-level knowledge structure humans have developed. If schemas are "genes" and scripts are "cells," then stories are three-dimensional "organs": visible to the naked eye and palpable to both the author and the reader in a way that schemas and scripts usually are not. Stories are bundles of *actualized* scripts. Some of these actualized scripts are more central to the story and thus are identified with the main plot, while others are subplots or loose ends, frilled offshoots that sprout from the story at various points but are given no space to develop beyond being mere schemas. The stress on offering *actualized* scripts sets stories apart from schemas and scripts as generalizable, fill-me-in categories. Stories are not generalizable. They are always specific: with these-and-not-other characters, this-and-not-that setting, plot, style, narrative pace, and so forth. The relationship between schemas and scripts on the one hand, and story on the other, is like that between being a human being, a human life in general, and a human life lived as a particular person. Another difference between story and schemas/scripts is that most stories are actualizations of numerous scripts. Even though one script may be dominant—Shrek and Fiona's falling in love in *Shrek* or their domestic problems in *Shrek 2*, for example—in most cases a story will employ at least a few scripts in different configurations: complementarity, parallelism, alternative, scalar relational, and so on. Finally, stories are not only, as Schank says, the *densest* knowledge structures. They are also, as socio-narratologist Arthur W. Frank insists, *alive*. They do things to the reader the way heart pumps blood for the body.

This, incidentally, is my other large premise. Any explanation about how human understanding works must be complemented with an account of how stories work. The two aspects are inseparable. In any engagement with literature, we bring in assumptions about the nature of stories. Mine draw from Judith A. Langer's take on literary understanding[11] and Ursula K. Le Guin's the carrier-bag theory of fiction[12]. Recently this position has been eloquently theorized through the lens of socio-narratology in Arthur W. Frank's *Letting Stories Breathe* (2010). Frank's argument echoes the claims of cognitive science about how stories are templates for experience and about the nature of narrative understanding. Yet, he goes beyond these two in several ways. First, Frank insists that the conventional understanding of stories as merely *imitating* experiences that have already happened is misleading. It denies the fact that stories also *prefigure* human action (10). This assertion is not merely a restatement of Schank's claim that knowledge *is* stories but gestures toward stories as a source of cognitive order. Stories work primarily as "people's *selection/evaluation* guidance system" (Frank 2010, 46) [13]. They not only guide the reader about what to notice and what to ignore, but also cue her about evaluating what has been selected. This guidance system, obviously, is not made up of rules or principles but is rather a "tacit system of associations." It processes a large portion of the reader's "candidate-experience" (47), some of which may later become actualized experience as our thinking unfolds *with* stories that scaffold our action and motivation. This perspective is foundational to my argument about the importance of justice scripts in stories read and watched by young people today.

Another overlap between Frank's account of how stories work and my focus is how stories make the unseen compelling. Just as Nussbaum approaches justice by discussing it in terms of actualized capacities, so too Frank theorizes stories through their specific capacities. Among the thirteen capacities he mentions, three are especially relevant to a script-based reading of fiction through the lens of justice. The first relevant capacity is that practically any story starts with an out-of-the-ordinary complication, what Frank calls "Trouble" (28). Both in their exemplification of what Trouble is and in their presenting model scenarios for dealing with Trouble, stories resort to scripts, which at some point involve questions of right and wrong or justice writ large. Thus, the fairy tale creatures' invasion of Shrek's swamp is the Trouble that sets the story in motion. Second, since story is character-driven and characters are typically dealing with Trouble, any story worth its salt will involve a test of the protagonist's character. Her choices and response to Trouble reveal who she is. This, Frank says, is the inevitable function of characters *being cast* into specific roles by the story in the same way real people are being cast into social and other roles by their birth, education, culture, and status. In the *Shrek* example, Shrek is cast into being a monster, then a savior of fairy tale creatures, then Lord Farquaad's champion knight, then Fiona's rescuer, and so forth. The story's interest, Frank claims, is what the protagonist does with that casting.

"All stories," he insists, "are about characters resisting or embracing or perhaps failing to recognize the character into which they have been cast" (30). In this sense, the work of stories is to make recognizable motivational schemes for characters cast into specific roles, requiring them to struggle with what is the right thing to do as they respond to Trouble. Third, by informing the readers' sense of what counts as good or bad—which Frank sees as gut feelings rather than moral principles (36)—stories elicit and structure a moral response from the reader.

Obviously, not all stories affect everyone and not in the same way. Those that do, however, remain "*resonant* even when they are not consciously remembered" (40, italics in original). This resonance is not a monological interpretation that fits merely the specific situation reflected in that particular story, but a resonance with interpretive openness. It provides evaluations of the world through affective experiences that allow multiple future uses, even in unanticipated situations. In this sense stories form symbiotic relationships with people, functioning as evolving and shape-shifting entities. As Frank puts it—echoing Haraway—stories are alive as humanity's "companion species" (43). They help each other be. This account, of course, is fully in line with the consensus in cognitive science that when dealing with a story—fictional or not—human understanding develops "through the same sort of complex, recurrent processes of encoding that [it employs] in ordinary life" (Hogan, *Cognitive* 42). In life and art we rely on the same cognitive architecture and on the same library of scripts.

My third large premise is that any discussion about the representationalist cognitive architecture of the human mind and how it can be examined in stories is lacking unless it involves a metalevel explanation, namely the consideration of human evolution. The software of the human cognitive architecture—based on schemas, scripts, and stories—is inextricably embedded in the hardware of the evolved MindBrain. "MindBrain" is a term Panksepp and Biven use to stress that the conventional distinction between the supposedly incorporeal mind and the physical brain misses the crucial point that in its operations it is "a unified entity lacking any boundary with the body" (xiii). Cognition and its processes—including imagination, learning, memory, thoughts, and so on—is a function of tertiary processes executed by the MindBrain complex. It is built on, and serves as an "emissary" to, the ancestral brain with its primary-process affects (490). In other words, neocortical regions responsible for uniquely human, higher cognitive capacities play a secondary role in generating our emotional responses, including our responses to literature. "The ancient MindBrain substrates for emotional affects," Panksepp and Biven argue, "are not only governors of how we behave, but … also prompt us to dwell on the complexities of our lives as we navigate social worlds" (490). In our lives and the stories we tell, we are inheritors of ancient, pre-cognitive affects that constitute the very ground of meaning within our minds.

If the above implies that literature is an evolutionary adaptation, one must be careful to specify in what way. Like all art, literature is not an adaptation in the strict biological sense—where adaptation is a term defined as yielding reproductive advantage in a specific environment. Nevertheless, fiction is a product of human evolution and—as demonstrated in Brian Boyd's *On the Origin of Stories* (2009)—elements of literature do have adaptive functions that are both cognitive and affective. They hone our specifically human abilities to detect social and agential patterns, direct attention, preselect information, and refine our understanding of events and intentions. When the human cognitive architecture is seen through the evolutionary lens, the picture that emerges is something along the following lines. Our narrative capacities and storytelling competence are highly specialized adaptations that developed with the big bang of human culture between 60 and 30 thousand years ago (Mithen 194). This period saw the emergence of full cognitive fluidity—"the defining property of the modern mind" (210)—which is the ability to project concepts across various, hitherto separated domains, such as language, social intelligence, natural history intelligence, technical intelligence, and other specialized modules. With the rise of cognitive fluidity and generalized intelligence, new types of activities such as art, religion, storytelling, and science became possible, forever separating Homo sapiens from other hominidae. Although we share with them and other mammals our primary affective systems, the hominidae's cognitive architecture has remained specialized and is "fundamentally different" from the human one (212).

Cognitive fluidity equipped us with at least three things that have direct bearing on our ability to tell stories and learn from them. First, it enabled us to understand others as intentional agents: entities animated by minds with intentions, beliefs, goals, and knowledge. In modern cognitive psychology this capacity, labeled in 1978 as Theory of Mind[14], is defined as the ability to attribute mental states to the self and others so that their behavior can be predicted and explained with reference to these mental states. The ability to understand these unseen factors has, over the years, been established as a uniquely human capacity (Gopnik np). A number of questions still need to be answered, but it is clear that ToM knowledge is constructed in social contexts, develops gradually, and is age-related, with major break occurring at age four (Heyes 122). Additionally, a number of mental disorders related to cognitive failures, such as autism and schizophrenia, are forms of a ToM deficiency[15]. Normal ToM, as measured by the standard competence displayed by ordinary human adults, enables people to read intentionality; grasp up to fourth-order false belief[16]; develop, detect, and enjoy humor, irony, sarcasm, and other emotions[17]. ToM enables the development of metacognition and metarepresentation[18], makes possible teleological, religious and existential reasoning[19], undergirds human social intelligence as well as intuitive epistemology and ontology—the so-called folk-psychology[20]. In short, ToM makes humans human in a multilevel, complex way that neither nonhuman animals nor machines can match.

Whereas the information-processing paradigm provided a conceptual model for the development of schema theories, so too ToM research has become an important framework for cultural and literary studies. Its most recognized proponent has been Lisa Zunshine, who theorizes literature as the most versatile form of calisthenics for the human mind-reading capacity. Works of fiction, Zunshine argues, "provide grist of the mills for our mind-reading adaptations that have evolved to deal with real people, even though on some level we do remember that literary characters are not real people at all" (Zunshine, *Why* 16–7). Narrative fiction, she insists, has a cognitive advantage over oral narratives because it can accommodate more complex and deeper levels of intentionality than oral texts, thus testing and stretching human cognitive systems to their limits (Zunshine, *Theory* 211). If literary simulations "tend to be more vivid and detailed, more emotionally compelling" than ordinary life experiences (Hogan, *How* 26), they are always metarepresentations. Reading about them, we do not necessarily experience the same emotion as the character, but we are able to understand these emotions through empathic identification and our emotional knowledge of what it means to experience such emotions.

The second benefit of cognitive fluidity has been the development of consciousness—a theme familiar to literary scholars, by default implicated in the human cognitive architecture, but only recently given full attention from the neurobiological perspective. The neural architecture that supports consciousness was first described by Antonio Damasio in *The Feeling of What Happens* (1999), some of whose findings shed light on the affective-cognitive importance of the story. First, Damasio demonstrated that consciousness and emotion are inseparable, with emotion inducing important changes in the mode of cognitive processing (80) and actually preceding thought[21]. Second, he discovered that the foundation of consciousness is a "*nonverbal, imaged* narrative" (186) in the form of a non-languaged map of logically related events akin to a movie. This nonverbal narrative of consciousness is primordial and prelinguistic, even though humans immediately convert it into language. Damasio also established what has since became a widely recognized fact in cognitive research, that the brain behaves in exactly the same way when it imagines an activity as when it perceives it "in reality" through the senses. The nonverbal narrative of consciousness is thus not merely a record of sensory impressions from the outside world, but a mind's own creation, reflecting the left hemisphere's propensity for "fabricating verbal narratives that do not necessarily accord with the truth" (187). In theorizing consciousness as a level of biological processing, and specifically in the study of "as-if body-loop" mechanisms, Damasio incidentally created a powerful argument for the epistemic nature of "what if" mechanisms that also happen to be central to narrative fiction. In literary studies, this direction has been developed by studies that build on affective psychology, such as Suzanne Keen's *Empathy and the Novel* (2007), Blakey Vermeule's *Why Do We Care about Literary Characters?* (2010), Hogan's *What Literature Teaches Us about Emotions* (2011), and recently Maria Nikolajeva's *Reading for Learning* (2014).

Finally, the third advantage of the human mind's cognitive fluidity has been the development of art. As outlined in the most comprehensive account to date, Boyd's *On the Origin of Stories*, art meets all criteria as a third-order Darwin machine: an evolutionary subsystem designed for creativity in unpredictable, fast-changing environments, in which challenges undergo transformations too fast to be tracked by the selection of genes (Boyd 123). Art, says Boyd, "develops in us habits of imaginative exploration" (124), which enables us to see the world as open to our shaping rather than closed and prearranged. It creates a conceptual space where the actual can be explored within the framework of the possible, the conditional, and the impossible. This habit of imaginative exploration is enhanced by two pre-linguistic and biologically programmed cognitive processes: first, the capacity for understanding and representing events in a complex way, referred to as narrative comprehension, and, second, the capacity to understand others as intentional agents, referred to as ToM. Both capacities operate unconsciously through multilevel elaborate inference systems that reinforce one another.

In this picture, narrative and event comprehension—which allow humans to make associations between events and to form and follow representations of these events or narratives—are not, Boyd insists, a matter of cultural convention. Rather, it is narrative conventions that "reflect the regularities most important to human lives and minds … our mode of understanding events, which appears largely … to be a generally mammalian mode of understanding" (131). This argument about narrative as our primary mode of understanding allows Boyd to claim that fiction is a specialized adaptation of the information-craving human cognitive system, whose biological function is to recalibrate minds so as to "extend and refine our capacity to process social information" (192). Perhaps the most crucial advantage of this mental refinement is the human capacity to metarepresent: to project social information—especially that related to characters and events, allies and enemies, goals and obstacles, actions and outcomes—as seen from the perspective of other individuals, places, circumstances, or times. Apart from this major mind-calibrating role, fiction also serves a number subsidiary functions, such as status building, clarifying and reinforcing communal values, contributing to solving problems of cooperation, encouraging the development of a moral sense, and enhancing creativity—a quality Boyd considers "the most important function of pure fiction" (199).

The three by-products of the human mind's cognitive fluidity listed above, groundbreaking as they have been, did not completely erase all traces of the modular Old Brain. I use the phrase "Old Brain" on purpose in the same way as Boyd evokes "mammalian" or "reptilian" understanding, because both terms ground our narrative competence in our tri-strata, bipolar brains. Making connections between ontogeny and phylogeny is always a risky undertaking, but here I draw on the recent work of Iain McGilchrist, Jaak Panksepp, and Lucy Biven, which demonstrates that the structure of the brain reflects its history as an evolving dynamic system, and on the work of Hugh Crago, who applied these findings to the telling and understanding of stories from childhood through old age.

The picture that emerges from these studies can best be described in the following terms. Human narrative and cognitive capacities are both mediated and limited by the nested structure of the brain. Its oldest core is the brain stem, or the reptile brain, whose functions include reproduction and survival (Crago 6). Enveloping this core is the limbic system, the mammal brain, which also controls survival and reproduction, but enhances these lower instincts with care, attachment, and other emotions that are shared by mammals. Taken together, these two ancient subcortical strata, which Crago refers to as the "Old Brain" (9), consist of at least seven emotional systems identified by Panksepp and Biven as seeking, fear, rage, lust, care, panic/grief, and play. Each of these primary systems controls specific instinctive and archaic types of behavior, their associated physiological changes, and affective consciousness (Panksepp and Biven 2). On top of these, and inextricably dependent upon their processes, is the lateralized cerebral cortex: the seat of secondary and tertiary processes of consciousness, where cognition occurs. Called by Crago the "New Brain," the cortex mediates the higher functions of the brain. In addition to its ability to negate or inhibit information that flows from the Old Brain (McGilchrist 198), the New Brain is able to reflect upon its own experience. These operations occur mostly in the frontal lobes of the cortex—the most recently evolved part of the brain that takes much more space in humans than in the brains of our animal relatives.

Each level in the Old-New Brain division is defined by structures that are not accounted for by laws at the lower level. The third level is additionally complicated by the fact that the cortex of the triune brain is lateralized. The account of this extremely complex polarity as offered in McGilchrist's *The Master and Its Emissary* boils down to the fact that each hemisphere processes information in a radically different way, offering us "two fundamentally opposed realities, two different modes of experience" (3). The right hemisphere, which McGilchrist sees as primary, is both the unreflective conduit for Old Brain functions (Crago 9), and "the first bringer into being of the world" (McGilchrist 196). It records the immediacy and completeness of human experience, offering us a sense of undivided whole of the world and us in it, with the awareness of relationships, emotions, and instant judgments. What this hemisphere experiences "seems intrinsically true and real" (Crago 9), largely because it is the right hemisphere that serves as the "handmaiden" for the ancestral brain's affective systems (Panksepp and Biven 490). The left hemisphere, by contrast, is secondary, even "parasitic" on the right one (McGilchrist 200). Yet, it is also the location of the observing self, which can represent and analyze what the right hemisphere sees. The left hemisphere is a place where distinctions are made into past versus present, self versus other; where complexity is built through categories, contrasts, and arrangement of whatever aspects of experience it receives from the right hemisphere

(197). The difference between hemispheric functions can best be understood as something akin to an unreflective experience versus awareness of the same experience after it occurs; something like showing versus telling.

The hardwired capacities of the triune lateralized brain bear upon our narrative competence in a fundamental way. In the nested MindBrain complex, higher or tertiary brain functions—the computational cognitive mind—participate in but must be distinguished from the subcortical, primary, and universally mammalian affective strata. Affective feelings—which reflective thought understands as ideas and literary studies often identify as "meanings"—are not functions of the cognitive apparatus but are merely filtered through it. They are the evolutionary givens of the MindBrain, the "foundation" for a variety of "learning and memory mechanisms" (Panksepp and Biven 9) without which "our higher cognitive minds could not work" (14).

Stories emerge in the right hemisphere, which is the place of convergence of our experience of the external world and a nonverbal, imaged narrative described by Damasio as the foundation of consciousness. When read, heard, or seen, stories offer "*a right hemisphere experience*" (Crago 13, italics in the original) of emotive immersion, either *recreating* actual/similar or *priming* possible/imaginable experiences, but always drawing on the primary affective system that provides a basic "value structure for higher mental activities" (Panksepp and Biven 12). Engaging with stories remains a right-hemisphere experience no matter how old we are, just as explaining/interpreting stories—the domain of literary criticism and the practice of education—is a quintessential left hemisphere activity. Because the right hemisphere does not distinguish between fact and fiction, we react to fiction and nonfiction just as strongly. This may explain the appeal of *Twilight* and some other fiction *despite* many readers' awareness of them being rather poor as literature. It also accounts for the common phenomenon of audiences confusing actors or authors with the fictional characters they play or describe. The affective grip that ignores the fact-fiction divide is especially powerful when stories evoke Old Brain primary affects like survival, self-preservation, violence, and sex.

Although the story impulse originates in the right hemisphere, it is the structuring activities of the left hemisphere that package stories into coherent units with beginnings and endings, specific settings in time and place, specific characters and implied audience, and a specific focus through the choice of what is narrated and how. This patterning of stories, a tertiary process in Panksepp and Biven's terminology, is a function of the left hemisphere that hones each person's capacity to tell and understand stories at various developmental stages. Although we are born equipped with the unlearned subcortical affective systems in place, our neocortical cognitive system is "largely a blank slate," even though primed for learning (Panksepp and Biven 10). Only with experience, it begins to fill with abilities, connections, knowledge, and skills. Thus, young children up to the age of 4 or so do not understand the concept of a beginning or an ending and so will rarely tell a well-formed story. The stories they tell and take in are timeless and immediate—an Old Brain narrative pattern that begins to break down only by the age of five (Crago 23) when the basic features of ToM are in place. Our storytelling capacities then develop further through childhood, adolescence, and young adulthood—each phase, according to Crago, defined by its own attention focus on what types and aspects of stories one finds especially compelling—until we achieve self-reflective storytelling maturity, which for most people happens around the age of thirty (126). Early adolescence is an especially crucial period, since it is then that our "capacity for deep reading" reaches its full potential and needs to be nurtured. "[U]nless maintained and enforced during the radical reconstruction of the adolescent brain, it is irretrievably lost" (Nikolajeva 226). This catching up of the left hemisphere, as Crago puts it, increases our abilities to tell and process complex stories, but even at its peak stories attract us not because they are "explained." Rather, they remain right-hemispheric "showings" that appeal through powerful resonance with Old Brain emotions. As people grow old and their left hemisphere capacities either falter or lose their glamour, even healthy seniors tend to return to stories that resemble "the Old Brain-dominated narratives of early childhood" (Crago 192).

Throughout this lifelong process, the New Brain—and specifically its left hemisphere—builds a reference library of schemas and scripts, most of which are packaged as stories inasmuch as stories are the easiest to remember and retrieve. Schemas, scripts, and stories are both processes and contents of memory, the basic blocks of our understanding. Upon encountering a new event, children "immediately assume it to be a script" (Schank and Abelson 225). New objects, in turn, become schemas. In other words, "first experiences with objects [or events] tend to define [them] by establishing an initial script [... which] serves as the basis of an adult script that evolves from it" (225). The same applies to structuring and storing interpersonal experiences, which are organized largely by prototypical cognitive representations. "To know the meaning of the word such as *fear*," explains the social psychologist Beverly Fehr, "is to know a script in which events unfold in a particular sequence," where the script "contains prototypical antecedents, physiological reactions, facial expressions, behaviors, and so on" related to a given concept (193). Concepts important for interpersonal experiences—love, forgiveness, respect, closeness, jealousy, and others—become linked to specific scripts. Often called relational/interpersonal schemas—"cognitive structures representing regularities in patterns of interpersonal relatedness" (Baldwin 461)—these scripts inform interaction between self and other, response patterns, and procedural knowledge we unreflectively rely upon. If scripts play such a foundational role in structuring human understanding, can they tell us something about our understanding of justice and the stories we tell about justice? I believe they can.

JUSTICE SCRIPTS IN LITERARY AND FILMIC NARRATIVES

The English word justice has at least three meanings: a just state of affairs, a motivational stance or the intention behind a given action, and a virtue or the state of character. When theorized through the lens of cognitive science, all three meanings—and possibly others—are part of a cognitive domain. Justice as a domain is a system of circuits clustered around the lexical entry for the term "justice" and including our ideas about justice, our beliefs about justice, attitudes to justice, examples of justice, correlates of justice such as "fairness," "rightness," "law," and other related information. This information, like in other lexical entries, can be accessed and activated in a number of ways—from a meaning, from referent, from part of meaning, from association, or from other lexical entries, with which justice is linked. Thus the lexical entry "justice" can lead one from "crime" to "court" and then to "punishment," but also, in a set of scalar links, it can lead from "crime" to "mistake." In a complementary link, it can lead from "victim" to "offender"and in an antithetical link, it can lead from "justice" to "injustice." In an idiosyncratic link, it can lead from "punishment" to "unfairness" and then on to "Aunt Chrissie." Theorizing justice as a domain has at least two advantages. First, it reflects the breadth and affective power of justice in ordinary human understanding, especially when contrasted with the extremely limited, bug-on-the-pin legal definition of justice. Justice as a domain is necessarily a polyphonic concept, dialogical in the Bakhtinian sense as shaped in interaction with its alternatives, but also multiple and open-ended. Second, like any domain, justice has an internal structure (Hogan, *Cognitive* 43) defined by a list of features including fairness, apology, violence, hurt, legitimacy, desert, and other properties as well as relations among them. Because these features and associations are not important on their own but only in the various structured/lexical complexes of which they are part, our understanding of justice is shaped by exemplary, even stereotypical, causal chains of events that involve beliefs, motivations, and other mental states. Justice, in other words, is processed by our minds as a script.

In its most general sense, a justice script is a sequence of causally linked events stretched between what the narrating agent identifies as harm or injustice and what she identifies as resolution or redress. Justice scripts include goal-directed actions and observable behaviors, but they also encompass expectations about the thoughts, feelings, goals, and actions of relative parties, as well as the roles in which characters in a script are cast. For example, the opening scenes of *Shrek* cast him as a monster beyond the reach of Lord Farquaad's new social order but also explain the invasion of fairy-tale creatures into Shrek's swamp, which they rightly see as a place of refuge. Grasping this, although acting on a selfish motivation to simply get everyone out of his swamp, Shrek becomes the spokesman for the persecuted and undertakes a quest, in return for which Lord Farquaad promises that fairy-tale creatures will be allowed to live in the kingdom, subsequently leaving

Shrek alone. The persecution introduced at the outset is to be redressed by a promise of coexistence—or at least tolerance of otherness—but Shrek and Fiona's falling in love complicates matters. The unforeseen solution turns out to be the elimination of Lord Farquaad, whose demise brings on the collapse of the racist social order he sought to introduce. With a number of unexpected twists, *Shrek's* plot is far from stereotypical as a romance, yet it is also an actualization of the social justice script: a sequence of events with participant-oriented patterns of effort aimed at overcoming the Trouble of social injustice and concluding with a resolution that re-establishes diversity, if not equality. The social justice script used in *Shrek* can thus be analyzed through its principal actions, involving participants cast in specific roles. Each role is defined by goals and character motivation, which positions actors in the story as advancing or thwarting the achievement of justice. In other words, scripts inform narratives on the level of plot, themes, and characterization.

One obvious problem evident in the above example is its reductionism. To say that *Shrek* is a story about social justice seems to overlook the fact that it is primarily a romance (see Attebery, *Stories* and Crago), which playfully subverts the many fairy-tale, romance, and quest conventions. Just as a description of human behavior in terms of scripts and schemata seems too vague and generalizable to an almost meaningless level, much like Campbell's monomyth, so too a justice script, at its basic level, seems not effectively different from a Proppian story function: it can be applied to so many different situations that its usefulness is doubtful. This is a valid objection but only if one claims that a close reading of narratives through the lens of scripts is supposed to exhaust their meanings. I do not think so. The recognition or focus on a particular script does not invalidate other layers of the story but merely brings the layer in question up for a critical scrutiny. Each story is a thick and complex entity that has many other layers of meaning above and beyond its embedded scripts. My point is that scripts are fundamental structures for patterns of meaning processed by the reader's mind (see Hogan, *How* 26). And if excavating a script in a text or film seems too violent an imposition on its richness, it is worth taking a look at a genre where scripts and schemata are exposed as a story's exoskeleton, namely computer games, collectible card games (CCG), and trading card games (TCG).

Take *Magic, the Gathering* (since 1993). For all of its immense complexity, the first CCG in the world uses only seven card types, five types of magic, and fifteen "ability words"—types of abilities. One of the most popular card games played by adolescents today, it uses these card types in combinations that, amplified by chance, generate an almost infinite range of actions and effects. This unpredictable game is, therefore, highly structured. Power, toughness, mana, and other qualities are computable and translatable: for instance, paying x mana inflicts x damage or conjures an x creature. *Magic, the Gathering* is also structured through a 5-step turn order that repeats preparation and confrontation sequences until the opponent's life total is reduced to zero. Playing the game, in other words, amounts to enacting ever-new constellations within the same "battle script," with slots for setting, characters, goals, and means to achieve them. Most computer games are also unabashedly script-based: players can choose among a handful of character types, customize them, then gather tokens, and quest in mission after mission. If dialogues appear in the game, in most cases they are reducible to simple information transfer: no matter how randomly you choose your character's answer—usually out of the three options available—you are going to end up with one mission or another.

Obvious as these formulas are, young people do not find them boring. On the contrary: they feel challenged and empowered by various reenactments of the same script just like a child feels empowered while listening to the same story time and time again (see Crago). The growing worldwide market for CCGs (Turkay et al. 3701) and other highly formulaic forms of play are a living testimony to the incredible appeal of script-based storytelling. In this light, some critics' tendency to downplay the importance of formulae in narrative practice seems to me the legacy of the Romantic "originality" imperative, where the word "derivative" was an anathema. Given our awareness of the cognitive advantages of repetition and pattern recognition and of how stories draw on stories (see Attebery, *Stories* and Frank), such a position smacks of anachronism. Familiarity with scripts is a prerequisite for altering them and for developing various higher-level cognitive skills. For example, research on card and computer games suggests that they help players develop analytical thinking, negotiation and social manipulation, iterative design, but most of all "learn[ing] from their mistakes and us[ing] this as a strength in the future" (Turkay et al. 3703).

In a larger picture, the charge of reductionism will never be conclusively resolved. There is as much resistance to the idea of scripts regulating human cognitive processes as there are arguments for script-based understanding and hardwiring of the human brain. Narrowing justice to a stereotypical sequence of either-or steps and roles, as is the formula used in the "due process of law," has been criticized, and rightly so, as not just making *New Justice* legal systems operate less fairly than they should, but actually creating a system that is incapable of operating fairly at all. This, I believe, is a valid point but one that must be addressed by legal scholars and moral philosophers. Given how the human cognitive apparatus functions, I am convinced that we are hardwired for script-based understanding—also when it comes to justice. At the same time, given our cultural adaptability and intelligence, we are not bound to follow any particular script. We modify and create alternative scripts, thus changing which scripts are culturally residual, dominant, or emergent. In light of what I said in the previous chapter, I find reasons to believe that the rise of pluralistic *Open Justice*—a reaction to the rigid formalism and monologism of *New Justice*—is a move toward recognizing more justice types and more justice scripts. The reductionism inherent in law has, of course, been addressed by literary critics. It yields a picture in which literature emerges as law's counterweight, a discourse recording, as Dimock put it in *Residues of Justice*: all of these specific details of messy human lives that the clean abstractions of the *New Justice* law are unable to accommodate, leaving them unresolved, unredressed, and often unarticulated. This, of course, is true, for literature has long been a record of the residues of justice. Yet, fiction is not only a sedimentation of life's processes but, as the *Shrek* example suggests, an envisionment of alternatives too. In this larger focus, the script-based study of representations of justice in literary and filmic narratives makes sense for a number of reasons.

First, because the nature of the mind is that it understands by objectifying—and it objectifies through patterning—we construct and apply justice scripts in the same way we use scripts for gender relations, work, recreation, and other spheres of life. Since our understanding of justice is script-based, any justice-seeking activity involves enacting any one, or the combination of, several justice scripts. The extent to which an individual participates in the creation of a new script, enacts or modifies an existing one, is a complex question with different answers that will be discussed in the following chapters. The important point is that scripts are foundational to our thinking and behavior. Largely subliminal, justice scripts are unreflective behavioral protocols that offer procedural knowledge of the if-then nature. Similar to story skeletons or plot structures, justice scripts can be filled with specific details and still remain structurally the same. We participate in them by virtue of living a particular human life, and they become visible only upon examination.

Second, there is not one but many types of justice scripts. In this study, I identify six scripts associated with six types of justice—poetic justice, retributive justice, restorative justice, environmental justice, social justice, and global justice—but, obviously, this taxonomy can be expanded. I also propose that each script comes with its own specialized tracks. For example, "endangered species," "preservation," and "sustainability" are identified as tracks of the environmental justice script. "Freedom" and "rights" are tracks in the social justice script, whereas "feudal" and "transcendentalist" are two tracks of the poetic justice script. Other tracks and scripts can also be singled out, but the ones I named seem to be the most statistically prevalent, especially in literature and film for young people. In life as in stories, scripts rarely function in their pure form. In most cases they interact with and feed into one another, creating practical hybrids that blend elements of two or more scripts. Justice scripts can be studied through enactments in real-life situations—for example, in a teacher's reaction to student bullying or the public reaction to an event such as the Russian annexation of Crimea. They can also be examined in narrative genres, oral and written, historic and contemporary. This, of course, is where my interest lies. My point is that in both literary representations and real-world actualizations, justice operates as a cognitive script.

Third, given the reciprocal influence of scripts on stories and stories on scripts, looking at justice scripts in narrative fiction offers a comparativist perspective that helps explain dominant narrative patterns and mindsets in specific historical periods. It also sheds light on the evolution of and relationship between genres. Hurt and compensation, threat and response to threat, crime and punishment, reward or reconciliation, rights and obligations are among the most ubiquitous problems posed by the social world. If this makes the concern with justice a statistical human universal, conceptualizations of justice have been different in different periods and cultures. Narrative fiction is a record of that evolving difference and a carrier of justice scripts, which spreads and reinforces expectations related to these scripts. At the same time, when justice scripts are studied in filmic or literary narratives, they are always unique actualizations, with specific casts of characters, plot twists, and settings—all situated against the stereotypical set of expectations and event-links provided by a given script as a fill-me-in category. Each actualization is different, but all instantiations of a specific justice script will also share its key constitutive elements related to the focus, goal, stages, actors, and means necessary for the achievement of a given type of justice.

Fourth, seeing justice as a script makes sense within an evolutionary approach to fiction. If literature is an evolutionary adaptation, as Boyd has it, the fact that justice issues lie at the heart of most literary narratives cannot be accidental. As an evolutionary adaptation, fiction should suggest solutions to specific and recurrent problems that humans encounter in their multifarious modes of life. Inasmuch as the fundamental problem of social life is "free-riding, taking advantage of the benefits of cooperation without paying a fair share of the costs" (Boyd 303), justice scripts provided by fiction—especially those related to a fair distribution of burdens and benefits—offer cognitive advantages for dealing with problems of cooperation central to human social existence. This, of course, is not to say that stories present ready formulas for achieving justice or identifying injustice. They do, however, provide actable models and cognitive scripting no less effective than real-life learning about justice.

Last but not least, what allows positing justice as a cognitive script is not only a script-based theory of understanding but the very structure of human cognitive architecture. I am pretty sure that humans have a cognitive-affective program for *detecting injustice*. This program is a specialized complex, which feeds on a bundle of mechanisms. These include inference procedures for detecting "violations of conditional rules when these can be interpreted as cheating on a social contract" (Cosmides and Tooby, "Cognitive" 205) and the emotion detector and the empathizing system within ToM, which enable identifying hurt and carry "the adaptive benefit of ensuring that organisms feel a drive to help each other" (Baron-Cohen, "Empathizing" 473). Helping us detect injustice are also payoff-biased and conformist transmission mechanisms, which ensure social cohesion by stabilizing "cooperation and punishment ... through human populations" (Henrich and Boyd 88) and finally a universally mammalian sense of fairness, self-righteous indignation at having been treated unjustly, and a sense of "rightness" that informs understanding of reciprocity (Boyd, Singer, Panksepp and Biven). Whether this program for detecting injustice is a kind of inter-modular interaction due to cognitive fluidity between various domains, as Mithen could theorize it, or one of evolved human motives, as Carroll might see it, it is an open behavioral program, whose openness to contingent factors accounts for why *reactions to injustice* are culturally varied. As I argued in the previous chapter, our understanding of justice has continued to evolve, its scope and purpose modeled and remodeled through time and contexts in which it was applied. In the process, what has changed and will likely change in the future is the form and content of specific justice scripts, established as they are through cultural practice. What has not changed and is unlikely to change is the cognitive-affective systems and processes through which the human mind makes the concept of justice available to consciousness.

I am also willing to risk a statement that *all humans* engage in justice-seeking activities, be it trying to resolve family conflicts or seeking justice on the local, global, or international level. When they seek justice, humans do so by drawing on a universal cognitive process that structures ideas about justice in their minds called scripting. Yet—because the content of justice scripts is culturally and historically varied—there is no universal script for *achieving justice*. Whether or not there are mental adaptations underlying the human capacity for engaging in justice-seeking activities, I find reasons to believe that there are psychological adaptations for participating in such activities, even though they may be side-effects of other mechanisms; that there are motivational systems for allocating effort and energy in those activities; and that there are cognitive mechanisms—such as scripts—for spelling out and grasping principles of justice, monitoring compliance with those principles, and punishing those who violate them.

A corollary to all this, and another advantage of the evolutionary lens, is that it allows seeing different justice scripts as structures differently embedded in our biological and cultural history. Thus, out of the six justice scripts I have examined, the poetic, the retributive, and the restorative ones are largely Old Brain creations that draw their sap from Panksepp's primary affective processes, especially fear, rage, and care. The fact that even modern people are irresistibly, if embarrassingly, drawn to descriptions of bloodshed—especially bemoaned in the field of computer games—may thus be best contextualized through the grounding of retributive justice impulses in the primary, subcortical affective systems of fear and rage. As I demonstrate in the chapters that follow, the poetic and retributive justice scripts reflect assumptions inscribed in *Old* and *New Justice*, and so these scripts dominated Western perceptions of justice throughout the nineteenth century. The marginalized restorative justice script—an Old Brain construct but relegated to the nursery and then to women's and children's fiction—was able to claim wider legitimacy only in the second part of the twentieth century. The social, environmental, and global justice scripts, in turn, are New Brain, left-hemispheric constructs that gained prominence only in the twentieth century.

Another way to visualize the relationship between these six scripts is to see them as positioned in different places on the *Old-New-Open Justice* continuum. Poetic and retributive justice are essentially the two faces of *Old Justice*—available, respectively, to the powerless and the powerful. The fact that social justice is often taken today as encompassing the entire field of justice can be explained by its being the pinnacle aspiration of *New Justice*. The rise of plural and situated restorative, environmental, and global justice, in turn, must be linked with the emergence of *Open Justice* thinking, which challenged the monologic and absolute claims of *New Justice*. Last but not least, the debate between *New* and *Open Justice* can well be considered as a facet of the larger conflict between an increasingly mechanistic, fragmented, and decontextualized world created by "the unopposed action of a dysfunctional left hemisphere" (McGilchrist 6), and the holistic, organic, and in-touch-with-reality perception of the world as synthesized into a usable whole by the right hemisphere (176). The left-hemispheric coherence imperative is another version of quantifiability and equivalence posited as the ideal by *New Justice*. Yet, because everything can perfectly cohere only in the virtual world of the left hemisphere, *Open Justice* is an argument on behalf of right-hemispheric synthesized perception. After all, the right hemisphere is not just "more" in touch with reality than the left one. It is in touch with reality in an absolute sense, inasmuch as the left hemisphere operating alone is a closed, virtual world with no links to reality—something that can also be said of the well-meant, but virtual project of *New Justice* as fine-tuned by Rawls. It is no wonder that this conflict between *New* and *Open Justice*, between lifeless absolute order and organic though messy diversity, has been a staple of young people's literature and film since the 1990s. From *Shrek* (2001) and Lord Farquaad's ambition to create perfect order through the recently released *The Lego Movie* (2014), where Lord Business seeks to glue the world together, replacing creative unpredictability with a static and hierarchically organized whole, young people today are presented with different versions of the same cultural debate that rages on in all fields, including education, with its push for testability, measurability, and universal standards.

When all these arguments are brought together they help explain why justice scripts are not just a ubiquitous feature of our everyday experience, but why they are especially important in literary and filmic narratives for young people. Perhaps the most illuminating study that supports this conclusion is Roberta Seelinger Trites's *Disturbing the Universe* (2000). According to Trites, issues of power are central to adolescent and YA novels, not just because the roots of the modern YA novel lie in the social unrest of the 1960s (9), but also because the psychological essence of YA and adolescent engagement with narratives is based in developmentally determined interest in seeking empowerment against or within institutional and social repression (20). If Trouble is foundational to any story, the keenly registered by the young audience injustice, wrongness or hurt are sources of the most genuine conflict. They build narrative tension and create expectations about the story's resolution. In other words, the Trouble of injustice evokes scripts and creates learning situations, in which the reader tests her script competence against the challenges the story brings, including making predictions and value judgments.

The higher sensitivity to justice issues and justice scripts is also a marker of the cultural moment we are in now: heading toward a more multicultural and diverse world that is already here but could definitely be more just and equitable. It has been pointed out (see Dresang, Winograd, and Hais) that today's young people are significantly more comfortable than their parents and grandparents with race and gender range of characters they embody in computer games, watch on TV, or read about. They are more used to diversity and do not perceive a stable identity as a key principle of engagement with the story. The case of computer games and other multimedia best exemplifies this trend, where one can create, adopt, and change one's avatar's identity at will. One day playing as an old male dwarf, another day playing as a young female dragon, yet another playing as a blue-skinned time-traveling alien, the millennial generation are experts in switching identity within the story. Due to this mental flexibility—multimodal and interactive "playspace" literacy (see Mackey, "Case" 106)—young people today are more likely to appreciate diverse and inescapably situated perspectives on what justice is and what doing the right thing may mean to different parties. In other words, Millennials react to fiction differently than is assumed by studies supporting the assumption that readers identify most with characters and situations that closely represent their own category of identity: age, gender, ethnicity, and so on. This is certainly true in many cases, but the counterintuitive position seems just as valid, especially for adolescent reading. Teens appreciate the differences, novelty, and strangeness of a story. These qualities are what most stimulates the cognitive and affective engagement with, and learning from, texts (see Nikolajeva). Empirical research suggests that even when they see their relative identity categories reflected in a text, adolescents "seek a primary relationship with the esthetic form" (Blackford 12), moving beyond identity politics to larger philosophical questions, and engaging primarily with theme and form. To put it differently, Katsa's story in *The Hunger Games* appeals to boys and girls alike not because of Katsa's gender or social position, but because the plot features the kind of Trouble young readers are especially sensitive to: disempowerment by an oppressive system and resistance to the many injustices it perpetuates.

Another reason the justice scripts lens provides a valuable perspective—possibly more important to scholars and teachers rather than to young readers—is that it enables us to *name* important justice themes in the story and *identify* how different justice scripts are interwoven in the story. This includes guiding our attention to the competition among scripts for legitimacy in the story, illuminating what characters are motivated by what script—and with what consequences—as well as shedding light on how different justice scripts project the achievement of justice. Identifying justice scripts can be empowering because it helps establish a certain new type of story as a desirable/emergent model and it helps liberate us from other, perhaps still dominant types we no longer want to be part of. The ethically ambiguous yet excellent example of this mechanism is the success and continuing appeal of Disney's *Little Mermaid*. By jettisoning the poetic justice script, where wrongs are not righted in this world and the reader is expected to accept the imperfect reality of human lives fraught with misunderstandings and misdirected loves in favor of a regressive but appealing retributive justice script—where wrongs are righted, dreams become realities, and the evil is summarily punished—Disney has created an Old Brain narrative: a wish-fulfilling romance ending that appeals to adolescents even though it tragically simplifies the literary original and reinforces an oppressive masculinist ideology. By the same token, the success of the *Shrek* and *Ice Age* movies can be largely explained by their creative blending of the emergent social justice and the environmental justice scripts. Both franchises advocate diversity and plurality against monologic homogeneity, and both are positive about the possibility of addressing remediable injustices such as power imbalance, social exclusion, and racism. As is almost standard in contemporary narratives, all these movies condemn violence as a solution and question the validity of retributive justice.

This argument could be extended but I want to wrap up this chapter by summing up my main points. Scripts are memory and knowledge structures that become visible through statistical causality as a feature of our thinking and actions. The script-based patterning is also a fundamental feature of narrative sequences, literary and filmic alike. Any story at once depends on and transgresses standard patterns of action and expectation encoded as scripts. The greater the complexity of the story, the more its scripts are obfuscated—thus an attempt to bring them to analytical scrutiny appears reductive and simplistic (Hogan, *How* 26). Script formulism, however, is not a limitation but the very foundation of creativity (see Hogan, *How* and Harman, *Story*), which occurs when the standard pattern is enhanced or transformed.

Justice scripts play an especially important role in young people's literature and film. Scripts structure narrative sequences just as they structure our understanding of them, but they are not consciously created by authors and not consciously registered by readers or viewers. They are automatic processes of memory and understanding grounded in our affective-cognitive architecture. At the same time, even if the story-writing or reading flows effortlessly as a purely right hemisphere non-analytical activity, the left hemisphere's equally effortless activity does not lag behind. The left hemisphere automatically processes event strings into patterned narratives, whose elements resonate—through contrast or analogy—with the already stored patterns and categories: scripts and schemata. Both processes happen at the same time, making scripts a Schrödinger cat of the literary experience. In this superposition state, a script comes to exist only when it is analyzed and recedes into indeterminacy during the literary experience itself. Scripts, however, have observable effects. They inform our understanding of chains of events or plots that involve characters moved by emotions and beliefs. Inasmuch as scripts are both a structure of memory—what readers bring into the reading process—and an affective-cognitive process by which readers modify scripts to accommodate newly introduced ideas about justice, narratives both reflect and challenge justice scripts predominant in a given society. In both capacities, stories are a form of cognitive scripting as effective as real-life stories about justice but usually more emotionally intense. In the following chapter, I illustrate this process by taking a closer look at two tracks of the poetic justice script that are used to inform folk and fairy tales of pre-modern Europe.

NOTES

1. According to MacIntyre, "theories of justice and practical rationality confront us as aspects of traditions, allegiance to which required the living out of some more or less systematically embodied form of human life, each with its own specific modes of social relationship, each with its own canons of interpretation and explanation in respect of the behavior of others, each with its own evaluative practices" (*Whose* 391).
2. For example, see the essays collected in *Un-Disciplining Literature: Literature, Law, and Culture* (1999).
3. It is no coincidence that AI and cognitive psychology were twins born in the same year, 1956. For cognitive psychology that year saw the dethronement of behaviorism and the adoption of the information-processing paradigm, which reintroduced the mind and mental states back into psychology and stimulated a still ongoing research on how people develop mental concepts. *A Study of Thinking* (1956) by Jerome Bruner, Jacqueline Goodnow, and George Austin—which posited cognitive activity as dependent "upon a prior placing of events in terms of their category membership" (231) within pre-existing conceptual frames—was the first book that seriously took the notion of cognitive strategies. Yet, it was not until the late 1970s that a consensus emerged about the information-processing paradigm as the best way to study human cognition.
4. In this book Schank and Abelson approach knowledge structures by asking questions about the organization of memory, and then reasoning inductively toward understanding. They claim that the basis of everyday knowledge and human memory organization is episodic memory, "organized around personal experiences or episodes rather than around abstract semantic categories" (17). One of the principal components of memory, they argue, is thus "a procedure for recognizing repeated or similar sequences" (18) called scripting. Scripting,

in this picture, is an economy measure in the storage of episodes that tends to eliminate details, but enables people to remember types of events "in terms of a standardized generalized episode" or script (19).

5. According to Schank, humans are set up to understand stories, because only stories provide context necessary "to help them relate what they have heard to what they already know" (Schank, *Tell* 15). In this picture, the major processes of memory are "the creation, storage, and retrieval of stories" (16). While scripts are indeed the basic nature of human thought and understanding—"They serve to tell us how to act without our being aware that we are using them. They serve to store knowledge that we have about certain situations. They serve as a kind of storehouse of old experiences of a certain type in terms of which new experiences of the same type are encoded" (8)—they are not the final answer. Because people reason from experience, and because the effective use of memory calls for effective indexing, scripts tend to get bundled into stories: extremely useful information carriers that come with many indices. Human memory, Schank concludes, is a cluster of experiences labeled in complex ways to allow the retrieval of relevant information at the right time. While not all experiences are stories, stories are especially interesting because they are multiply-indexable.
6. These limitations—at least limitations from a literary scholar's perspective—can best be seen in two areas, which are the late offshoots of the AI phase: evolutionary cognitive psychology, focused on how human experience is represented in the mind, and cognitive linguistics, whose central concern has been on the relationship between the mind and language. Evolutionary cognitive psychology—exemplified in its flagship collection *The Adapted Mind* (1992)—argued that culture arises from evolved psychological mechanisms, that the human "adapted" mind functions through highly specialized mental modules (Tooby and Cosmides, "Psychological" 113), and the human cognitive architecture is primarily biological rather than cultural—claims that were made *post hoc* without specific scientific evidence and have largely been discarded. Cognitive linguistics, in turn, especially as represented in Narrative Comprehension Framework Theory (Emmott) and Text World Theory (Gavins), seems to me of limited use to literature scholars. Although both NCFT and TWT are indebted in schema-theoretic view of cognition (Emmott 15–36, 97–8) and recognize how contexts—textual and extra-textual—structure understanding of any discourse-based communication, their use of literary examples works on the level of single sentences, becomes wobbly when applied to paragraphs, and caves in under the weight of larger narrative units. For literary analysis the kind of information that NCFT seeks—for example, about at exactly which point in a sentence a particular inference is made—is between irrelevant and ridiculous. Also, cognitive linguistics' compulsion for elaborate categorizing—for example, in TWT a sentence "I didn't need to look [1], but I wondered what would happen next [2]" amounts to constructing of embedded enactor-accessible modal-worlds, first a negative deontic modal-world [1], followed by a an epistemic modal-world (Gavins 137)—is at best tangential and usually positively narcoleptic for literary analysis. Linguistic analyses of literary texts do illuminate some of the mechanisms underlying the creation of meaning, but they are usually not, in Schank's terms, good stories.
7. Although I will use the term script in psychology, education, and linguistics of the period, the preferred term was schemata. The term gained currency through

the pioneering studies in reading comprehension by the cognitive psychologist David E. Rumelhart in the mid-1970s but is traceable to studies of memory carried out by British psychologist Frederic E. Bartlett in the 1930s. Schema-theories contemporaneous to Schank's work explored scripts/schemas as concepts crucial for research in the fields of language and narrative comprehension, early education, discourse processing, studies of communicative or narrative competence, and related areas. As Rumelhart puts it, "all knowledge is packaged into units ... the schemata. Embedded in these packets of knowledge is, in addition to the knowledge itself, information about how this knowledge is to be used" (Rumelhart 34). Thus, schema-theoretic models have stressed the basic characteristics of understanding also posited by Schank: its multilevel and interactive character—meaning that various levels of knowledge structures are activated in the process and that each level can contribute input, in a heterahierarchical fashion, at various points of the understanding process—and its hypothesis-based processuality—meaning that in the process of constructing the most plausible interpretation of events, clues, or what Schank calls indices, are picked up, tested, and then rejected or accepted. For all that, schema-theorists focused mostly on text comprehension and only indirectly on how schemata influence the work of memory, goal setting, decision making, intelligence, and action in the physical world. These aspects, however, have been central to AI research, in which script has been posited as a cognitive protocol applied by individuals to structure understanding and guide action. For examples of schema-theoretic studies, see Roy E. Freedle's *New Directions in Discourse Processing* (1979), Rand J. Spiro, Bertram C. Bruce, and William F. Brewer's *Theoretical Issues in Reading Comprehension* (1980), or Alyssa McCabe and Carole Peterson's *Developing Narrative Structure* (1991).

8. Whereas schema-theories have been concerned with how preexisting knowledge is formed, stored, and applied in the understanding tasks, theories of narration developed in communication studies and rhetoric have drawn attention to how central storytelling is in generating knowledge and social reality. Two theories that have stressed the epistemic nature of storytelling and emerged during the AI phase include Ernest Bormann's Fantasy Theme Analysis and Walter R. Fischer's Narrative Paradigm Theory. Bormann's FTA offers a descriptive model of what communication activities, and how they do it, contribute to creating a common social reality that frames processes of understanding and sense making for its members. Although he does not use the terms script or schema, Bormann's communicative activities like "rhetorical visions" and "fantasy types" are clearly scripts with "dramatis personae and typical plot lines" (Bormann, "Vision" 398), and "stock scenario[s] repeated again and again by the same ... or ... similar characters" (Bormann, "Theme" 434). The epistemic dimension of narration has also been demonstrated in another ambitious proposal developed in the field of rhetoric, Walter A. Fischer's Narrative Paradigm Theory. Rather than taking storytelling as a way to construct social reality, as Bormann would have it, Fischer's NPT asserts that narrative reflects a more primal cognitive mechanism. For humans, Fischer says, the species-specific way of constructing, interpreting, and understanding events is narrative logic. It is through this logic that people come to believe and act. Humans for Fischer are *homo narrans* (Fischer 62)—a storytelling species—whose understanding and communication is possible only as a narration. "*[B]ehind* any structure that is *given to* human

communication," he asserts, "the perceptual framework of narration will always also be constraining and projecting meaning" (193, italics in the original).

9. To savor the nuances and terms that describe the various stages of the cognitive process, see especially Schank's *Tell Me a Story* (1990), Turner's *The Literary Mind* (1996), Herman's *Story Logic* (2002) and *Storytelling and the Sciences of Mind* (2013), as well as Hogan's *Cognitive Science, Literature, and the Arts* (2003), *Affective Narratology* (2011), and *How Authors' Minds Make Stories* (2013).
10. Memory, of course, is a complex superstructure with several subsystems: short-term versus long-term memory, semantic versus episodic memory, procedural, working, explicit, implicit, and other types of memory. For my purposes, however, these divisions are not important. For discussion of memory, see Schank, *Tell* 118–21, Hogan, *Cognitive* 13–30, 34–42, Panksepp and Biven 214–22.
11. See Langer's *Envisioning Literature: Literary Understanding and Literary Instruction.* (2011).
12. See Le Guin's collection *Dancing at the Edge of the World: Thoughts on Words, Women, Places* (1997), especially 165–170.
13. This claim has been articulated by many scholars. One argument is that "humans are not really set up to understand logic" (Schank, *Tell* 15) but stories. A less radical position is that the basic form of human logic is narrative rationality, "a storied context" that provides empirical, actable knowledge (Fischer 48). Yet another version of this is a claim that stories are "a chief means through which humans organize their complex motivational dispositions into a functional program of behavior" (Carroll, "Revolution" 43).
14. In D. G. Premack and G. Woodruff's "Does the chimpanzee have a theory of mind?"
15. In 1995 Simon Baron-Cohen's *Mindblindness: An Essay on Autism and Theory of Mind* demonstrated that autistic patients do not develop a ToM and thus established autism as a ToM disorder; recent studies have suggested similar connections between ToM malfunctioning and schizophrenia, as schizophrenics are unable to coordinate various types of information that are required for normal ToM processing. See, for example, Leiser and Bonshtein's "Theory of mind in schizophrenia: Damaged module or deficit in cognitive coordination?"
16. The capacity to realize that someone may have a different idea about something than what actually is the case develops gradually. By early adolescence humans can navigate through fourth-order constructions such as what A thought that B thought that C believed that D understood about E's thoughts. Fifth-order tasks exceed processing capacities of most adults and are suggested as the human processing limit. See, for example, Simon Baron-Cohen, O'Riordan, Stone, Jones, and Plaisted's "Recognition of faux pas by normally developing children and children with Aspenger syndrome or high-functioning autism" or Kinderman, Dunbar, and Bentall's "Theory of Mind deficits and causal attributions."
17. See especially Baron-Cohen's "The Emphathizing System: A Revision of the 1994 Model of the Mindreading System" in which he expanded his model of mindreading into the eponymous Empathizing System by adding to it two new components, The Emotion Detector, TED, and The Emphathizing SyStem, TESS. In this new model, TESS "is the real jewel in the crown" in that "it carries with it the adaptive benefit of ensuring that organisms feel a drive to help each other" (473). The modular neurocognitive Empathizing System is foundational for

social interaction and consists of six integrated subsystems: ID, intentionality detector which represents behavior in terms of volitional states; EDD, eye direction detector that detects the direction of the gaze and appraises what the eyes are looking at; SAM, a shared attention mechanism, which allows an individual to be aware of whether they attend to the same object or event as others; TED, The Emotion Detector, that provides information about affective states; TESS, The Empathizing SyStem, that allows an empathic reaction to another's emotional state; and finally ToMM, Theory of Mind Mechanism, that integrates input from the other components and develops a coherent mental model.

18. For the discussion of the capacity for cognition of cognition, see, for example, Smith, Shields, and Washburn's "The comparative psychology of uncertainty monitoring and metacognition." For an extended discussion of the relationship between ToM and metarepresentations, their levels, and role, see Josef Perner's *Understanding the Representational Mind.*
19. See Evans and Wellman's "A case of stunted development? Existential reasoning is contingent on a developing theory of mind."
20. See, for example, Jesse M. Bering's "The folk psychology of souls."
21. For a concise discussion of emotion processing as it applies to perception, in life and literature, see Hogan, *Cognitive* 168–190. Damasio's results have also been confirmed and significantly expanded by Panksepp and Biven.

WORKS CITED

Baldwin, Mark W. "Relational schemas and the processing of social information." *Psychological Bulletin* 112.3 (1992): 461–484.

Baron-Cohen, Simon, Michelle O'Riordan, Valerie Stone, Rosie Jones, and Kate Plaisted. "Recognition of faux pas by normally developing children and children with Aspenger syndrome or high-functioning autism." *Journal of Autism and Developmental Disorders* 29 (1999): 407–18.

Baron-Cohen, Simon. *Mindblindness: An Essay on Autism and Theory of Mind.* Cambridge, MA: MIT P, 1995.

Baron-Cohen, Simon. "An Empathizing System: A Revision of the 1994 Model of the Mindreading System." *Origins of the Social Mind: Evolutionary Psychology and Child Development.* Eds. Bruce J. Ellis and David F. Bjorklund. New York: Guilford 2005: 468–492.

Bering, Jesse M. "The folk psychology of souls." *Behavioral and Brain Sciences* 29.5 (2006): 453–498.

Blackford, Holly Virginia. *Out of this World: Why Literature Matters to Girls.* New York: Teacher's College Press, 2004.

Bormann, Ernest. "Fantasy Theme Analysis and Rhetorical Theory." *The Rhetoric of Western Thought.* Eds. Golden, James L., Goodwin F. Berquist, and William E. Coleman. Dubuque, IA: Kendall-Hunt Publishing, 1983: 433–448.

Bormann, Ernest. "Fantasy and Rhetorical Vision: The Rhetorical Criticism of Social Reality," *Quarterly Journal of Speech* 58 (1972): 396–407.

Boyd, Brian. *On the Origin of Stories: Evolution, Cognition, and Fiction.* Cambridge, MA: Belknap Press, 2009.

Bruner, Jerome, Jacqueline Goodnow, and George Austin. *A Study of Thinking.* New York: John Wiley & Sons, 1956.

Carroll, Joseph. "The Human Revolution and the Adaptive Function of Literature." *Philosophy and Literature* 30.1 (2006): 33–49.

Cosmides, Leda and John Tooby. "The Psychological Foundations of Culture." *The Adapted Mind: Evolutionary Psychology and the Generation of Culture.* Eds. Jerome H. Barkow, Leda Cosmides, and John Tooby. New York: Oxford UP, 1992: 19–136.

Cosmides, Leda and John Tooby. "Cognitive Adaptations for Social Exchange." *The Adapted Mind: Evolutionary Psychology and the Generation of Culture.* Eds. Jerome H. Barkow, Leda Cosmides, and John Tooby. New York: Oxford UP, 1992: 163–228.

Crago, Hugh. *Entranced by Story: Brain, Tale and Teller from Infancy to Old Age.* New York: Routledge, 2014.

Damasio, Antonio R. *The Feeling of What Happens: Body and Emotion in the Making of Consciousness.* New York: Harcourt Brace, 1999.

Dimock, Wei Chi. *Residues of Justice: Literature, Law, Philosophy.* Berkeley and Los Angeles: U. of California P., 1996.

Dresang, Eliza. *Radical Change: Books for Youth in a Digital Age.* New York: Wilson, 1999.

Emmott, Catherine. *Narrative Comprehension: A Discourse* Perspective. New York: Oxford UP, 1996.

Evans, E. Margaret and Henry M. Wellman. "A case of stunted development? Existential reasoning is contingent on a developing theory of mind." *Behavioral and Brain Sciences* 29.5 (2006): 471.

Fehr, Beverly. "The Role of Prototypes in Interpersonal Cognition." *Interpersonal Cognition.* Ed. Mark W. Baldwin. New York: Guilford Press, 2005: 180–205.

Fischer, Walter A. *Human Communication as Narration: Toward a Philosophy of Reason, Value, and Action.* Columbia, SC: U. of South Carolina P., 1987.

Frank, Arthur W. *Letting Stories Breathe: A Socio-Narratology.* Chicago: The U. of Chicago P., 2010.

Gavins, Joanna. *Text World Theory: An Introduction.* Edinburgh: Edinburgh UP, 2007.

Gopnik, Alison. "Theory of Mind." *The MIT Encyclopedia of the Cognitive Sciences* (MITECS). Eds. Robert A. Wilson and Frank Keil. Cambridge, MA: MIT Press 1999. https://cognet-mit-edu.ezaccess.libraries.psu.edu/library/erefs/mitecs/gopnik.html. March 12, 2014.

Haraway, Donna. "Situated Knowledges: The Science Question in Feminism and the Privilege of Partial Perspective." *Feminist Studies* 14.3 (1988): 575–599.

Henrich, Joseph and Robert Boyd. "Why People Punish Defectors: Weak Conformist Transmission Can Stabilize Costly Enforcement of Norms in Cooperative Dilemmas." *Journal of Theoretical Biology* 208.1 (2001): 79–89.

Herman, David. *Story Logic: Problems and Possibilities of Narrative.* Lincoln, NE: U. of Nebraska P., 2002.

Herman, David. *Storytelling and the Sciences of Mind.* Cambridge, MA: MIT Press, 2013.

Heyes, Cecilia M. "Theory of Mind in Nonhuman Primates" (and commentators' discussion). *Behavioral and Brain Sciences* 21.1 (1998): 101–48.

Hogan, Patrick Colm. *How Authors' Minds Make Stories.* New York: Cambridge UP, 2013.

Hogan, Patrick Colm. *Cognitive Science, Literature, and the Arts: A Guide for Humanists*. New York: Routledge, 2003.

Hogan, Patrick Colm. "Fictive Tales, Real Lives: Problems with Reading Law as Literature." *Un-Disciplining Literature: Literature, Law, and Culture*. Eds. Kostas Myrciades and Linda Myrciades. New York: Peter Lang, 1999: 271–290.

Kinderman, Peter, Robin Dunbar, and Richard P. Bentall. "Theory of Mind deficits and causal attributions." *British Journal of Psychology* 89.2 (1998): 191–204.

Langer, Judith A. *Envisioning Literature: Literary Understanding and Literary Instruction*. New York: Teacher's College Press, 2011.

Le Guin, Ursula K. *Dancing at the Edge of the World: Thoughts on Words, Women, Places*. New York: Grove Press, 1997.

Leiser, David and Udi Bonshtein. "Theory of mind in schizophrenia: Damaged module or deficit in cognitive coordination?" *Behavioral and Brain Sciences* 26.1 (2003): 95–6.

MacIntyre, Alasdair. *Whose Justice? Which Rationality?* Notre Dame, IN: U. of Notre Dame P., 1988.

Mackey, Margaret. "The Case of Flat Rectangles: Children's Literature on Page and Screen." *International Research in Children's Literature* 4.1 (2011): 99–114.

McGilchrist, Iain. *The Master and His Emissary: The Divided Brain and the Making of the Western World*. London: Yale UP, 2009.

Mithen, Steven. *The Prehistory of the Mind: The Cognitive Origins of Art, Religion, and Science*. New York: Thames and Hudson, 1996.

Nikolajeva, Maria. *Reading for Learning: Cognitive Approaches to Children's Literature*. Amsterdam and Philadelphia: John Benjamin's Publishing Company, 2014.

Nussbaum, Martha. *Poetic Justice: The Literary Imagination and Public Life*. Boston: Beacon Press, 1995.

Panksepp, Jaak and Lucy Biven. *The Archeology of Mind: Neuroevolutionary Origins of Human Emotions*. New York: W.W. Norton, 2012.

Perner, Josef. *Understanding the Representational Mind*. Cambridge, MA: MIT Press, 1991.

Premack, D. G. and G. Woodruff. "Does the chimpanzee have a theory of mind?" *Behavioral and Brain Sciences* 1 (1978): 515–526.

Rumelhart, David E. "Schemata: The Building Blocks of Cognition." *Theoretical Issues in Reading Comprehension*. Eds. Rand J. Spiro, Bertram C. Bruce, and William F. Brewer. Hillsdale, NJ: Lawrence Erlbaum Associates, 1980: 33–58.

Schank, Roger. "Conceptual Dependency: A Theory of Natural Language Understanding." *Cognitive Psychology* 3/4 (1972): 552–631.

Schank, Roger. *Conceptual Information Processing*. New York: Elsevier 1975.

Schank, Roger and Robert Abelson. *Scripts, Plans, Goals, and Understanding: An Inquiry into Human Knowledge Structures*. Hillsdale, NJ: Lawrence Earlbaum Associates, 1977.

Schank, Roger. *Dynamic Memory: A Theory of Reminding and Learning in Computers and People*. New York: Cambridge UP, 1983.

Schank, Roger C. *Tell Me a Story: Narrative and Intelligence*. Evanston, IL: Northwestern UP, 1995.

Singer, Peter. *The Expanding Circle: Ethics, Evolution, and Moral Progress*. 2nd ed. Princeton, MA: Princeton UP, 2011.

Slingerland, Edward. *What Science Offers the Humanities*. New York: Cambridge UP, 2008.

Smith, J. David, Wendy E. Shields, David A. Washburn "The comparative psychology of uncertainty monitoring and metacognition." *Behavioral and Brain Sciences* 26.3 (2003): 317–73.

Trites, Roberta Seelinger. *Disturbing the Universe: Power and Repression in Adolescent Literature*. Iowa City: U. of Iowa P., 2000.

Turner, Mark. *The Literary Mind*. New York: Oxford UP, 1996.

Turkay, Selen, Sonam Adinolf, and Deveyani Tirthali. "Collectible Card Games as Learning Tools." *Procedia - Social and Behavioral Sciences* 46 (2012): 3701–705.

Wilson, Edward O. *Consilience: The Unity of Knowledge*. New York: Alfred A. Knopf, 1998.

Winograd, Mortley and Michael D. Hais. *Millenial Momentum: How a New Generation Is Remaking America*. Piscataway, NJ: Rutgers UP, 2011.

Zunshine, Lisa. *Why We Read Fiction: Theory of Mind and the Novel*. Columbus, OH: The Ohio State UP, 2006.

Zunshine, Lisa. "Theory of Mind and Experimental Representations of Fictional Consciousness." *Introduction to Cognitive Cultural Studies*. Ed. Lisa Zunshine. Baltimore, MD: The John Hopkins UP, 2010: 193–213.

3 The World Is Not Fair

Poetic Justice Scripts

Sometime in the mid-1730s a workers' revolt occurred in a printing shop in Paris. The apprentices felt they had suffered abuse. They lived in a room that was freezing cold, ran errands from dawn to dusk, and received nothing but slop to eat. The apprentices' lives were also made miserable by cats. The master's household was full of pampered cats, and the neighborhood teemed with wild alley cats. These would howl on the roof above the boys' bedroom, making it impossible to get a full night's sleep. Seeking to right this inequitable state of affairs, the apprentices took to meowing over the master's bedroom, until he told them to get rid of the cats. Using their master's order as an excuse, the boys quietly killed the household cats. Then they openly massacred dozens of the alley cats, subjecting them to a mock trial and finally hanging them at a makeshift gallows (Darnton 77). The revolt and the cat massacre were described in a number of pamphlets that circulated around France and were reenacted across the country as the best of jokes.

Describing this event in *The Great Cat Massacre* (1984), the social historian Robert Darnton speculates about what made this story so hilarious to the eighteenth-century French. He concludes that our inability to see the cat massacre as a joke suggests the mental distance that separates us from the workers of preindustrial Europe. I believe it also does something more. It sheds light on the affective-cognitive perception of justice, structured within the *Old Justice* framework, that was a standard action protocol for most interpersonal relations across the social strata in pre-modern Europe. This standard protocol was the poetic justice script: a reaction to injustice or abuse at the hands of one's superiors or of the social system that dulled or rechanneled the outrage onto something other than the actual cause or agents of the abuse. The apprentices enacted the poetic justice script, because it was the only justice script available to those whose lower place in the social hierarchy denied them any claims against higherpositioned offenders, such as their master or even his cats. Although the poetic justice script seems to be an indirect form of retributive justice, it differs from the retributive justice script in its acknowledged inability to end the injustice. Although the "revolt" did take the boys' minds off the everyday abuse they were enduring, it did not end or *even hope* to end the mistreatment. Freezing sleeping quarters, long working hours, sloppy food, abuse at the hands of journeymen and the master, even the eventual sleepless nights as the cats returned to the roofs and to the master's household—all of these continued unchallenged. In the world of Jerome and Léveillé, mistreatment resulting from inequality was the galling default and justice was unclaimable.

This unclaimability is the key marker of poetic justice. In my use of the term, poetic justice is justice denied, where the victim's harm is not recognized as such but as a natural state of things. This definition differs from Nussbaum's theorization of poetic justice as justice mediated through literature (see Nussbaum, *Poetic*) and from the traditional definition of poetic justice as a feature of the narrative, in which virtue is ultimately rewarded and vice is ultimately punished—a structure I identify as the transcendentalist track of the poetic justice script. Rather, I see poetic justice as an inherent feature of the *Old Justice* socio-economic complex in which might makes right and justice is endorsed as hierarchical, absolutist, and gradational; to be had, if at all, by the "deserving," depending on where one is placed on the social hierarchy. The poetic justice script in this understanding was the product of the social realities and power relations that defined pre-modern and early-modern Europe. It was the action- and thought-protocol for the powerless, for whom justice was unachievable. This script was a cognitive programming for everyday, existential confrontations with people and institutions that had absolute power over one's life. In the *Old Justice* paradigm, the mirror image of the poetic justice script was the retributive justice script. It was available for the powerful as the right to punish those below them and will be examined in the following chapter.

My proposal in this chapter is that the poetic justice script comes in the feudal and transcendentalist tracks. The feudal track is the pre-Enlightenment, traditional form of poetic justice that denies injustice; the transcendentalist track is its post-Enlightenment manifestation that acknowledges injustice but defers redress. Their effects are the same. Both support the *Old Justice* idea of justice as the exercise of power by those who wield it, the kind of Hobbesian-Benthamite-Nietzschean line of reasoning that explains justice away as "it is what they say it is." Since the *Old Justice* paradigm is largely extinct today, so too is the poetic justice script. Evidence of the poetic justice script, however, can be found in pre-and early-modern folk and fairy tales, which I examine as taproot texts for contemporary YA speculative fiction.

OLD JUSTICE REALITY, OLD WIVES' TALES, AND THE POETIC JUSTICE SCRIPT

The poetic justice script was the basic action-protocol for most Europeans through the late eighteenth century and in some parts of Europe throughout the early twentieth century; in some parts of the world, where the hierarchical *Old Justice* mindset is still a dominant paradigm, poetic justice remains a sad reality even today. In Europe the ubiquitous influence of this script was derived from the many centuries of social, economic, and political realities shaped by feudalism and then absolutism. In a world shaped by the *Old Justice* mindset, justice was an elusive and highly arbitrary concept not only for the masses but even for the nobility[1]. Although they thrived by exploiting the socially inferior, the nobility also lived under constant fear of being victimized by the arbitrary decisions of ruthless monarchs. While codified law was becoming more elaborate, in all too many cases justice was an expedient, authoritarian device used to suit the political and economic goals of the powerful. This perception was reflected in Hobbes' reasoning that violence is the human norm, and the only way to avoid a total war is an authoritarian government.

In the religious and social realms, justice was likewise unachievable by being largely unclaimable. What are considered abuses today were then seen as unavoidable facts of life. To confirm that perception was the survivalist function of the poetic justice script: a function scholars recognize today by acknowledging that traditional folk and fairy tales are records of "psychological defense [mechanisms] and means of emotional survival" (Haase 362). The therapeutic function of the poetic justice script, in turn, was activated when outrage against mistreatment could no longer be contained. It consisted of re-channeling simmering rage into an outburst against something other than the actual cause or agents of the injustice. This function, grounded in the fear and anger circuitry of the Old Brain, is illustrated by the great cat massacre and provided a template cathartic mechanism for the mass displacement of rage.

The dawn of the Enlightenment changed this paradigm. Although *New Justice* failed to immediately dislodge the cluster of ideas about justice as either unachievable or hierarchical, it exposed the poetic justice script as obsolete. Soon after the American and French revolutions—themselves products of the *New Justice* thinking—the pre-Enlightenment, feudal track of the poetic justice script was jettisoned in favor of the transcendentalist track, which became an upgraded though uneasy norm for most of the nineteenth century. The entire poetic justice script, however, was doomed once the practices that had long been assumed sad but natural contingencies of life were suddenly exposed as socially constructed injustices. Identified by the late eighteenth-century philosophers and revolutionaries as denials of inalienable rights to life, freedom, and the pursuit of happiness, the practices and ideological foundations of *Old Justice* continued throughout the nineteenth century, fuelling such oppositional ideologies as Chartism, Marxism, socialism, abolitionism, and emerging nationalisms. Resistance to the injustice of arbitrary power had, of course, its antecedents. By the late eighteenth century, for example, it had long been enshrined as the cornerstone of Anglo-American constitutionalism[2]. In many of its applications, though—for example, the enfranchisement of women or decolonization—*New Justice* was a new mindset and remained a contested one at least until the adoption of the Universal Declaration of the Human Rights in 1948. By the time modern fantasy took shape in the works of Lewis and Tolkien, the resistance to arbitrary power had become one of the key themes of Anglo-American speculative fiction. In twentieth-century fantasy, dystopia, and science fiction the plots have tended to be tales of resistance against some arbitrary power. The protagonists are the rebels, never the empire, and they actively seek the achievement of justice.

This was not the case in most folk and fairy tales throughout the late nineteenth century. In these stories, ontologically concerned "with exploitation, hunger, and injustice familiar to the lower classes in pre-capitalist [and early-capitalist] societies" (Zipes, *Breaking* 6), there was no justice to be achieved. The ethics promoted in these narratives is dubious, in as much as both good and evil actions or characters are so not because of their moral character, but because of the roles in which they are cast by the plot. Since folk and fairy tale protagonists usually achieve their goals through immoral means, there is "very little justice" in these narratives (Nikolajeva 183). In folktales there is, at best, a consolation of survival and some occasional triumphs through trickstery. In fairy tales there are miraculous deliverances that redress injustices, usually for the royally-deserving, but the magic nature of those occurrences only confirms that they are not to be counted upon in real life. Although *New Justice* did introduce a non-arbitrary and non-hierarchical understanding of justice, it took a century before the idea and its practical implications seeped down into the collective consciousness of ordinary people, began to shape their dreams, and influence the stories they told. The bulk of pre-twentieth-century folk and fairy tales continued to reflect one or the other track of the poetic justice script.

In his discussion of the universal genres, which he defines by plot prototypes, Hogan claims that each prototypical plot type is defined by the goals characters pursue, where goals are identified with emotional and motivational systems that underlie them (Hogan, *How* 29). This focus applies to all justice scripts but is especially helpful in distinguishing tracks within each script. The feudal and transcendentalist tracks reflect different emotional and motivational systems that constitute distinct compositional stages of the poetic justice script as it existed before the arrival of *New Justice* thinking and then in the first century thereafter. Partly due to this chronology, the older, feudal track is to be found mostly in pre-modern folktales. Alternated occasionally with elements of retributive justice, it is the continuator of the hierarchical-absolutist conceptualizing of justice that also informs myths and epics. In these earlier forms, poetic and retributive justice scripts coexisted in varying degrees, always according to the principle that retribution belongs with the nobility, whereas poetic justice—the suck-it-up-and-move-on pattern—is the fate of a common person, an ethnic or religious alien, a woman, a child, or an outcast.

The transcendentalist track, in turn, emerges as an underlying structure in most literary fairy tales written in the nineteenth century. In cognitive-evolutionary terms, this track emerged as a result of the process Fauconnier and Turner call conceptual blending (17–38), when the mental space or lexical content of the feudal track was blended with two new and mutually reinforcing mental spaces: Christian evangelical ideas of salvation in the next world, and the Enlightenment promise of salvation in this world. The emergent quality—one that was absent in the feudal track—is that the transcendentalist track decidedly cues the audience to interpret various kinds of inequality *as* injustices. However, whereas the feudal track denies redress, the transcendentalist track defers it, keeping justice *de facto* unachievable. It offers instead a promise that injustice suffered in the present will be made right in the future. It will be made right when the rational society will eventually create perfect institutions or when those who abuse others will undergo a change of heart. Alternatively, the injustice will be made right when death removes the sufferer to a better world in heaven. In effect, the transcendentalist track is one of resigned acceptance of the ongoing inequality that affirms justice as unachievable. If, as Fauconnier and Turner claim, cultural practices "offer us methods for setting up a blend" (72), the late eighteenth-century emergence of speculative fiction and then nineteenth-century explosion of the literary fairy tale can surely be seen as a space where the new blend of the transcendentalist justice could be created and run for specific cases. In terms of cultural space, both tracks were tied to specific social and political contexts: the feudal track deriving from the feudal social conditions; the transcendentalist one, from the bourgeois world of nineteenth-century capitalism.

To claim that the feudal track informs the folktale and the transcendentalist track underlies the fairy tale begs the question how the two genres differ. The distinction between the folktale and the fairy tale is at the same time rather obvious and deeply problematic. It is obvious in as much as there is a marked difference between an oral story that circulates in its multiple versions, transforming constantly from one telling to another, and a story that is written down in one specific version, stamped by particular social and cultural circumstances of its creation and the personality of its author. The distinction is also problematic, though, inasmuch as not all oral tales are folktales and not all literary tales are fairy tales. There are problems with all existing taxonomies—Aarne's index, Propp's system, even with the traditional folkloristic division between myth, legend, and folktale[3]. What does seem certain is that oral folktales and literary fairy tales coexisted for centuries and interacted in complex ways (see Zipes, *Tradition*). Since the great majority of people until the mid-nineteenth century were illiterate, pretty much everyone participated in oral traditions, either as a listener or as a teller. Motifs and plots from folktales were absorbed into fairy tales but the transmission went both ways. Against what Ruth B. Bottigheimer sensationally claims in her *Fairy Tales: A New History* (2009), there *was* a tradition of oral fairy tales[4]. Even those literary fairy tales that have not been adaptations of oral folktales tended to dribble down to the illiterate and semi-literate audiences and melt into the pool of oral tradition—a process well documented in folkloristic, anthropological, and literary studies (see Dégh, Zipes, *Spells*; *Tradition*; and *Why* 47–8). Literary fairy tales were also transmogrified through abridgments into versions distributed in chapbooks and were thusly appropriated by oral storytellers, the literary becoming the source for an oral tradition (Zipes, *Tradition* xxi). This complex interaction has continued until the present, although in the last decades it has been more of an intertextual dialogue between texts—reworkings, revisions, subversions—rather than between literary texts and an oral tradition.

The many problems of definition explain why a large body of folk and fairy tale studies exists where the distinction between the two categories is seen as superfluous. Semioticians (Propp), social historians (Darnton), psychologists (Bettelheim), folklorists (Lüthi)[5], even literary scholars such as Bottigheimer have collapsed the two categories, referring to "the folk or fairy tale, the *Märchen*" as one broad genre (Bottigheimer, *Bad* 8)[6]. There is, on the other hand, another tradition that upholds a distinction between folk and fairy tales, the line between the two genres being a function of their media and the projected audience's social class. This argument posits that the folktale "is part of a *pre-capitalist people's oral tradition*" (Zipes, *Breaking* 25) that records the struggles of their everyday lives but also dreams of a better future. The fairy tale, by contrast, "is of *bourgeois coinage* and indicates the advent of a new literary form which appropriates elements of folklore to address and criticize the aspirations and needs of an emerging bourgeois audience" (25).

Grasping the intended audience's social class—with its living conditions, cultural modes, and dominant concerns determining the focus and main themes of folk and fairy tales—is crucial for understanding the difference between stories that employ the poetic justice script in its feudal or transcendentalist tracks. Across Continental Europe until WWI, class membership was the key determinant of one's capabilities in the sense of the term used by Sen and Nussbaum: ways of doing or being that people managed to achieve. In other words, for about a century after the emergence of the *New Justice* egalitarian thinking, the *Old Justice* hierarchical social structures were still firmly in place, especially outside of urban areas. Class and other hierarchies were, of course, foundational scaffolds of social and cultural life before the nineteenth century. Thus folktales, told among the illiterate or semi-literate common people, reflected life circumstances and emotional-motivational systems of the mostly non-urban people who recounted them. Fairy tales, originating in the literate world of the urban middle class, registered dreams and aspirations of the bourgeois city-dwellers, including their romantic visions of the pastoral countryside. Although there was much narrative cross-breeding, each category retained elements of the belief system and social realities of their primary audience.

Obviously, not all folk and fairy tales contain justice issues. Most of them, however, pose ethical questions and focus on pivotal moments in protagonists' lives. The protagonists' choices invite or cue the audience to evaluate these choices in terms of right or wrong. Exploring practical but also ethical consequences of characters' choices makes the reader confront issues of practical justice. Is it acceptable to steal from a giant or lie to someone who means you harm? Will the protagonist regret her choices? Are the characters aware of the consequences of their actions? In many cases, folk and fairy tales revolve around certain social impossibilities, such as the trope of a commoner marrying a princess, or around "unreachables" such as justice. When the folktale ends happily, whatever triumphs the folk hero may have achieved are always seen as temporary and isolated. The happy endings in fairy tales, in turn, capture the dreams of a more upwardly mobile middle class, but again these magical happy endings are so attractive precisely because they embody the intersection of specific, real-life impossibilities and their achievement in fantasy[7]. In this sense, folk and fairy tales through the late nineteenth century were structured on the poetic justice script that projected justice as unachievable. The feudal track framed most folktales through the early nineteenth century; the transcendentalist track exploded in fairy tales in the 1830s and remained dominant until the end of the century.

THE POETIC JUSTICE SCRIPT IN FOLK AND FAIRY TALES

For a long time the only form of the poetic justice script was the feudal track. This version of the poetic justice script was the thought- and action-protocol for many generations of commoners in pre- and early-industrial Europe. Its function was survivalist and therapeutic. This action-protocol prompted one to either ignore one's perception of injustice or to convert the resultant outrage into outbursts against something or someone other than the actual cause or agents of the injustice. Arbitrary feudal justice was not only part of commoners' everyday experience but was also reinforced as a cognitive script in all European folktales that register and take for granted what today is identified as social inequality. Informed by the feudal track of the poetic justice script, folktales project justice as unachievable because it is unclaimable. In this sense they can be seen as "historical documents" (Darnton 13) that exhibit the terrifying irrationality of the early modern peasantry and the terrifyingly oppressive world of the feudal social system the peasants inhabited[8].

The folktale was a product of those conditions. French and German villages were still very much politically feudal until the end of the eighteenth century; economic feudalism thrived in most of Europe well into the nineteenth century (Ganshof 168; Strayer 68). As late as 1789, the most advanced country of Continental Europe, France, was parceled out into manors where peasants were weighed down with multiple, onerous burdens. Although there were differences among provinces regarding the breadth and extent of seigneurial rights, most peasants could not change their dwelling-places, occupations, or nearly anything else about their lives. They were subjected to never-ending lists of dues and fees, most of which were outright extortions (Sydney 10–31) and some of which included other forms of social obeisance. In some provinces, even after the *formariage* fee was paid, the bride was "obliged to dance for the amusement of the *seigneur* or his agents" (10). Additionally, seigneurs had monopolies on the means of production—the mills, the bakehouses, the presses for wine and oil, and so on. They enforced those monopolies, often cheating villagers of their grain, oil, and wine. The peasants could not fish, hunt, keep pigeons, or kill any game that ravaged their crops. If this was not enough, seigneurs also had tremendous judicial powers. Their courts handled about nine-tenths of the total number of cases (34). Peasants thus lived their lives subject to courts of law that were private property and enforced the rights of seigneurs as well as the mechanisms of agrarian feudalism. Attempts to carry peasant grievances against the seigneur to the royal court were suicidal and ended in the utter ruin of those who dared claim any rights (51).

These interlocking levels of oppression, augmented by chronic poverty, made peasants' lives very grim. Men and women labored from dawn to dusk, yet much of their harvest was drained from them. The great majority lived in a state of chronic malnutrition; as demonstrated in Piero Camporesi's *The Magic Harvest* (1998), in Italy and some other regions of Europe malnutrition of the lower classes was the norm throughout the late nineteenth century. Children had to work almost as soon as they could walk (Darnton 29). Due to spousal deaths, marriages lasted on average fifteen years, so families were broken and patched (27). Children and infant deaths were common: one out of two children born in eighteenth-century France died before the age of ten. Countless infants were smothered by their sleeping parents, as whole families slept together in one bed or a single room, often surrounded by livestock to keep warm (28). As Darnton comments, "[t]he human condition has changed so much since then that we can hardly imagine the way it appeared to people whose lives really were nasty, brutish and short" (29).

The social realities of this extremely brutal world—where inexorable, unending toil, degrading hunger, angry stepmothers, and abandoned children were an everyday reality—can also be found in European folktales. Far from representing—as Jungian and Freudian psychologists would have it—immutable inner realities of human life, the folktale described the grim conditions of pre-modern life and the resultant mentality that life produced[9]. The gritty reality of these stories, repulsively stimulating to a twenty-first-century suburbanite, was a realistic background for the miraculous triumphs of luck or cunning that held its original audience in rapture. On the cognitive-affective level, the folktale helped peasants navigate their world. Folktales reflected the essence of how their lives were being lived and captured their dreams of full bellies, prosperity, and safety. Yet, folktales also demonstrated the folly of expecting anything more than cruelty from a harsh social order. The essence of this social order was competition for power among the nobility and competition between the nobility and the peasantry. This never-ending struggle was the reality of late-feudal European social order and the dominant theme of pre-modern folktales. This focus on "might makes right" has been defined by Zipes as a set of assumptions that power is everything, that class and power determine social relations, that morals are nonexistent or irrelevant, and that human life is confined only to this world (Zipes, *Breaking* 29–30). These assumptions inform the feudal track of the poetic justice script. They reflect the folktale worldview with its representation of heinous crimes and gruesome punishments as "everyday features" (Lüthi, *European* 35), and its belief "that the world is truly the way [the folktale] perceives and portrays it to be" (88).

While many tales can be quoted, examples drawn from French and German folktales recorded by Perrault and the Brothers Grimm are sufficient to illustrate actualizations of the poetic justice script in its feudal track. Neither Perrault's nor Grimms' tales are anywhere close to being the authentic rendition of an oral folk tradition. However, if mapped on the continuum linking folklore and literature, Perrault's and the Grimms' versions, especially in the first two editions of *Tales*, can be placed closer to the folktale rather than the fairy tale on the spectrum. Perrault's eight tales in his 1697 *Histoires ou Contes du temps passé* were all slightly polished versions of popular folktales (Zipes, *Happily* 33; *Why* 72–3). Despite adjustments in tone and language, they retained almost all elements of their folk originals (Zipes, *Happily* 34; Darnton 62). The Grimms' collection, in turn, began as a scholarly project[10]. The first two editions of the *Nursery and Household Tales* (1812–14 and 1819–22), therefore, consist of clearly identifiable folktales, retold with only mild alterations. As Maria Tatar has shown, with each subsequent edition—up to the seventh in 1857 that included a total of 211 tales—the Grimms "veered more sharply away from the rough-hewn simplicity of their first editions to a sanitized and stylized literary form that proved attractive to both parents and children" (204). By abandoning scholarly folkloristic ambitions and instead embracing the needs of the market for respectable children's literature, the Grimms reworked their folk originals until, in the final edition, they appear as highly self-conscious literary creations. As I argue later in this chapter, at least some of the Grimms' alterations illustrate the transition from the feudal to the transcendentalist track of the poetic justice script. Inexorably indebted to its folk tale origins yet forever drawn—through an ever-finer grinding in the cultural mill—toward the socially acceptable, bourgeois, literary fairy tale, Perrault's and Grimms' tales retained enough of the folktale flavor to provide clear examples of how the feudal track of the poetic justice script operated in folktales recounted in pre-modern Europe.

In both Perrault and the Grimms, the feudal track informs tales in which characters are abused or meet a bad end, but the abuse is either unnoticed—by the characters and the narrator—or is passed over in silence as business as usual. Perrault's version of "Little Red Riding Hood"[11] is perhaps the best example in his collection of this early justice script, both in what it omits and what it adds. Perrault worked on the basis of a popular folktale that survived well into the nineteenth century and contained several elements he excised. These included cannibalism, striptease, and finally the girl's escape under the pretext of needing to relieve herself. At each of these steps, except for the scatological happy ending, the girl's unperturbed acceptance of what befalls her bespeaks of unclaimable justice. She demurely accepts the wolf's declaration that the "meat and wine" she just ate are the flesh and blood of her grandmother. She then obeys the wolf to undress and get into bed with him. When she asks what to do with every piece of clothing she takes off, the girl is told to throw it on the fire for she will not need it anymore. Eventually, stark naked, she gets into bed with the wolf, notices how hairy he is, and the familiar exchange of questions and answers begins (Darnton 9–10; Zipes, *Trials* 21–3). What is especially puzzling is how the girl remains unperturbed. There is no resistance. There is no physical violence. The girl obediently does as she is told. Her passivity, coupled with the cool, calculating, and absolute power of the wolf, captures the essence of the relationship between the powerless and the powerful that defined the feudal *Old Justice* world of premodern Europe[12].

Although he eliminated cannibalism and striptease, Perrault did retain the basic folk structure of this tale by replacing the scatological happy ending with a terse equivalent of the motifs he expurgated, the girl's demise in the beast's belly. The killings of the grandmother and Little Red Riding Hood are presented matter-of-factly, and the wolf is not blamed for slaying them. His predatory behavior is natural and expected; whatever blame can be assigned lies with the victims for exposing themselves. Irrespective of whether the most common pre-Perrault version of this tale ended on a tragic or hopeful note (see Darnton 9 and Zipes, *Trials* 4), Perrault's version became the standard reference both for the Grimms and through the early twentieth century. His Little Red Riding Hood is more helpless than in some folk versions. She will not be saved by moral outrage, by her own survival instincts, or external assistance. The girl enacts the feudal track of the poetic justice script, her social position and gender casting her as a victim in the *Old Justice* hierarchical social order. When, in their prudish version, the Grimms added a happy ending grafted from "The Wolf and the Kids" (Darnton 12), they managed to both punish and save the protagonist. First they saved Little Riding Hood's chastity and then—after she was appropriately punished by being devoured—they had her rescued by the hunter. This timely salvation reverses the otherwise doomed end, cuts the tale's mooring in the feudal track, and alters it far more than the elimination of cannibalism or sexuality. Torn between two happy endings, the Grimms eventually retained both. In each version the wolf dies, either by falling down on account of heavy stones in his stomach or by slipping off the roof and drowning in a trough. Both versions of his death are projected as *deserved* punishment, and the whole tale is clad in Christian moralizing that was absent from its earlier versions.

The introduction of the notion of moral desert—suggestively linked to a set of unexpected happy circumstances that reverse an otherwise unhappy end—is a powerful new idea that strikes at the heart of the poetic justice script. This idea was alien to the naturalistic peasant culture and trickled down into folktales from the mental world of the nobility. Perrault's collection may have been critical in effecting this change, for his greatest alteration was introducing various levels of moral desert in every one of his eight tales. He did so by grafting on his folktale originals the connection between physical beauty and good character and then, by extension, associating the conceptual blend of physical beauty and good character with good fortune implied to be a form of moral desert[13]. In cognitive terms, Perrault created a "megablend" (Turner and Fauconnier 153) where the blend of beauty and goodness was mapped onto the blend of good fortune and moral desert, integrating these different-domain elements within a single frame. This megablend of beauty-goodness-moral desert has become a prominent feature of Western mentality since then and is the hallmark of Walt Disney productions even today. For all these alterations, vestiges of the feudal track shine through in as many as five of Perrault's tales[14].

Most expressive in this respect is "Little Thumb." A French version of "Hansel and Gretel," it tells of parental abandonment at the time of famine. The poor parents, whose "seven children inconvenienced them greatly, because not one of them was able to earn his own way" (Perrault np), decide to get rid of the boys by losing them in the woods. This happens not once but twice. On the second occasion the boys wind up at the house of an ogre, who wants to eat them. Tricking the ogre into slaughtering his own daughters, the brothers return home, where the family is happily reunited. The boys do not bear any grudge against their parents or the ogre. Neither abandonment in the woods nor an attempted murder is seen as abuse. In fact, the behavior of the ogre couple—referred to in the tale as "a good woman" and "a good husband" (np)—is strikingly similar to that of Little Thumb's parents. The fact that the ogre woman saves the boys, whereas their own biological mother urges her husband to abandon them, actually tilts the moral comparison in favor of the ogre couple. The ogres are also devastated by the death of their children, whereas the human parents are devastated that the boys return. Scared as they may be, Little Thumb and his brothers do not expect fair treatment nor do they act fairly toward the only person who treats them well. Although the ogre's wife takes them in, feeds them, and saves them from death, Little Thumb has no qualms about inducing the murder of her daughters, and then, when he finds the Ogress grieving, he has no qualms about inveigling her to give him all of their family savings. His is a world where might is right.

Little Thumb's cunning agency—like that of The Master Cat—are exceptions, however. Other characters in Perrault's tales never resist maltreatment at the hands of those socially superior. The passivity of daughters in "The Fairies," the abuse suffered silently by Cinderella—who "bore it all patiently, and dared not tell her father, who would have scolded her" (Perrault np)—the imprisonment and barely avoided cannibalistic murder of the Sleeping Beauty and her children at the hands of her husband's mother, as well as the demure manner in which the local peasants follow the Master Cat's threat to lie about the ownership of the land unless they "shall be chopped up like mincemeat" (Perrault np) are all behaviors supported by the poetic justice script. They reflect a world in which it is unwise to have any expectations beyond survival, where might is right, and where justice is unclaimable.

The feudal track of the poetic justice script also entered the Grimms' collection. Although its representation of abuse as natural and random clashed with the Grimms' teleological worldview, the poetic justice script was still supported by the social and economic realities of life in the first decades of the 1800s. For the Grimms, as for their audience, the basic facts of fairy-tale family life—with its child abandonment, infanticide, hostile stepparents or siblings, and everyday abuse—squared with the facts of everyday life. Because of this proximity, many of the Grimms' tales continued to reflect the poetic justice script in its feudal track. This can be seen with particular clarity in the tales where the incomprehensible randomness of injustice was so jarring that the Grimms expurgated them from the second edition onwards. These tales—for example, "The Hand with the Knife," "The Mother-in-Law," and "The Starving Children"—reveal a world of arbitrary suffering, in which ideas of justice or moral desert are nonexistent. Each of these stories recounts one or more episodes of unexplained abuse. In "The Hand with the Knife" the mother heaps all the chores on a daughter so that her three lazy sons do not have to work. One of the hardest things the girl protagonist does daily is digging peat. Obtaining this cooking fuel with a stick takes long hours and is exhausting. As it happens, though, an elf who lives in a nearby hill develops sympathy for the girl. Whenever she passes a particular boulder, he stretches out his hand and gives her a magic knife, which does the peat-cutting job in minutes. On her way home, the girl taps the boulder twice and returns the knife to a hand that appears from it. As soon as the mother notices how quickly the work is done, she tells her sons to follow the sister. The boys end up wrenching the knife from her and cutting off the elf's hand to make sure he will not help the girl again (qtd. in Tatar 195–6). This bizarre tale not only showcases the nastiness of particular family but evokes a number of puzzling questions. No reason for the girl's abuse is provided. Given that she performs chores that benefit the entire family, it is baffling to see why the mother is troubled by the work being done efficiently and fast. It is equally odd to see how eagerly the girl's brothers get rid of the elf. With the magical helper out of the way, the family continues to enjoy abusing the girl and things are back to normal.

A similar example of gratuitous cruelty can be found in "The Mother-in-Law"—a truncated version of the second half of Perrault's "Sleeping

Beauty." The tale describes how, after the king goes away to war, his cannibal mother imprisons the queen and her two little sons. As in "The Hand with the Knife," no reason is given for the old woman's hatred of her own grandchildren. The intended murder and cannibalism cannot be explained in rational or practical terms, serving thus as an icon of arbitrary violence that will not meet with any retribution (qtd. in Tatar 201–2). "The Starving Children," finally, is a tale of a starving family, in which a widow-mother decides to kill one of her two daughters to feed herself and the other child. Every time she is about to do so, however, each daughter begs the mother to spare her life. When both working for food and stealing it prove insufficient, the daughters suggest a solution: let all of us fall asleep and not awake until the Day of Judgment. The girls do so—in a thinly veiled metaphor for death—whereas the mother, apparently unable to devour the bodies of her dead children, leaves the house and is never seen again (qtd. in Tatar 202). Although structurally different than the other two tales, "The Starving Children" points to the naturalistic bottom line between starvation and survival that informs "Hansel and Gretel" and other tales, in which children are abandoned on account of famine. It affirms the arbitrariness of suffering that strikes people irrespective of what their personal merits may be.

As committed Christians whose teleological worldview demanded reasons for everything that happens, the Grimms were uneasy with the arbitrary, hierarchical justice and the idea of a random universe it implied. Their resistance shows in two types of editorial changes: first, by grafting the idea of moral desert onto the largely amoral tales, they created a *New Justice* foundation for a deserved reward or punishment; second, by grafting magical happy endings, they performed Frankensteinian surgery on tales whose plots were largely geared toward an impending doom. The contrast between improbable happy endings and the otherwise believable plots involving cannibalism, incest, infanticide, abuse, and abandonment that fill the worlds of Grimms' fairy tales has been noted. However, it has been discussed mostly in the context of the most revolutionary operation the Grimms spearheaded: the adoption of the folk and fairy tale for socializing children in middle-class values (see Zipes, Bottigheimer, Tatar, and Cashdan). My suggestion is that the Grimms' textual alterations also reflected their creative resistance to the poetic justice script embedded in the folktale material they had collected. By registering various criminal motivations and linking them with all kinds of severe punishments, the Grimms were able to rework some folktales informed by poetic justice into narratives with retributive justice conclusions[15]. At the same time, the fact that the Grimms were unable to completely erase the feudal track of the poetic justice script from many of the tales they extensively reworked indicates how deeply their folk originals were rooted in patterns of poetic justice. Even in the final edition of the *Nursery and Household Tales* many tales are structured on two key markers of the feudal track-based stories: the unscrupulous struggle for survival in an unjust world and the characters' passive acceptance of inequality and its attendant miseries.

A brutal struggle for resources is the dominant theme of, among others, "Little Farmer," "The Thief and His Master," "The Three Little Birds," and "Puss in Boots." The first tale exemplifies ruthless competition among the peasants in the village, the second represents competition among strangers in the countryside, the third typifies competition among the sisters in a royal household, and the fourth epitomizes competition across the social strata: between peasants and the nobility. "Little Farmer" (qtd. in Zipes, *Tradition* 21–25) tells the story of a swindler-farmer who takes over the entire village after tricking everybody else into committing mass suicide. Preposterous as this plot sounds, its moral evaluation implied by the upbeat conclusion is even more so, unless seen as bespeaking of an *Old Justice* mentality where might makes right. How else can one interpret the narrative's happy ending about the entire village wiped out and Little Farmer, a sole survivor, becoming a rich man? "The Thief and His Master"—a story of how a poor father apprentices the son to a master thief and then tricks the master out of his fee (qtd. in Zipes, *Tradition* 359–60)—likewise conveys a worldview particular to folk narratives structured on poetic justice. This tale implies that it is better to prosper by any means rather than to have qualms about how one comes into wealth. Cunning and shrewdness are also suggested as the way to make it in the unjust world in "The Three Little Birds" and "Puss in Boots" (qtd. in Zipes, *Tradition* 302–5 and 402–5). "The Three Little Birds" is a tale of sibling rivalry, with evil sisters, a king who never talks to his wife, and a queen who never protests the abuse she endures from her sisters and husband. The happy ending in this tale—when the supposedly murdered children survive, the wicked aunts' crimes are punished, and the bone-weary mother is released from prison—is the reverse of what happens in "Puss in Boots." There, the shrewd cat and his master actually succeed in a similar course of action as that taken by the evil sisters in "Birds." Besides evoking a convenient moral code of privileging human over non-human—one that allows murdering ogres but not people—"Puss" and similar tales suggest that while tricksterism may not always work, it is as efficient as any other means to secure one's future in an unjust world. These tales affirm that in a world of might makes right, anyone who can act in cruel and unjust ways will do so. As long, of course, as they can get away with it.

The dominant marker of the feudal track of the poetic justice script in the Grimms' tales is the protagonists' passive acceptance of the ongoing injustice. It is a rule in these tales that females persecuted by stepmothers or mothers-in-law never complain to their fathers or husbands, as is the case in "The Three Little Men in the Wood," "Cinderella," and "The Mother-in-Law" (qtd. in Grimm, *Household* 54–9 and 93–100; Tatar 201–2). Children abandoned, abused, or threatened with death by their own parents never bear any grudge and passively accept whatever happens to them. This dynamic is especially evident in "The Twelve Brothers," "Hansel and Gretel," "The

Starving Children," and "The Magic Table, the Golden Donkey, and the Club in the Sack" (qtd. in Grimm, *Household* 37–41 and 62–9; Tatar 202; Zipes, *Tradition* 427–34). All these tales rationalize abandonment and abuse as normative defaults. In "The Magic Table," for example, each of the three sons is expelled by the father on the basis of goat's accusations and, without protesting, each takes to the road. In "Iron Hans," like in a closely related tale "The Wild Man" (qtd. in Zipes, *Tradition* 329–34 and 323–25), a young prince is kidnapped by the eponymous Iron Hans/Wild Man but accepts his fate and never seeks his family again. The protagonists' resigned acceptance goes so far that even characters who are freed from injustice return to live with their former tormentors, as if no other option was conceivable. Hansel and Gretel are abandoned not once but twice, Snow White—in the Ölenberg version—faces two expulsions (Joosen), and the triple sequence of repeated expulsions in other tales, such as "The Magic Table, the Golden Donkey, and the Club in the Sack," also speaks of abandonment as a recurring phenomenon. The fact that in most cases passive endurance is rewarded by a miraculous deliverance suggests moral desert—a concept whose grafting on folktale originals reflects the Grimms' attempts to replace the poetic justice script with a more rational or ethically comprehensible vision better suited to their *New Justice* sensibilities. The addition of moral desert was just one among many revisions the Grimms introduced, but I find it intriguing that practically all of those revisions have to do, in one way or another, with the Grimms' resistance to the poetic justice script.

When it comes to punishing evildoers, the Grimms were torn. On the one hand, they were eager to see evil punished. In a retributive justice remake of these poetic justice narratives, they added a number of graphic details about righteous punishments of evil sisters, stepmothers, and other troublemakers. On the other hand, they accepted the notion that human justice is imperfect and featured a number of tales about wrongfully accused protagonists, positing that human judgments can sometimes be misjudgments. A derivative of this awareness about the imperfect character of human justice was the Grimms' belief, also projected into the tales, about the perfect character of divine justice. This idea finds expression in "A Tailor in Heaven" (qtd. in Grimm, *Household* 141–43), which tells of how, acting in righteous anger, a poor tailor in Heaven punishes a thief on Earth but is reprimanded by God. When he tells God that he knocked the thief with God's own footstool, the Lord replies:

> "Oh, thou knave, were I to judge as thou judgest, how dost thou think thou couldst have escaped so long? I should long ago have had no chairs, benches, seats, nay, not even an oven-fork, but should have thrown everything down at the sinners. … No one shall give the punishment here, but I alone, the Lord." (143)

The assumptions that all humans are sinners and that punishment must be left to God are important components of the transcendentalist track of the poetic justice script, which the Grimms seemed to have embraced, yet were unable to fully impress on their tales. Leaning more toward the earthly fulfillment that marks folktales, they made most of their happy endings into resolutions in the here and now. These resolutions, however, did not deny larger patterns of injustice in the world. As members of the *New Justice* generation, the Grimms were ideologically pulled toward the transcendentalist understanding of the poetic justice script. Yet their tales remained framed by the feudal track, even while they were already moving away from it through the introduction of moral desert and happy endings.

THE TRANSCENDENTALIST TRACK OF THE POETIC JUSTICE SCRIPT

The emergence of the upgraded poetic justice script in its transcendentalist track was concurrent with the reforms enacted under the banner of *New Justice* and the rise of the middle class between the 1830s and 1900. This period saw three main developments in the literary fairy tale. First, fairy tales for young audiences came into their own only in the 1830s. Second, from 1835, the tales of Hans Christian Andersen introduced a new standard for fairy tales as crossover fiction, with individual tales "that could be grasped readily by children and adults alike but with a different understanding" (Zipes, *Spells* xxiv). Third, a great flowering of the fairy tale in Europe and America further diversified the genre's role and potential—not least in its interaction with the emergent genres of fantasy and science fiction (see Ashley 331–333).

The first signs of the transition within the poetic justice script from the feudal to the transcendentalist track come about in the rebellious fairy tales written by German Romantic authors: Wilhelm Heinrich Wackenroder, Ludwig Tieck, Novalis, Adalbert von Chamisso, Friedrich de La Motte-Fouqué, and E.T.A. Hoffman. Their works lambasted the *Old Justice*-based hierarchical order and offered a critique of the worst aspects of absolutism as well as the abuses of the arrogant Enlightenment. In these tales the forces of injustice are no longer allegorical representations but assume a social hue associated with the bourgeois society or the decadent aristocracy. These forces are so ubiquitous, though, that the majority of the German Romantic tales end with a disaster, where the protagonists either go insane or die. In this sense these tales registered the reality of social injustice but did not see a solution to it. It was up to a master of transcendentalist justice, Hans Christian Andersen, to soothe the consciousness of nineteenth-century Europe by affirming that all injustices are noted and will be redressed in due time.

A unique combination of personal, cultural, and ideological reasons made Andersen the spokesman for the transcendentalist track of the poetic justice script. First, he was a social climber. Unlike Perrault, the Grimms, or the German Romantics, he came from the dirt-poor, small town proletariat. His life was haunted by embarrassment over his origins: no amount of success could erase the fact that he had a cobbler father and a washerwoman mother. Second, Andersen was "one of the greatest mythomaniacs, hypochondriacs, and narcissists" of the nineteenth century (Zipes, *Dreams*, 2nd ed. 109). He invented for himself a quasi-biographical fairy tale about a poor but talented boy who, by God's will, conquers the world of the upper classes. Using his tales therapeutically to account for the traumas and tensions of his own life, Andersen helped spread the "wait, you will be rewarded" message of poetic justice in its transcendentalist track. Third, his tales were published at a time of powerful simmering across the European Continent forced to comply, through the Congress of Powers, with the backward-looking *Old Justice* political structures. Andersen's Europe was a world of regressive social order: placid on the surface but hiding a volcanic buildup of rebellion from the increasingly numerous socialist, nationalist, Chartist, and other reformers who subscribed to the *New Justice* rationality and protested against what they saw as the remediable injustices of their time.

As a result of these factors, Andersen's tales came to reflect the widely felt and deeply frustrated aspirations of the underprivileged in the rigidly classed Europe. Immensely influential in Europe and America, Andersen's tales canonized the transcendentalist track of the poetic justice script. They told of innocent victims quietly suffering abuse and injustice—each victim's acceptance projected as the measure of their moral worth—until the injustice is ended by a divine intervention or by the removal of the victims to an afterworld in which they are amply compensated. So defined, transcendentalist justice became a social pacifier—an alternative to the retributive justice script lurking under the word "revolution"—and a culturally promoted response to the widespread injustice these tales registered and their audiences experienced in their everyday lives. The poetic justice script broadcast in Andersen's fairy tales derived primarily from the fact that Andersen saw his own life in terms of a poetic justice script with a transcendentalist promise. The unfairness of the fact that he was born in poverty and had to struggle for what was the given for the bourgeois was bound to be redressed, Andersen believed, in one way or another: preferably through his fame and the recognition that would follow. He waited for this moment all his life, but it was always deferred, always incomplete. This tension between Andersen's servility and rebellion, between injustice and its deferred redressing, largely accounts for the appeal of his tales (see Zipes, *Hans*). A man who "hated to be dominated though he loved the dominant class" (Zipes, *Dreams*, 2nd ed. 82), Andersen registered through his tales, a burgeoning class conflict resulting from an unjust social order. Yet, he legitimized that same oppressive order by subsuming it under a higher, absolute order of Divine Providence.

This blending operation of mapping current injustice onto God's larger plan is exactly what is at stake in the transcendentalist track. This track recognizes the injustice of social inequality and affirms that it must be made right. However, by promoting a resigned acceptance of the ongoing injustice as a test whether or not one is worthy of the reward of justice in the future, it defers justice and makes it *de facto* unachievable. The acceptance of present injustice in the name of future redress is, of course, closely aligned to the Christian ideal of suffering in this world in the name of salvation in the next. The affirmation of the necessity of suffering in the present is the cornerstone of transcendentalist justice and explains why the majority of Andersen's tales reveal a strong belief in God's design and moralistic evangelicalism.

The class divide that prevents the diligent and deserving commoners from enjoying equality with the upper classes was seen in folktales as the way of the world. In Andersen's fairy tales, by contrast, it was clearly registered as injustice. This is evident in most of his 156 tales but especially in those of love frustrated by class differences—such as "The Little Mermaid" and "The Steadfast Tin Soldier"—and in those about criminal or social injustice—such as "The Rose Elf" and "The Little Match Girl." In various degrees of explicitness, the transcendentalist track of the poetic justice script also informs other tales: "The Traveling Companion," "The Snow Queen," "The Red Shoes," "The Girl Who Trod on the Loaf," "There Is a Difference," "A Vision of the Last Day," "The Cripple," and others. All these tales script justice on the assumption of trust in God's design, even if the reward would come after death.

A good case in point is "The Little Mermaid." In this tale, the injustice of social inequality that bars one from achieving fulfillment is first mitigated by the mermaid's gamble of breaking through the species barrier. When this fails, the resolution to the unfairness of her soulless birth is projected into a postmortem existence three hundred years in the future. The fact that even the postmortem existence of the mermaid as "the daughter of the air" Andersen still calls a "time of trial" (Andersen, *Tales* 89) makes "The Little Mermaid" an eye-opening illustration of how indefinitely the transcendentalist track defers redress. The mermaid's entire life is one patient enactment of the poetic justice script with a transcendentalist promise. It affirms the future reward of justice but requires her to continuously prove being worthy of it. Having first proven her love to the prince by saving his life, she then proves it through a physical mutilation, then through becoming his lap dog—"[t]he Prince said that she should always remain with him, and she received permission to sleep on a velvet cushion before his door" (82). The trial does not end there. The Little Mermaid must then prove her love by refusing to kill the Prince, then by dying for his love, and eventually through centuries of trial as the daughter of the air. She "had suffered unheard of pains every day, while he was utterly unconscious of all" (86). Her suffering is presented as necessary for the very possibility of redress in the future.

"The Steadfast Tin Soldier" is a similar case of love frustrated through an unbreachable class divide with the promise of fulfillment in some vague afterlife. In this tale both protagonists are mute, and the character cast to enact the transcendentalist track is the tin soldier. Immobile but forever hoping, he is the ultimate model of proper behavior for someone troubled by undeserved inequality but locked within the transcendentalist promise of the poetic justice script. Expelled from and then miraculously returned to his cozy yet class-divided home, the tin soldier soon falls to arbitrary injustice again when one of the boys throws him into the fire "without rhyme or reason" (Andersen, *Tales* 127). Dying and melting, the soldier is joined by the little dancer who, carried by the breeze, perishes in the same fire. A small tin heart that the maid finds in the ashes the next day leaves the reader with a vague consolation that the two are joined in death—a conclusion that confirms the appropriateness of their unfulfilled lives and the promise of justice meted out in the afterlife.

While the theme of unfulfilled love runs through numerous tales, Andersen also authored tales that were more blunt about other kinds of injustice. "The Rose Elf," for example, tells of how an evil brother murders his sister's lover and buries him in the woods. The murder is witnessed by the Rose Elf, who then reveals it to the maiden. The girl digs up the body of her lover, plants his head—with a little spring of jasmine—in a flowerpot, and eventually dies of sorrow to find herself "in heaven with him whom she loved" (Andersen, *Tales* 351). Since the girl's suffering ends with a reunion in heaven, in this tale Andersen allows himself to sprinkle a bit of righteous revenge. When the evil brother inherits the beautifully blooming jasmine, the flower-beings execute justice by pricking his tongue with poisonous spears. The tale ends with the Queen Bee singing about how, behind every smallest leaf, "there dwells One who can bring the evil to light, and repay it" (352). This ending transforms the revenge into God's will and affirms that all evil deeds will be punished. At the same time, it removes justice from human hands and locates it either in the afterlife or in God's providence. The girl does not even move a finger to expose the murder. The elf, flower-souls, and the bees actively pursue justice only because they are instruments of God's larger plan.

Such execution of justice is lacking in what is perhaps the single most glaring example of the poetic justice script in Andersen's oeuvre, "The Little Match Girl." This tale, known virtually throughout the whole Western world, targets the injustice of social inequality that tolerates grinding poverty and leads to incidents such as a poor girl freezing to death on New Year's Eve. "Shivering with cold and hunger … a picture of misery" (Andersen, *Tales* 187), the match girl dies staring, from the windy street corner, at brightly lit windows where the rich enjoy their sumptuous meals. The culprit in this tale is social inequality and its attendant abuse. The redress Andersen offers, however, is happiness in heaven. Unable to make herself return home to receive a beating from her father—and hallucinating about food, warmth, Christmas gifts, and a loving family—the girl is embraced by the grandma-angel. Both fly "in brightness and joy above the earth, very, very high, and up there was neither cold, nor hunger, nor care—they were with God" (189). Even though this tale has been read as an implicit accusation against the wealthy for the suffering of the poor, its core message is that social injustice is unavoidable and that the suffering poor will be compensated only after death. In this and other tales, Andersen defined the poetic justice script in its transcendentalist track. The way to salvation, for him, was a path filled with suffering, humiliation, and torture, but only such a path can lead to compensation in the future. Regardless of how much one may suffer, the protagonist of "A Vision of the Last Day" is explicitly told in his dream that one must leave justice to God. According to the Angel, those who seek to persecute evil themselves are Mohammedans and those who seek revenge are Jews. A good Christian, the Angel continues, must "dwell in dust" and "copy Him who with patience endured the contradiction of sinners" (Andersen, *Tales* 209).

Articulated in Andersen's tales, transcendentalist justice and its attendant script spread through Europe like wildfire. By the time the fairy tale renascence occurred in England, it was no longer peculiar to Andersen's work alone. Starting in the 1840s and continuing to the end of the century, most authors were using fairy tales to raise consciousness about the suffering caused by social inequalities. These authors registered patent social injustices and voiced passionate concerns about the fate of the poor and exploited. Like Dickens in "A Christmas Carol" (1843) or Kingsley in *The Water Babies* (1863), however, they continued to present social injustice as inextricably woven into the fabric of human existence. Kingsley's chimney sweep protagonist Tom, for example, takes "being hungry, and being beaten ... for the way of the world, like the rain and snow and thunder." He also dreams of becoming a master himself, who would "bully [his own apprentices], and knock them about, just as his master did to him" (Kingsley 6). The solution to the injustice of early capitalism advocated by these authors was moral education that would lead to a voluntary transformation of the heart, as in case of Ebenezer Scrooge, or an education that would confirm the expectation of justice only after death, as in case of Kingsley's Tom. Although Tom is eventually resurrected, he returns to life with the knowledge that the administration of justice belongs with supernatural forces. The same message is stressed in George MacDonald's *At the Back of the North Wind* (1871). MacDonald's protagonist, Diamond, rides the North Wind to learn that suffering and death eventually lead to something good because everything happens according to God's will.

The transcendentalist version of the poetic justice script continued to be dominant into the second part of the nineteenth century, contributing to the rise of fairy tale utopias that questioned existing injustices. While their authors sought to oppose class domination and exploitation, even the most progressive artists found it hard to find alternatives to the poetic justice script. Oscar Wilde's two fairy tale collections, *The Happy Prince and Other Tales* (1888) and *The House of Pomegranates* (1891), offer good examples. Both are harsh critiques of social injustice and of the hypocrisy of the prosperous. Both were also structured by the transcendentalist track. In what was the hallmark of the first collection and remains the best known of Wilde's fairy tales, the "happy" Prince is dead. After enjoying a life of careless fun he has now awakened to the misery and suffering he had ignored when he was alive. "[N]ow that I'm dead," he says through the mouth of a golden sculpture of himself, "they have set me up here so high that I can see all the ugliness and all the misery of my city, and though my heart is made of lead yet I cannot choose but weep" (Wilde 255). The remorseful Prince asks a swallow to serve as his messenger. He asks the bird to pluck three precious stones that adorn the sculpture—two of them his eyes—and requests that they be delivered to three people: a poor widow whose son is dying for lack of proper medication, a poor poet who is numb with hunger and cold in his freezing garret, and—in an echo of Andersen—to a poor match girl who had lost all of her matches and faces a harsh punishment by an abusive father. After the swallow returns from his mission, the Prince reflects that "more marvelous than anything is the suffering of men and of women. There is no Mystery so great as Misery" (259). The Prince asks the swallow to fly over the city and then hears a tale of woe. Realizing the scale of suffering, the Prince orders the Swallow to pick off, leaf by leaf, the gold that covers his statue and distribute it to the poor of the city. This improves the situation for many, but the statue of the happy Prince becomes dull and the Swallow dies of cold. As the city council pulls down the sculpture, the Prince's broken heart ends up on a pile of trash alongside the body of a dead Swallow. The leaden heart and the dead bird are then declared by God "the two most precious things in the city" and brought to Paradise, where the bird and the Prince will live happily for ever (260).

Transcendentalist justice also underlies Wilde's other tales, most of which—such as "The Devoted Friend" and "The Selfish Giant"—reveal a poignant awareness of what today would be called social injustice. At the same time, all of these tales assume that injustice is the way of the world. "The Happy Prince" offers no solution, seeing misery as a "mystery" (260). It suggests that small acts of kindness, although noble, are a drop in the ocean. They will be rewarded in heaven, but on earth they barely change a thing and tend to exhaust one to the point of demise. In "The Selfish Giant," while the Giant learns the lesson of sharing *before* he dies, his contribution to the community is a drop in the ocean too. The social injustice in the world remains, and only death removes one to the realm where justice is possible and will reign forever. As was the case for many other nineteenth-century authors, Wilde's fairy tales reveal a sharp awareness of social injustice, but they lack a vision—or a script—of how such injustice may be addressed and countered. Those yearning for justice must die and find fulfillment in a transcendental beyond.

When specifically the poetic justice script in its transcendentalist track phased out is hard to say, but two things seem clear. First, it survived longer on the European Continent than in the Anglo-American world, as many Continental fairy tales and novels for the young, like Coloddi's *Pinocchio* (1882), Amicis' *Heart* (1886), and even Korczak's *King Matt the First* (1926), continued to favor the transcendentalist track of the poetic justice script well into the post-WWI years. Transcendentalist justice retained its strongest appeal in the genre of fantasy, and in the 1950s resurfaced powerfully in its nostalgic Christian underpinnings. The thinning and "all-beauty-will-fade" world of Tolkien's *The Lord of the Rings* is largely structured on the transcendentalist understanding of justice. For Tolkien the achievement of justice in this world is at best temporary: genuine fulfillment can be found only by sailing to the west. This vision of justice also underpins the "testing ground" universe of Lewis' Chronicles, in which the dream of justice is fulfilled in the realm of eternal Narnia among all the deserving dead. Yet, even as Tolkien and Lewis were composing their classics, the transcendentalist version of the poetic justice script was well on its way to go down in a blaze of glory. Within one generation after Tolkien and Lewis, even this late version of unachievable justice was largely a thing of the past, at least in the Western world.

Second, the foundational assumptions of transcendentalist justice first collapsed in the Anglo-American world. This came about as authors began to revise their perceptions of social inequalities and widespread misery from being inevitable to being redressable. As gradual reforms opened the way to enfranchisement in England, studies of mechanisms of the labor market redefined poverty as a product of specific socio-economic conditions rather than the natural state of things. Immense public interest in investigations such as Charles Booth's *Life and Labour of the People in London* (1889–1903) or Seebohm Rowntree's *Poverty: A Study of Town Life* (1901) created a reformist climate in which British fairy tale authors such as Nesbit, Grahame, Kipling and others began moving away from the transcendentalist, unachievable justice toward non-utopian visions of achievable justice. In the United States the process had begun earlier, as evidenced in Frank Stockton's collection *The Floating Prince and Other Fairy Tales* (1881). Free from the heavy Christian moralizing that was a standard in children's literature of the time, Stockton's tales replaced unachievable, absolute justice with its provisional and achievable alternatives. Because he abhorred war, Stockton's fairy tales were among the first ever to advocate peaceful conflict resolution. Instead of punishing oppressors, Stockton's narratives made them look ridiculous and pitiful. Unencumbered by the poetic justice script and denouncing any violations of human rights, Stockton's fairy tales were groping toward a humorous version of restorative justice—a direction even more pronounced in the work of L. Frank Baum. Baum's Oz books not only deplored violence and elevated compassion for others, but they also stressed positive and workable redress of injustice in the physical present. Also Baum's fairy tales, such as "The Queen of Quok" (1901), reflected his cheerful resoluteness and respect for basic human rights in the here and now rather than in some transcendent afterlife. In "The Queen," a ten-year-old king is sold in marriage but refuses to either marry or execute an old woman who buys his hand for three million dollars. In their hopeful visions of how the oppressed can resist force and transform their weaknesses into strengths, Stockton's and Baum's works rejected poetic justice in favor of visions of justice that is achievable in the present.

Because poetic justice and its supporting script were products of the *Old Justice* social order, they began losing appeal when that social order was fading. As the world of feudal oppression, illiteracy, and incessant toil was gradually replaced by arrangements that eventually became the bedrock of modern Western democracies, the feudal and then the transcendentalist versions of this script were discarded in favor of scripts that projected justice as a viable choice in the present. The last vestiges of transcendentalist justice were swept away, albeit with much resistance, by the civil and human rights movements of the 1950s, as "We Shall Overcome ... *someday*" was no longer found to be a satisfying message. When in 1963 Martin Luther King, Jr. demanded the immediate cashing of the check that would give African Americans "the riches of freedom and the security of justice" (np), he was challenging poetic justice in its transcendentalist track. The time for justice, he asserted, is now. The place for justice is here.

These sentiments have found expression in literary expectations. One reason why traditional folk and fairy tales have been so often reworked in the second part of the twentieth century is that they are informed by the poetic justice script. In depicting patent injustices that are tolerated or passed over in silence, this script is jarring to modern sensibilities. Disney's *Little Mermaid*, for all its controversial simplifications (see Hastings), is at heart a radical rejection of the assumptions of poetic justice in favor of achievable justice. Disney opts for the retributive justice script, where Ariel's suffering is projected as "harm" inflicted by the evil sea witch Ursula as part of Ursula's larger scheme of taking over Triton's underwater kingdom. To undo the "harm"—including Ursula's spells on other characters—is to restore justice, which happens when Ursula is killed by a sinking ship. Unlike in Andersen's original, in Disney's version Ariel's suffering pays off in this life. Justice—understood as restoration of order and the punishment of evildoers—is achieved rather than delayed. Today, when it surfaces in literary fiction, poetic justice is almost always the argument of the oppressors. In Cressida Cowell's *How to Train Your Dragon* (2003), the protagonist Hiccup confronts Green Death exclaiming, "[b]ut it's so unfair ... Why do YOU get to eat everybody, just because you're bigger than everybody else?" The dragon's answer, "[i]t's the way of the world" (152), is the *Old Justice* claim that not only fails to convince, but evokes resistance, resulting in the downfall of those who embrace the "might makes right" position.

NOTES

1. If ordinary people did not have much say in the matter, the nobility did negotiate, seeking advantageous justice through agreements with the rulers. By the

seventeenth century many European countries had some form of a parliament. England was arguably ahead of other nations. As the only country in Europe conquered and permanently repossessed by foreigners, already since the tenth century, England had become the arena of fierce legal negotiations among the conquerors' elite. For all that, Norman justice, as Robin Hood would agree, was that of a thieves' den. Its goal was to ensure enough spoils and guarantees of these to keep the barons and the royalty relatively unified, thus giving the Anglo-Saxons no chance to reclaim their country by rising against divided Norman occupants. In the long run, however, intra-Norman negotiations about justice spilled over to the rest of the society. Imperceptibly at first, they coalesced into a tradition of resistance to arbitrary power that eventually became the jewel and cornerstone of Anglo-American political pride: the tradition of constitutionalism.

2. From the codification of the English common law under Henry II in the twelfth century; through the Magna Carta (1215) and other documents in the thirteenth century—developments that gave Parliament an almost permanent place in subsequent English political history; through the Glorious Revolution of 1688 that established parliamentary monarchy; to the American Revolution of 1775 that, in the Constitution of 1788, established the first in the world constitutional republic and, by adding to constitutionalism the argument about inalienable natural rights, formulated claims central to modern democracies—all these milestones are part of the tradition that resists the injustice of enserfing people to the whims of arbitrary power.
3. The division between myth, legend, and folktale fails to be helpful in cases—as with *The Illiad*, *The Odyssey*—when these categories overlap to an extent that the distinction seems rather artificial. Tales that combine the elements of the three are impossible to classify in this system. Aarne's index, as Propp notes, is helpful as a practical reference but also is "dangerous for … [i]t suggests notions which are essentially incorrect." According to Propp, "[c]lear-cut division into types does not actually exist; very often it is a fiction. If types do exist, they exist not on the level indicated by Aarne but on the level of the structural features of similar tales" (Propp 11). This assessment is also shared by Zipes in whose opinion using tale types is limited and usually misleading (see Zipes, *Why* 129). Propp's approach, finally, ignores verbal aspects of the folktale and its changeable particular details. His focus, rather, is on the dynamics of a plot in several versions of one the same tale.
4. Bottigheimer succumbs to logical positivism in claiming that the study of the fairy tale should be limited only to published fairy tale collections and that the existence of oral fairy tales "among any *folk* before the nineteenth century cannot be demonstrated" (Bottigheimer, *Fairy* 7). This approach disregards the multiple parallel flows of stories from the written to oral and from the oral to written versions.
5. Lüthi's work—despite its high regard in some circles—is rather disappointing for its hopeless confusion of the two categories. This is glaringly made clear by the fact that the English translators of his *Es war einmal. Vom Wesen des Volksmärchens* (1964) and *Das Volksmärchen als Dichtung. Ästhetik und Anthropologie* (1975)—published in English as *Once Upon a Time: On the Nature of Fairy Tales* (1970) and *The Fairy Tale as Art Form and Portrait of Man* (1984), respectively—felt compelled to replace Lüthi's key term in the

titles, the folktale, with the term more appropriate to what Lüthi actually talks about, the fairy tale. The only study where Lüthi's use of the term the folktale seems legitimate, and so retained in its translation, is *The European Folktale: Form and Nature* (original in 1947, English in 1982)—a study that despite its occasional generalizations offers one of the best insights into the form and nature of the folktale. Lüthi was, of course, a child of his time through no fault of his own. What today appears as a confusion of terms was partly the result of the fact that Lüthi's output "stands in a tradition of folktale scholarship that in Germany and other European countries reaches back to the time of Jacob and Wilhelm Grimm" (Niles xviii)—a tradition based on a series of unquestioned assumptions largely discredited today.

6. In *Fairy Tales: A New History* Bottigheimer returned to this distinction and, in fact, saw it as central to the discussion of fairy tales, which she defined as a specifically urban and literary phenomenon developed independently of the necessarily oral folklore (4).
7. This pattern goes back to the first fairy tales to appear in Europe in the 1550s. For example, Straparola's fairy tales about a poor commoner marrying, with the help of magic, into royalty were written in Venice at the time when Venetian laws explicitly forbade such interclass marriages (see Bottigheimer, *Fairy* 21). The combination of the improbable and the illegal made those fairy tales especially appealing to its non-noble urban readership.
8. Obviously, seeing folktales as historical documents does not mean treating them as accounts of events that actually happened, or happened to specific people, and least of taking *all* the events they describe as mirroring the social conditions of pre-modern Europe. For all that, many events, personal relations, and plot elements that aim at verisimilitude—for example, grinding poverty, frequent starvation, child abandonment, widespread violence, and so on—do suggest a lot about pre-modern social realities. Some discussions of differences in the national character of European folktales (see Darnton 49–62) seem to suggest that these differences—ironic detachment in French tales, gory punishments in German ones, buffoonery and Machiavellianism in Italian ones—can be seen as different responses to the patent injustice of the world they register and take for granted. English folktales are on average gayer—i.e., more whimsical—that their continental counterparts partly due to the fact that British peasants were the first in Europe to escape the world of the Malthusian trap, which remained the reality for their Continental counterparts for another century or two. For example, the last peacetime famine in England occurred in 1623–24, in Germany in 1771–72, in France in 1778, in Ireland in 1845–49, in Finland 1866–68, and in Russia as late as in 1932–33.
9. This, of course, does not mean that psychological interpretations of folk and fairy tales are invalid. The plots of these tales invite multiple readings, and psychological-symbolic perspective can yield interesting interpretations. Nevertheless, the very idea that folk or fairy tales concern themselves with *inner* realities strikes me as a distinctly modern concept that goes back to the invention of psychology—Freud, Jung, Fromm, and Bettelheim, with antecedents in the Romantic thought. The idea that folk and fairy tales are symbolic representations—through concrete images—of psychic realities is a strictly twentieth-century perception that is at odds with conclusions of folkloristic and historical studies, most of which offer a compelling case for the psychological

shallowness and physical immediacy of folk and fairy tales as they functioned before the mid-nineteenth century.

10. As put by Wilhelm in the preface to the first edition in 1812, the brothers' ambition was "to collect these tales in as pure a form as possible" so as to produce a collection in which "[n]o details have been added or embellished or changed" (qtd. in Tatar 210).
11. Perrault's tales are known under various English titles, but the ones I use, as well as their translations, are taken from Perrault's *Mother Goose Tales* as listed on D. L. Ashliman's e-database *Folktexts: a Library of Folktales, Folklore, Fairy Tales, and Mythology*.
12. It also reflects traditional cultural notions of sex roles and male domination. For a detailed discussion, see Zipes, *Trials*.
13. Beauty is rewarded and ugliness is punished in "The Sleeping Beauty in the Wood," "Cinderella, or The Little Glass Slipper," "Little Red Riding Hood," "The Master Cat, or Puss in Boots," "Blue Beard," and even by the end of otherwise quite unusual "Ricky of the Turf." In each of these tales beauty/good character is also implied as evoking moral desert that works to the advantage of protagonists. Beautiful protagonists in "The Fairies," "The Sleeping Beauty in the Wood," "Cinderella, or The Little Glass Slipper," "The Master Cat, or Puss in Boots," and "Ricky of the Turf" end up with a fortune and a royal spouse; in the remaining two tales they just end up with fortune. In all tales the protagonists are also saved from certain doom by a pattern of happy coincidences augmented, as in "Little Thumb" and "The Master Cat, or Puss in Boots," by their own shrewdness.
14. Besides "Little Red Riding Hood," these include "Little Thumb," "The Fairies," "The Sleeping Beauty in the Wood," "Cinderella; or, The Little Glass Slipper," and "The Master Cat, or Puss in Boots."
15. This accounts for why there is also a tradition of examining the Grimms' tales as reflecting a preliterate legal tradition and as a major contribution to the history of crime and punishment. See, for example, Mueller, Tatar, and Bottigheimer, *Bad*.

WORKS CITED

Andersen, Hans Christian. "The Little Match Girl." *Fairy Tales of Hans Christian Andersen*. New York: The Orion Press, 1958: 187–89.

Andersen, Hans Christian. "The Little Mermaid." *Fairy Tales of Hans Christian Andersen*. New York: The Orion Press, 1958: 65–89.

Andersen, Hans Christian. "The Steadfast Tin Soldier." *Fairy Tales of Hans Christian Andersen*. New York: The Orion Press, 1958: 123–27.

Andersen, Hans Christian. "The Rose Elf." *Fairy Tales of Hans Christian Andersen*. New York: The Orion Press, 1958: 347–52.

Andersen, Hans Christian. "The Traveling Companion." *Fairy Tales of Hans Christian Andersen*. New York: The Orion Press, 1958: 13–34.

Andersen, Hans Christian. "A Vision of the Last Day." *Fairy Tales of Hans Christian Andersen*. New York: The Orion Press, 1958: 205–10.

Andersen, Hans Christian. "The Red Shoes." *Fairy Tales of Hans Christian Andersen*. New York: The Orion Press, 1958: 195–202.

Andersen, Hans Christian. "There Is a Difference." *Fairy Tales of Hans Christian Andersen*. New York: The Orion Press, 1958: 225–30.

Andersen, Hans Christian. "The Snow Queen." *Hans Christian Andersen's Life and Works: Research, Texts and Information*. 2011. http://www.andersen.sdu.dk/vaerk/hersholt/TheSnowQueen_e.html. July 11, 2014.

Andersen, Hans Christian. "The Girl Who Tread on the Loaf." *Hans Christian Andersen's Life and Works: Research, Texts and Information*. 2011. http://www.andersen.sdu.dk/vaerk/hersholt/TheGirlWhoTrodOnTheLoaf_e.html. July 11, 2014.

Ashley, Mike. "Fairy Tale." *The Encyclopedia of Fantasy*. Eds. John Clute and John Grant. New York: St Martin's Griffin, 1999: 330–33.

Bettelheim, Bruno. *The Uses of Enchantment: The Meaning and Importance of Fairy Tales*. New York: Vintage Books, 1977.

Bottigheimer, Ruth B. *Grimm's Bad Girls and Bold Boys: The Moral and Social Vision of the Tales*. New Haven, CT: Yale UP, 1987.

Bottigheimer, Ruth B. *Fairy Tales: A New History*. Albany, NY: SUNY Press 2009.

Camporesi, Piero. *The Magic Harvest: Food, Folklore, and Society*. Malden, MA: Blackwell Publishing, 1998.

Cashdan, Sheldon. *The Witch Must Die: How Fairy Tales Shape Our Lives*. New York: Basic Books, 1999.

Cowell, Cressida. *How to Train Your Dragon*. New York: Little, Brown and Company, 2004.

Darnton, Robert. *The Great Cat Massacre and Other Episodes in French Cultural History*. New York: Basic Books, 1984.

Dégh, Linda. *Folktales and Society: Story-telling in a Hungarian Peasant Community*. Transl. Emily M. Schossberger. Bloomington, IN: Indiana UP, 1969.

Ganshof, François L. *Feudalism*. Transl. Philip Grierson. New York: Harper and Row, 1964.

Grimm, Jacob and Wilhelm. "A Tailor in Heaven." *Grimm's Household Tales*. Vol. I. Transl. and ed. by Margaret Hunt. Detroit, MI: Singing Tree Press, 1968: 141–43.

Grimm, Jacob and Wilhelm. "Mother Holle." *Grimm's Household Tales*. Vol. I. Transl. and ed. by Margaret Hunt. Detroit, MI: Singing Tree Press, 1968: 104–7.

Grimm, Jacob and Wilhelm. "Frau Trude." *Grimm's Household Tales*. Vol. I. Transl. and ed. by Margaret Hunt. Detroit, MI: Singing Tree Press, 1968: 170.

Grimm, Jacob and Wilhelm. "The Three Little Men in the Wood." *Grimm's Household Tales*. Vol. I. Transl. and ed. by Margaret Hunt. Detroit, MI: Singing Tree Press, 1968: 54–9.

Grimm, Jacob and Wilhelm. "Cinderella." *Grimm's Household Tales*. Vol. I. Transl. and ed. by Margaret Hunt. Detroit, MI: Singing Tree Press, 1968: 93–100.

Grimm, Jacob and Wilhelm. "The Twelve Brothers." *Grimm's Household Tales*. Vol. I. Transl. and ed. by Margaret Hunt. Detroit, MI: Singing Tree Press, 1968: 37–41.

Grimm, Jacob and Wilhelm. "Hansel and Gretel." *Grimm's Household Tales*. Vol. I. Transl. and ed. by Margaret Hunt. Detroit, MI: Singing Tree Press, 1968: 62–9.

Grimm, Jacob and Wilhelm. "The Magic Table, the Golden Donkey, and the Club in the Sack." *The Great Fairy Tale Tradition: From Straparola and Basile to the Brothers Grimm*. Ed. Jack Zipes. New York: W.W. Norton, 2001: 427–34.

Grimm, Jacob and Wilhelm. "The Wild Man." *The Great Fairy Tale Tradition: From Straparola and Basile to the Brothers Grimm*. Ed. Jack Zipes. New York: W.W. Norton, 2001: 323–25.

Grimm, Jacob and Wilhelm. "Iron Hans." *The Great Fairy Tale Tradition: From Straparola and Basile to the Brothers Grimm*. Ed. Jack Zipes. New York: W.W. Norton, 2001: 329–34.

Grimm, Jacob and Wilhelm. "The Thief and His Master." *The Great Fairy Tale Tradition: From Straparola and Basile to the Brothers Grimm*. Ed. Jack Zipes. New York: W.W. Norton, 2001: 359–60.

Grimm, Jacob and Wilhelm. "The Three Little Birds." *The Great Fairy Tale Tradition: From Straparola and Basile to the Brothers Grimm*. Ed. Jack Zipes. New York: W.W. Norton, 2001: 302–5.

Grimm, Jacob and Wilhelm. "Puss in Boots." *The Great Fairy Tale Tradition: From Straparola and Basile to the Brothers Grimm*. Ed. Jack Zipes. New York: W.W. Norton, 2001: 402–5.

Grimm, Jacob and Wilhelm. "Little Farmer." *The Great Fairy Tale Tradition: From Straparola and Basile to the Brothers Grimm*. Ed. Jack Zipes. New York: W.W. Norton, 2001: 21–5.

Grimm, Jacob and Wilhelm. "The Hand with the Knife." *The Hard Facts of the Grimms' Fairy Tales*. Maria Tatar. Princeton, NJ: Princeton UP, 1987: 195–6.

Grimm, Jacob and Wilhelm. "The Mother-in-Law." *The Hard Facts of the Grimms' Fairy Tales*. Maria Tatar. Princeton, NJ: Princeton UP, 1987: 201–2.

Grimm, Jacob and Wilhelm. "The Starving Children." *The Hard Facts of the Grimms' Fairy Tales*. Maria Tatar. Princeton, NJ: Princeton UP, 1987: 202.

Haase, Donald. "Children, War, and the Imaginative Space of the Fairy Tales." *The Lion and The Unicorn* 24 (2000): 360–377.

Hastings, Waller. "Moral Simplification in Disney's The Little Mermaid." *The Lion and the Unicorn 17* (1993): 83–92.

Hogan, Patrick Colm. *How Authors' Minds Make Stories*. New York: Cambridge UP, 2013.

Joosen, Vanessa. "Back to Ölenberg: An Intertextual Dialogue between Fairy-Tale Retellings and the Sociohistorical Study of the Grimm Tales." *Marvels & Tales: Journal of Fairy-Tale Studies* 24.1 (2010): 99–115.

King, Martin Luther, Jr. "I Have a Dream." Aug. 28, 1963. http://www.americanrhetoric.com/speeches/mlkihaveadream.htm. May 10, 2014.

Kingsley, Charles. *The Water-Babies: A Fairy-Tale for a Land-Baby*. New York: Hurst and Company, 1899.

Lüthi, Max. *Once Upon a Time: On the Nature of Fairy Tales*. Transl. by Lee Chadeayne and Paul Gottwald. New York: Frederick Ungar Publishing, 1970.

Lüthi, Max. *The European Folktale: Form and Nature*. Transl. by John D. Niles. Philadelphia: Institute for the Study of Human Issues, 1982.

Lüthi, Max. *The Fairy Tale as Art Form and Portrait of Man*. Transl. by Jon Erickson. Bloomington, IN: Indiana UP, 1984.

Mueller, Gerhard O. W. "The Criminological Significance of the Grimms' Fairy Tales" *Fairy Tales and Society: Illusion, Allusion and Paradigm*. Ed. Ruth Bottigheimer. Philadelphia: U. of Pennsylvania P., 1987: 217–27.

Niles, John. D. "Translator's Preface." *The European Folktale: Form and Nature*. Max Lüthi. Philadelphia: Institute for the Study of Human Issues, 1982.

Nussbaum, Martha C. *Poetic Justice*. Boston, MA: Beacon Press, 1995.

Perrault, Charles. "Little Red Riding Hood." *Folklore and Mythology Electronic Texts*. Transl. and ed. by D. L. Ashliman. 2011. http://www.pitt.edu/~dash/perrault02.html. July 9, 2014.

Perrault, Charles. "The Sleeping Beauty in the Wood." *Folklore and Mythology Electronic Texts*. Transl. and ed. by D. L. Ashliman. 2011. http://www.pitt.edu/~dash/perrault01.html. July 9, 2014.

Perrault, Charles. "The Master Cat; or, Puss in Boots." *Folklore and Mythology Electronic Texts*. Transl. and ed. by D. L. Ashliman. 2011. http://www.pitt.edu/~dash/perrault04.html. July 9, 2014.

Perrault, Charles. "The Fairies." *Folklore and Mythology Electronic Texts*. Transl. and ed. by D. L. Ashliman. 2011. http://www.pitt.edu/~dash/perrault05.html. July 9, 2014.

Perrault, Charles. "Little Thumb." *Folklore and Mythology Electronic Texts*. Transl. and ed. by D. L. Ashliman. 2011. http://www.pitt.edu/~dash/perrault08.html. July 9, 2014.

Perrault, Charles. "Cinderella; or, Little Glass Slipper." *Folklore and Mythology Electronic Texts*. Transl. and ed. by D. L. Ashliman. 2011. http://www.pitt.edu/~dash/perrault06.html. July 9, 2014.

Propp, Vladimir. *Morphology of the Folktale*. 2nd ed. Rev. and ed. by Louis A. Wagner. Austin, TX: The U. of Texas P.: 1973.

Strayer, Joseph R. *Feudalism*. New York: D. van Nostrand Company, 1965.

Sydney, Herbert. *The Fall of Feudalism in France*. New York: F.A. Stokes, 1921.

Tatar, Maria. *The Hard Facts of the Grimms' Fairy Tales*. Princeton, NJ: Princeton UP, 1987.

Turner, Mark and Gilles Fauconnier. *The Way We Think: Conceptual Blending and the Mind's Hidden Complexities*. New York: Basic Books, 2002.

Wilde, Oscar. "The Happy Prince." *The Classic Fairy Tales: A Norton Classical Edition*. Ed. Maria Tatar. New York: W.W. Norton & Company, 1999: 253–60.

Wilde, Oscar. "The Selfish Giant." *The Classic Fairy Tales: A Norton Classical Edition*. Ed. Maria Tatar. New York: W.W. Norton & Company, 1999: 250–53.

Zipes, Jack. *Breaking the Magic Spell: Radical Theories of Folk and Fairy Tales*. Austin: U. of Texas P., 1979.

Zipes, Jack, "Introduction." Spells of Enchantment: The Wondrous Fairy Tales of Western Culture. New York: Penguin Books, 1991.

Zipes, Jack, Ed. The Trials and Tribulations of Little Red Riding Hood. New York: Routledge, 1993.

Zipes, Jack. *Happily Ever After: Fairy Tales, Children, and the Culture Industry*. New York: Routledge, 1997.

Zipes, Jack. *When Dreams Came True: Classical Fairy Tales and Their Tradition*. New York: Routledge, 1999.

Zipes, Jack. "Cross-Cultural Connections and the Contamination of the Classical Fairy Tale." *The Great Fairy Tale Tradition: From Straparola and Basile to the Brothers Grimm*. Gen. Ed. Jack Zipes. New York: W.W. Norton, 2001: 845–869.

Zipes, Jack. Hans Christian Andersen: The Misunderstood Storyteller. New York: Routledge, 2005.

Zipes, Jack. Why Fairy Tales Stick: The Evolution and Relevance of a Genre. New York: Routledge, *2006*.

Zipes, Jack. *When Dreams Came True: Classical Fairy Tales and Their Tradition*. 2nd Ed. New York: Routledge, 2007.

4 Find Them and Kill Them

Retributive Justice Scripts

The previous chapter suggested that the "might makes right" conception of justice was a product of the *Old Justice* world, which supported cultural adaptations, where those on the lower rung of the social ladder were almost powerless against abuse or harm in the hands of their superiors. The resulting mental model called poetic justice, in neurocognitive terms, was an adjunctive complex—a survival-motivated thought- and action-protocol of redirecting an unachievable need into habituated reactions that ignore or negate it, yet make one feel as if they were doing something real about that need (Paanksep and Biven 111). In its feudal and then transcendentalist versions, poetic justice was a mental construct and a knowledge structure, which, as a script, operated in real-life situations but also became a structuring element of oral and literary narratives including folk and fairy tales. Although elements of retributive justice can be found in these genres, they were created for and recounted by the unpowerful, non-artistocratic audience, featured unpowerful protagonists, and were in effect dominated by the poetic justice scripts. However, if poetic justice was a bottom-up scripting within the *Old Justice* paradigm, the top-down protocol—available to the powerful—was retributive justice.

Retributive justice has not passed from cultural and social practice the way poetic justice did. Instead, it was adopted and modified to the *New Justice* paradigm through limiting its scope, regulating its application, and rationalizing its procedures. This complex transformation that began with Beccaria's 1764 call for curtailing the arbitrariness, randomness, and cruelty of retributive justice is far from complete, since objections are raised against retributive justice even today. Although the Western world has seen a tremendous advance in this field—especially through the delegalization of torture, private vengeance, slavery, ethnic, race, gender, and other forms of formerly quotidian oppression—retributive justice continues to be the foundation of our criminal law. This is due partially to the fact that retribution has enjoyed a long and bloody cultural history. With notable exceptions, it was a moral, religious, and legal norm in societies throughout the pre-modern period. Rather, it is primarily because as a response to harm or punishment for an infraction, retribution is hardwired into our affective/cognitive architecture. This does not mean that we are biologically doomed to violence. Far from it, as the following chapter will argue, under certain conditions retribution offers fewer evolutionary advantages than forgiveness and can be culturally contained. The impulse for retributive violence, however, is ineradicably embedded in our Old Brain, mammalian and primary-process affective functions.

My proposal in this chapter is to take a look at the evolution of the retributive justice script, which—over the nineteenth and twentieth centuries—has been shedding the license for unrestrained violence and gravitating toward violence in its "unavoidable minimum"—whatever this may mean. I contrast the pre-modern form of the retributive justice script with its three most common tracks to be found in modern YA speculative fiction: defensive war, Darwinist survival, and rebellion. These are distinguished by their enactor agents and the scope of violence projected as necessary to achieve justice. In all three tracks, the retributive justice script is activated by what is perceived as a harm or violation and it projects the achievement of justice as retribution against the offender. Within this script, retributive redress is seen as a justified and unavoidable response to harm, even though it is also acknowledged as possibly begetting further retribution in a self-reinforcing feedback loop. Despite this tragic flaw, in life and as in fiction, the retributive justice script often seems the only script available to deal with certain kinds of injustice. The apparent unavoidability of this script, compounded by the painful awareness of its inherent limitations, lies at the root of the current dissatisfaction with the practice of the criminal law in liberal democracies. It also sheds light on the appeal of literary and filmic representations of justice, where the killing of antagonists—sometimes humans, but often zombies, orcs, or other monsters—is projected as a resolution of the conflict. The fact that the pool of circumstances in which the protagonist can legitimately kill another character is rapidly shrinking is the effect of the rise of the restorative justice thinking that will be examined in the following chapter.

RETRIBUTION, VIOLENCE, AND THE RETRIBUTIVE JUSTICE SCRIPT

The word "retribution," meaning a type of punishment that is morally right, deserved, and extracted in recompense for a thing done or suffered, has a history going back to the Bible and other ancient writings. In its current usage it describes an ideal of *New Justice* when it comes to appropriate response to harm. As defined by Beccaria, Kant, Bentham and other *New Justice* theorists, retributive or criminal justice was constructed as a set of rational objective principles to replace the ill-defined retribution, or vengeance, that was the substance of *Old Justice* laws and everyday practice. Although much has been achieved in this respect, two and half centuries later we still find it difficult to distinguish between the two. Empirical research demonstrates that while retribution remains the main motivation underpinning punitive attitudes in life and in courts, it has retained two distinct dimensions: retribution as revenge, a reaction predicated on the desire of "getting even and making the offender suffer," and retribution as just deserts, the domain of procedural justice in the criminal courts, which seeks the restoration of justice through proportional compensation from the offender (Gerber and Jackson 72). Interestingly, the correlations between the two dimensions of retribution and their two different motivational structures support my contention that these two understandings of retribution are grounded in, respectively, *Old* and *New Justice* mindsets. Thus, embracing retribution as revenge has been revealed as tied to such objectives as endorsing status boundaries, wanting to dominate out-groups, supporting "harsh punitive measures" including "the denial of procedural fairness," and a desire to "re-establish a position of dominance" (76)—all these being *Old Justice* motivations. When retribution is conceived as just deserts, however, the leading motivation has been "a more constructive desire to communicate good moral values to society" and a reduced desire to "deny due process and respect to criminal offenders" (77): in other words, the equality imperative central to the *New Justice* thinking. Thus, *New Justice* ideas of proportionality, equality, and compensation work as a buffer against revenge.

One reason why we find it so hard to distinguish retribution as just deserts from retribution as revenge is that retributive justice is more deeply embedded in the Old Brain systems any other type of justice. In the taxonomy used by Panksepp and Biven, it builds on reptilian rage, a primary-process affect that does not need an intentional object and can be aroused by a number of stimuli, and then on mammalian anger, a secondary-process affect that combines the attribution of blame for what we identify as the wrong and an impulse to correct it. Drawing on these Old Brain emotions, retributive justice, such as hatred and jealousy, is a tertiary and thus cognitive process (Panksepp and Biven 145–7), where the affects of rage and anger are structured in a left hemispheric operation of conceptual blending that integrates the concepts of blame, responsibility, intentionality, offender, victim, hurt, wrong, punishment, and restoration, leading to a plan or determination to set things right. In other words, the self-conscious cognitive component of retributive justice is merely a surface coating to an unfolding of the reptilian brain and mammalian brain's affective arousal that is part of our capacity for violence. Because this capacity is "a universal human trait" (McCullough, *Beyond* 11)—and a generally mammalian evolutionary adaptation crafted to deal with three specific adaptive problems: to deter the recurrence of harm from the same group or individual, to deter third-party would-be harm doers, and to force free-riding members of one's own group to cooperate (48–61)—it has developed its own organs in the brain and its own cognitive programs. On the hardware side, human cognitive architecture includes at least four systems that trigger violence: the seeking system, the dominance system, the fear circuit, and the rage circuit. Since they are organized hierarchically, these systems easily interact. Fear, pain, rage, craving, dominance, and other triggers for violence tend to reinforce one another (Pinker 497–502). What gives us the edge over other mammals, however, is that the Old Brain levels—the physical organs associated with violence—occupy much less space in our brains than in animal brains and are enveloped by the much larger bihemispheric New Brain, whose sections are associated with self-control, morality, empathy, sensitivity to norms, and sociality (502–6). Such neuroanatomy implies that our Old Brain impulses for violence and revenge have to contend with the powerful systems of New Brain cerebral restraint.

On the software side, violence operates through a number of scripts. Pinker, for example, identifies these scripts by categories of and incentives for violence: in his description, the retribution-as-revenge script is one of the three main violence scripts humans employ. Its underlying incentive is safety and deterrence, its violence is predominantly instrumental, and it is chosen as an action script only in the absence of inhibiting factors (36–47). My focus here is only on the retributive justice script, not violence scripts *per se*, but I believe that the Old Brain grounding of revenge explains its cross-cultural ubiquity. Revenge has been a major cause of violence in all human societies up to the present (529). It differs from other types of violence in that it is in certain circumstances, neurologically pleasant: offering the brain stimuli identical to those activated by craving and consummating pleasures such as nicotine or cocaine (531).

Other reasons for the currency of retributive violence are biological and cultural. For example, in humans and other mammals aggressiveness and male sexuality tend to go together, making violence a heavily gendered evolutionary endowment. This, at least, is suggested by statistics: as perpetrators and victims, males outstrip females by a wide margin in all serious forms of violence. In all human societies, past and present, the levels of lethal violence committed by men and women are widely disproportionate, on average 20 to 1 (Barash 146–7). Another powerful legacy to contend with is the fact that many forms of violence have been culturally ritualized and emotionally sacralized. This has been the case with domestic violence but also with revenge and large-scale intrahuman violence called war. That, at least, is the argument advanced by the military historian John Keegan in his much-celebrated *A History of Warfare* (1993) but also by the feminist political writer Barbara Ehrenreich in her *Blood Rites* (1997). For Keegan war is the expression of culture. For Ehrenreich the violence of war is a psychological-evolutionary adaptation that includes but goes deeper than explanations of innate male aggressiveness. "[O]ur peculiar and ambivalent relationship to violence," she claims, "is rooted in a primordial experience that we have managed, as a species, to almost entirely repress: … the experience of being preyed on by animals that were initially far more skillful hunters than ourselves." If the violence of war reenacts "the human transition from prey to predator" (22), the human fascination with and ambivalence toward violence is psycho-biologically programmed.

None of this denies the fact that the capacity for retributive violence can be culturally contained. There is even evidence, to be examined in the following chapter, that it is no more a biological default than cooperation. This is not just wishful thinking. Contrary to the impression created by modern media preoccupied with violence and danger; contrary to human moral psychology, which thrives on a sense of danger; and contrary to social theories from Durkheim to Foucault that have emphasized the criminogenic nature of modern society, an increasing number of data-based studies demonstrate that violence has been diminishing in human societies over long stretches of time[1]. What Norbert Elias called the civilizing process has brought radical changes in our toleration of violence, and this trend is also notable in young people's literature. The cognitive architecture of the human brain is set to run programs that are adaptable, and the threshold cue for revenge—recently defined as "empathy erosion" (Baron-Cohen, *Evil* 6)—is getting steeper. The pacification and civilizing processes that Pinker and others have spoken about are not always hypocritical, chauvinistic arrogance[2]. As the circumstances in which humans function change, alternatives to violence become more pronounced.

This, too, is the case with revenge, which the *New Justice* project has undertaken to sharply distinguish from retributive justice: "a response to harm that includes retribution" (Capeheart and Milovanovic 45) and is carried out by the impartial state. In the *New Justice* framework—which we all mentally inhabit—revenge differs from punishment not only because it is a private, emotional response versus an impartial, state-sanctioned one. It also differs because it seeks a tit-for-tat response to harm rather than a response that *translates* harm into the most appropriate form of punishment to serve a desired effect: deterrence, rehabilitation, social defense, or something else. This ambitious project of rationalizing and humanizing the retributive justice system has been successful in many ways. For example, it did put an end to the pre-modern *Old Justice* perception where a proper response to harm—both personal and official—was envisioned along tit-for-tat lines[3]. For centuries much of Europe "lived under the sign of private vengeance," which was imposed on individuals "as the most sacred of duties—to be pursued even beyond the grave" (Bloch 123). Except in England, where the delegalization of revenge happened the earliest but was long seen as one of the aspects of the royal "tyranny," governments in Europe throughout the late seventeenth century "confined themselves to moderating the more extreme manifestations of [revenge] practices which they were unable or perhaps unwilling to stop altogether" (126). The rise of *New Justice* changed the picture. Within a century punishment became more regulated and gradually moved away from the idea of punishments as mirror images of crimes to punishments as a translation of harm. All these have been positive developments, when compared to *Old Justice*, and only in the second part of the twentieth century have the absolutist aspects of these claims been contested[4]. For all this criticism, the state-executed retributive justice is considered today a legitimate form of punishment, whereas individually executed retribution is not. The fact that the state-sanctioned criminal law is a form of retributive justice is a testimony to how nearly universal is the perception of retribution as an indispensable element of justice[5].

Retributive justice has a long history. Framed, on the one hand, by the Old Testament Israelites who justified their atrocities as Yahweh's retribution for the gentiles' evil ways and, on the other, by the Homeric Greeks, whose "rape of Troy"—to use Gottschall's phrase—was a retribution for the abduction of Helen, the history of the Western world, and much of its literature, is one of never-ending retributions. Within the *Old Justice* framework retribution and justice marched hand in hand. For much of the premodern period there was no limit to the amount of violence that could be committed in the name of a "just" retribution. For example, when Odysseus returns home from his wanderings—one episode of which was the massacre at Troy—he slaughters the 108 suitors, having first rejected their excuses and offers of compensation payments worth more than 2160 oxen; he then has 12 slave women brutally executed, kills not-at-all-guilty Leodos alongside the unknown number of the suitors' attendants, and orders a slow torture and dismemberment of Melanthius (Gottschall 148). All this is nothing unusual by the standards of Homeric violence, yet the whole episode is unique as not being a violence of war or a raid but one of retribution for a domestic offense. Homer does not go to particular lengths in explaining why the carnage is necessary in the first place. He merely assumes its unavoidability. And it is exactly this assumption that lies at the heart of the retributive justice script.

Evidenced in works of heroic literature—from *The Epic of Gilgamesh*, *Mahabharata*, *the Bible*, *Iliad*, *Odyssey*, *Aeneid*, *Beowulf* to other grand works of the mythic age—this basic form of retributive justice script is defined not so much by its violence as by its tragic inevitability. There is no alternative to the retributive justice script for men caught and dying in the endless cycles of retributive violence depicted in heroic literature. Retribution is meted out on behalf of Gods, family, kin or tribe, but it is never the last word. Those who win at a given point are doomed to lose at another. The unfathomable rage of never-ending wars and raids that lay waste to entire regions and bring mass genocide is acknowledged with grim resignation as inescapable, imposed on humanity by God(s), fate, or nature acting through men's passions. As Gottschall has shown, the shadow of ever-present violence that hangs over protagonists in myths and epics is responsible for the dominant mood of suffering and futility that permeates these works (141). The inevitability of violence is a vicious circle that helps create, reinforces, and sustains specific patterns of behavior, gender roles, and value systems. It locks people in a thought- and action-protocol of enacting retributive violence called the retributive justice script.

The two signature phenomena of the retributive justice script are the retribution imperative, no matter what the cost, and the license for unrestrained violence—not merely against the aggressor but against anyone that may be associated with them. This script can be reconstructed from heroic and epic literature but also from social and political practice throughout the late seventeenth century, especially in large-scale bursts of violence that accompanied pogroms, crusades, wars, and colonial conquests. The retributive justice script was also a staple of literature. It informs the spectrum of violent punishments described in the *magnum opus* of retributive equivalence, Dante's *Divine Comedy* (1321). It accounts for much of the tragic appeal of Shakespeare's tragedies, from *King Lear* to *Macbeth*, and it frames Milton's *Paradise Lost* (1667). Most of these pre-modern masterpieces do not set limits on the amount of justified violence. They reflect an *Old Justice* world, in which attempts to contain people's predilection for violence over those less powerful were largely unsuccessful. Only since the late seventeenth century, with Europe reeling from the religious wars, were the first questions posed about a viable rationale and acceptable scope of retribution.

The rise of *New Justice* brought a major change in the retributive justice script. In the course of a complex cluster of processes discussed, among others, by Ruff and Pinker, retribution has become milder. The imperative for retribution has survived—in human cognitive architecture as well as in law—but over the past two centuries the violence it entails has been significantly curtailed. For example, a recent evolutionary interpretation of Odysseus' retributive rampage, Gottschall's argument in *The Rape of Troy*, accounts for it by the shortage of women in Homeric society and by Odysseus' "specific sense of violated sexual ownership" (66). This may well be so. Yet, if a modern Odysseus razed the entire neighborhood of someone who made advances to his wife, no explanation would even come close to legitimizing it. What Gottschall's study makes clear in the first place is that the *New Justice* sensitivity requires incomparably stronger reasons to accept retribution than an *Old Justice* one. Unrestrained violence is no longer seen as a default reaction, in part because we have more justice scripts to choose from and in part because the retributive justice script has evolved to require convincing explanations for the use of violence in its unavoidable minimum. The essence of the change is captured by the late twentieth-century popularity of sentiments such as "we are not like them" or "we are better than that" voiced in response to violent harm. In the *New Justice* framework, the function of law, and the role of punishment, is to reduce "recourse to the use of force in the settling of disputes" (Cragg 8).

THE RETRIBUTIVE JUSTICE SCRIPT IN MODERN YA SPECULATIVE FICTION

Although examples of the pre-modern retributive justice script can be found in many nineteenth-century fairy tales—like the stepmother's dogged persecution of Snow White—I want to illustrate it with an account from a modern YA novel that retains all the qualities of this script as it informed the actions of Odysseus and other pre-modern characters. A group of armed men stalks along the shore. Each carries two shields and a sword or spear. The leader has a helmet with a ridge across the top, two chin-protecting panels on the sides, and a hawk's mask visor that makes him weird and otherworldly. Reaching a cove, the men set a small fire under a metal pot. When the brew is ready, they drink it. The leader smears the leaves from the pot into the face and clothes of an 11-year-old boy who accompanies them. For the men drunk with the wolf-brew, only smell allows distinguishing a friend from an enemy. After a stealthy run, the panting and frenzied warriors stop at the edge of a bluff overlooking a village.

Pandemonium starts when the group descends on the village. Torches are thrown on thatched roofs, and every living being is put to the sword. Villagers are felled by rocks, clubbed or speared, or run through or brained with axes. They are thrown on the ground and chopped up. Their bodies' contents pour forth in viscous torrents: brains, eyes, and intestines splash on the ground, mixing with severed heads, arms, and legs. Blood pours everywhere. Men, women, and children are trapped in burning houses, which collapse on their heads. Some try to escape but in vain. The boy who watches from the bluff, unable to move or look away, sees the leader behead a young woman and throw her baby back into the flames. When no humans are left alive, cattle are driven from the barns and likewise slaughtered. By the time the sky turns pink, what used to be a village is a collection of smoldering heaps strewn with parts of human and animal corpses. The traumatized boy wanders back to the beach and wades in, looking for the consolation of death. When he comes to his senses in broad daylight, standing above him is the leader of the group. As he flops down to clean the blood from his sword, the giant Viking says: "I hope you're thinking of nice things to say about me" (Farmer, *Sea* 161).

The above is a description of a berserking raid in Nancy Farmer's *The Sea of Trolls* (2004), rated grades 6–9. Set in the extremely violent eighth-century England, Scotland, and Scandinavia, the book follows the adventures of an Anglo-Saxon bard-apprentice in a world torn by a clash of three cultures: Anglo-Saxon Christians, Viking Norsemen, and Celtic as well as Pictish polytheists. Carried away by the rampaging Northmen, Jack—the reader's alter ego in the story—is made to witness the raid so that he can compose a saga in praise of the berserkers' leader. He learns that in the Norse worldview, the raid counts as a justified retribution—a noble deed that shows respect to an oath-breaker chieftain who owns the village. The fact that the slaughtered people are innocent, or that the oath-breaker Gizur Thumb-Crusher may not even be there, is irrelevant. The inhabitants fall to what the Vikings take as a justified retribution. Like Odysseus, so too Olaf One-Brow is bound to defend his honor, violated by Gizur's unnamed offense. The only way it can be done is by enacting the retributive justice script against Gizur or, in his absence, unleashing death and destruction on anything or anyone associated with him, such as one of his villages. When Jack hears that Olaf's men will first poison the village dogs, he asks the Viking why he cannot then take what he wants without fighting. Olaf is offended. "What kind of honorless brute do you think I am?" he asks. "If I took Gizur's wealth without engaging in battle, I would be no better than a thief. I would show him no respect" (156). The understanding of honor, of what is offensive or not, and of what is the appropriate response to harm are as puzzling to Jack as they are to the reader. They place Olaf's actions within the *Old Justice* thinking, where the causes for retribution are many and the limits on it are few.

The rise of *New Justice* made such logic unacceptable but did not eliminate retributive justice altogether. Instead, it rationalized it by linking its legitimacy to specific sets of circumstances. In YA speculative fiction the credit or the blame for reintroducing the two *New Justice* versions of the retributive justice script goes to two authors: the British socialist fantasist William Morris and the American cross-genre author Robert E. Howard. In terms of their childhoods and education Williams and Howard hailed from two different worlds. Morris was born into a well-to-do bourgeois family and enjoyed a pampered childhood in the 1840s, including his own suit of armor and a pony he rode through a forest near his family estate. Howard, the only child of a travelling physician, grew up surrounded by violence in pre-WWI Texas and lived a frontier lifestyle with few possessions. Both authors, however, were drawn to retributive justice. Each drew from the revival of interest in myth that made available to Western readership the world of *wyrd* and retributive justice depicted in Norse, Icelandic, and Celtic myths.

As a co-translator of a number of Icelandic sagas and epics, Morris was soaked right through with the Norse worldview[6]. Irresistibly drawn to tales of slighted honor, conflicting loyalties, and bloody revenge, Morris was also a committed socialist. He abhorred the devastation brought by the Industrial Revolution and was fascinated with a stage in the development of civilization called "upper barbarism." In his eyes, "the upper barbarian was a noble, vigorous, and naturally moral human being, living in the heroic community of the gens or clan and dwelling in political, social, and economic cooperation with his fellows" (Silver 121). In line with his own declarations that the leading passion of his life, except for art, was the "hatred of modern civilization," and that he looked forward to "barbarism once more flooding the world" (qtd. in Silver 121), Morris identified upper barbarism with ideal communism and made its defense the theme of the first two among his eight fantasy romances. Before his interests drifted toward somewhat mystical and psychological quests, in *The House of the Wolfings* (1889) and in *The Roots of the Mountains* (1890), Morris described how the noble kindreds of the Mark resisted the invasions of powerful enemies. In doing this, he not only reimagined the history of Northern Europe but also forged the first *New Justice* version of the retributive justice script: its defensive war track. In *Wolfings* the rotten civilization of Rome is defeated in Germania; in *Roots* the repelled enemy is the Huns. In both novels a powerful enemy is routed through heroic resistance. Both novels are tales of retributive justice and follow the same structural pattern.

There are four signature markers of the defensive war track of the retributive justice script. One is that the acting agent is the entire community coalesced in the face of a threat. Two is that the acting agent is presented as innocent of any previous harm against the attacker, with the current conflict being an unprovoked aggression and a violent response being the only means of survival. Three is that the aggressor meets a justified retribution, often involving the killing of the aggressors to the last man. Four is that the retribution is asserted as ensuring the resumption of a happy, peaceful life from before the aggression.

The House of the Wolfings is a model actualization of this script. In the first part of the book Morris goes to lengths to describe the Wolfings as peace-loving people living in democratic community with the neighboring tribes. "Merry was the folk with that fair tide, and the promise of the harvest, and the joy of life, and there was no weapon among them so close to the houses, save here and there the boar-spear of some herd-man or herd-woman late come from the meadow" (Morris, *Wolfings* 7). This idyllic existence is then threatened by the Roman invasion; from Chapter 2 through the final Chapter 31, the novel tells of how the people of the Mark unite and fight a series of lost battles in an attempt to stop the advance of the Roman army. In the course of the campaign, the Romans commit a number of atrocities, and many of the Wolfing warriors die. Often they do so, when they bravely respond to tragic dilemmas, such as that faced at one point by a small party led by the war-duke Otter. If they attack right away, they may prevent the slaughter of women and children in defenseless Wolfstead, but since they are badly outnumbered, they will likely be wiped out. If they wait for the arrival of Thiodolf's reinforcements, however, the Romans would have slayed the civilians. Having made up his mind to attack, the band rides on against the Romans. Otter heroically dies and only a handful are still alive when Thiodolf's force arrives and allows them to withdraw.

Before the final battle a couple of days later, and in line with the goal of justice projected in the retributive justice script, Thiodolf declares that "no Roman man must be left alive" (166). Warming his warriors with declamations of heroic poetry, Thiodolf then leads the Markmen against the enemy. The description of the battle takes three chapters. It includes episodes of a berserker rage that carries Arinbiorn to his death, a description of the Roman attempt to burn alive a number of the Wolfing captives, and extended passages about the eventual massacre of the Romans in the Wolfing Stead. Inevitably, the battle ends in mass slaughter.

> Then fell the first rank of the Romans before those stark men and mighty warriors; ... And the Romans, they had had no mercy, and

> now looked for none: and they remembered their dealings with the Goths, and saw before them, as it were, once more, yea, as in a picture, their slayings and quellings, and lashings, and cold mockings which they had dealt out to the conquered foemen without mercy, and now they longed sore for the quiet of the dark, when their hard lives should be over, and all these deeds forgotten, and they and their bitter foes should be at rest for ever. (181)

When the ramparts are breached, the advancing warriors are joined by a frenzied crowd of "old men and lads and women" (181). No longer able to restrain themselves, they fall on the thinning Romans and tear them to pieces. As the remnants of the Romans are hewn down, Thiodolf saves the Wolfing captives but is mortally wounded and dies a heroic death. He does not die in vain, though. As a result of this victory, the Wolfings and other tribes return to their peaceful lifestyle and "make good all that the Romans had undone" (198). The novel ends with a declaration that "the Wolfings throve in field and fold, … [growing] more glorious year by tear" and that no Roman army "ever fell on the Mark again" (199). In this way the plot of *The House of the Wolfings* is structured on the four key markers of the defensive war track.

Like Morris in his romances, Howard in his barbarian fiction also drew on mythic patterns, especially Irish-Celtic history, myth, and legend (Eng 32). What attracted Howard to the mythic blend of history and legend was the clash of barbarism and civilization—a variation on the theme of Darwinist survival evident already in his first professional sale, "Spear and Fang" (1924), a story in which a Cro-Magnon rescues his mate from a Neanderthal. Howard's interest in things mythic was then strengthened by his acquaintance with H. P. Lovecraft. In the course of their correspondence, Howard admitted he "wished that he had been born a barbarian or on the frontier of the previous century" (de Camp 147) and passionately debated the merits of barbarism over civilization (Finn 145–75).

Howard's ideas about barbarism were different than those of Morris, though. He did not believe in an idyllic barbarian community any more than in an uncorrupt civilization. As he understood it, barbarism was prowess, survival skills, and simplicity; civilization, by contrast, meant indolence, decadence, and limitations on individual freedom (Eng 35, 53). These ideas sat well both with his personality and with his experiences of growing up in Texas boomtowns. His father being a traveling physician, Howard saw injury and violence and heard firsthand stories about their effects (Finn 35); often a new kid at school, he saw and confronted bullies, both among teachers and schoolmates (de Camp 137). As a result, Howard developed an appreciation for physical strength and a "natural inclination toward fiction of violence" (Knight 123). His tales all reflect "a gloomy conviction … of the ultimate failure of man's best efforts toward peace: 'Barbarism is the natural state of mankind. Civilization is unnatural. It is a whim of circumstance. And barbarism must always ultimately triumph'" (Leiber 7).

Conan the Cimmerian was the last one and the most fleshed out of Howard's literary characters. He grew out of Howard's earlier work, especially the Kull and Kane tales, but also reflected Howard's familiarity with the Texas country roughnecks (Knight 121). All these characters blended into Conan and made him the most recognizable sword and sorcery character ever. But there was more. In the twenty-one tales of the lone barbarian Howard completed, he gave the fullest expression to the Darwinist survival track of the retributive justice script. As shaped by Howard, the Darwinist survival track is characterized by three markers. First, the acting agent is a lone figure, usually a barbarian outcast or mercenary, motivated by his own survival. Second, the acting agent is largely amoral, lashing out violently against any obstacle, danger, conspiracy, or threat. Third, the retribution, usually death, that the acting agent's opponents meet with does not bring the end to violence. In the Darwinist survival track, the world is assumed to be a place of never-ending conflict. Retribution is seen as the natural, default action-protocol.

These elements are pronounced, albeit to differing degrees, in each of the original Conan tales. For example, "The Phoenix and the Sword" (1932), the first of the Conan tales, tells the story of a failed assassination coup against Conan, a middle-aged usurper king of Aquilonia. The story opens with a prologue from *The Namedian Chronicles,* which depicts Conan as a barbarian conqueror: "black-haired, sullen-eyed, sword in hand, a thief, a reaver, a slayer, with gigantic melancholies and gigantic mirth, [coming] to thread the jeweled thrones of the Earth under his sandalled feet" (Howard 7). The first section of the story describes the nest of vipers that his palace is. On that particular night, a noblemen outcast Ascalante and his mage-slave discuss the upcoming assassination of Conan in the hands of other noblemen who think Ascalante is working for them. Obviously, he is planning to use and then dispose of them to become king himself. In section two, which depicts the king's evening conversation with his councilor, dejected Conan is presented as an unhappy king. As his grumblings make clear, Conan became king by accident, merely because one person he had slayed happened to be the last king of Aquilonia. All he ever wanted, however, "was a sharp sword and a straight path to my enemies" (11). In sections three and four the upcoming coup acquires an additional supernatural twist. The mage-slave recovers his long-lost ring of power and sends a demon to slay Ascalante and any witnesses. Conan, in turn, has a dream in which he is called across the gulfs of time and space to meet a long dead mage who transforms his broadsword into a magic weapon. In section five, taking more than half of the story, Conan single-handedly massacres the assailants and then slays the demon too.

This paradigmatic story highlights all three elements of the Darwinist survival track. Conan is a lone figure who fights for his own survival even though he is now a king. In defeating his would-be assassins, Conan administers retributive justice, but as an assassin himself, he does so in the name of Darwinist survival rather than moral desert. In his world, all conflicts are solved by force, be it the scramble for the crown as in this story, a brawl in a tavern as in "The Tower of the Elephant" (1933), or a clash of riding parties as in "The Frost-Giant's Daughter" (1976). In line with Howard's view of barbarism, Conan as "the natural killer" embodies "the unconquerable primordial" (Howard 22). This primeval barbarism saves him not only from the assassins but from the demon as well. When the horror paralyzes the civilized men in the room, it rouses in Conan a berserking fury that allows him to slay the demon. A lone survivor of the massacre, Conan reconnects with his barbarian self, but the fun ends too soon. His nature is war, not peace, and he will go on being a grumpy king who misses everyday brawls. Fortunately, his triumph that night is a provisional solution. It will not prevent other plots against Conan in the future, like the one described in "The Scarlet Citadel" (1933).

The three signature phenomena of the Darwinist survival track recur in other tales as well. First, whatever alliances Conan strikes, as a king, lover, mercenary, pirate, thief, or in any other capacity, are always temporary. He always acts for himself, even though he may be serving a community that pays him. Second, all his triumphs are the triumphs of force in an amoral and inherently violent world. Usually, like in "The Tower of the Elephant" (1933), "Queen of the Black Coast" (1934), or "Red Nails" (1936), they involve the death of everyone but Conan. Third, no retribution is a final solution; instead, each marks another episode of Darwinist survival and confirms the logic of the retributive justice script.

Between themselves, Morris and Howard shaped the two dominant tracks of the retributive justice script in modern YA speculative fiction. Morris' defensive war track set the tone for mythopoeic and heroic fantasy and other genres, in which the hero's triumph is morally desirable—a victory in the name of a community that defends its values. Howard's Darwinist survival became a model for sword and sorcery fantasy and its generic hybrids, in which the hero functions in a kill-or-be-killed world and wins because he—it was always he—is the toughest. The two tracks then spilled over to other genres and gave rise to a modern superhero concept. For Morris and Howard the retributive justice script was attractive not only because of their love of myth. It was attractive—and set them at odds with earlier authors who embraced the poetic justice script—because of their secularism. Morris and Howard constructed worlds where religion, if any, is a force of oppression and where justice must be actively sought in the material present. Their preference for retributive justice reflected a rejection of the complacency of civilized life in the late Victorian England and the post-frontier Texas respectively. It was also supported by a belief they shared about living at the time of civilizational collapse: a time when the violence of retributive justice is fully legitimate.

Like the poetic justice script described in the previous chapter, the retributive justice script coined in Morris' and Howard's works was a product of specific circumstances. Morris offered an anti-imperialist script of resistance to an armed invasion that sought to make a free nation into a colony. Published in the same year that saw the beginning of the second Anglo-Boer War (1899–1902)—in which the British eventually managed to convert the Boer republics into British colonies—*The House of the Wolfings* expressed Morris' spirited defense of independent ways of life. It projected the retributive justice script as a legitimate form of resistance to colonialism and did so at the time when the British Empire reached the zenith of its colonial expansion. Howard, in turn, offered a spirited praise of individual Darwinist survival at the time of civilizational upheavals. The first Conan story was published at the height of the worst economic meltdown in American history, when the collapse of market mechanisms left people to their own devices. Politically too, the Conan years 1932–36 were far from mellow. Preceded by the rise of communism in Russia and of gangsterism in the United States, they were contemporaneous with the rise of totalitarianisms in Europe and Japan—all of which bespoke of civilization soon crumbling under the tides of barbarism. Although Howard did not live to see the carnage of WWII, the track of the retributive justice script he canonized in his stories paved the ground for the emergence of a long list of lone superhero vigilantes, starting with Superman in 1938, Batman in 1939, and Captain America in 1941.

In this sense, Howard's fiction was part of the axial age of the shaping of the retributive "American monomyth." As described by John Shelton Lawrence and Robert Jewett in their excellent yet overlooked *The American Monomyth* (1977) and then *The Myth of the American Superhero* (2002), this archetypal plot pattern had become the driving force of American popular culture by 1941. Derived from secularized Judeo-Christian dramas of community redemption, the American monomyth crystallized as a script, in which a community in a harmonious paradise is threatened by evil. When societal institutions fail to contend with it, a selfless superhero emerges to "carry out the redemptive task" (Lawrence and Jewett, *Myth* 6). He then rides out into the sunset and withdraws from the society. Although Howard did not invent the superhero concept, he penned an adaptable script for stories about naturalistically legitimate retributive justice administered by a lone outsider who prowls the fringes of civilization. It does not come as a surprise that Jerry Siegel and Joe Shuster, the creators of Superman, were avid readers of *Weird Tales*, the magazine that published most of the Conan stories (Bowers 11–25).

The dangerous years of WWII and then of the early Cold War conflicts provided a fertile soil for these two tracks of the retributive justice script to strike deep root in narrative fiction. The spectacular boom of fantasy and science fiction in the 1950s and 1960s spread the retributive justice script through many genres, its memetic cross-pollination resulting in the creation of numerous hybridic versions, often peculiar for specific genres. As represented by the works of Lewis and Tolkien, for example, mythopoeic fantasy continued to draw on the poetic justice script, but it also relied on the defensive war track of the retributive justice script. Writing fantasy predicated on the Christian worldview, Lewis and Tolkien were nevertheless uneasy with the retributive justice script. As a result, they toned it down with forgiveness and attempts at reconciliation—a fact largely overlooked in most current interpretations of their work and completely erased from their filmic adaptations. In other genres too the retributive justice script was undergoing various adaptations. By the late twentieth century, however, three tracks of the retributive justice script in YA speculative fiction seemed dominant: two evolved from those established by Morris and Howard, and a recent track coalesced from blending their elements.

The Darwinist survival track functions as a spectrum of actualization across various generic media, but its appeal seems more limited in literature than in movies and computer games. In its hard version, it remains the hallmark of sword and sorcery fantasy, military science fiction, dark epic fantasy, and post-apocalyptic horror. The fact that the first two of these are almost extinct by now testifies to how problematic this track is for extended narrative fiction. Post-Howardian sword and sorcery fantasy enjoyed a revival in the *Flashing Swords!* anthologies (1973–81) edited by Lin Carter but with the demise of the series, it largely disappeared from the publishing market. Military SF, like Heinlein's *Starship Troopers* (1959), fell out of favor by the end of the Cold War and, with the shift to the defensive war track, became part of apocalyptic SF. The Darwinist survival track has fared well in dark epic fantasy, like George R. R. Martin's A Song of Ice and Fire series (1996–in progress), but even better in dystopian superhero fiction, like the Watchmen series (1986–7), and post-apocalyptic horror, like The Walking Dead franchise. In its focus on graphic violence, the Darwinist survival track works best in short narratives and visual arts rather than in novels. Martin's series is an exception, but even so its appeal rests largely on masking rather than exposing the Darwinist survival track it employs. When examined across all genres of YA speculative fiction, the Darwinist survival track is most frequently found as a subplot in narratives that employ the trope of the revenge of the dead, like Vera Brogsol's *Anya's Ghost* (2011) or Maggie Stiefvater's *The Raven Boys* (2012).

Where it fails in narrative fiction, the Darwinist survival track is a foundational structuring for movies about unavoidable revenge, from *Desperado* (1995), *Kill Bill* (2003), *Taken* (2008), *Inglorious Bastards* (2009), and *Safe* (2012), and in filmic/comic stories of post-apocalyptic horror, like *World War Z* (2013) and *The Walking Dead* (since 2010). Both categories are addressed to adults and glamorize ultra-violence directed against ruthless criminals or zombieized monsters. When it comes to the YA audience, the Darwinist survival track remains the dominant script for the genres of first person shooter and real time strategy video games, which came to prominence with *Wolfenstein* (1992) and *Warcraft: Orcs and Humans* (1994), respectively. Both genres foreground the joy of retributive violence over contextual explanations for it. Each has generated its spin-off novels. In its indisputable predilection for violence, however, this script seems to work better as a computer game than a literary narrative. Unless, of course, violence is made into joyful play.

This is the case in the superhero genres, where lone vigilantes or, more often now, superhero gangs, execute toned-down retributive justice, acting *both* for the fun of action it brings and in the name of some greater good. If this latter aspect is highlighted, the retributive justice script is usually couched within the social justice script, with retributive measures serving as a means to redress social injustice. All of these genres employ the Darwinist survival track in its softened and more communal form. The tendency for superheroes to proliferate and join superhero gangs can be traced to the 1960 premiere of DC Comics' *the Justice League of America*, followed in 1961 by its competitor Marvel Comics' the *Fantastic Four*; this trend is so strong today that superhero gangs seem to have almost led to the extinction of solitary superheroes. The multiplication of superheroes—also of women, children, and animal superheroes—is fundamentally at odds with the idea of a mythic or even Conanic hero, who had always been a solitary male, a class of his own. Whether or not this trend may be leading toward the extinction of the superhero concept itself, the popularity of superhero gangs is symptomatic of modern democratic models of social action and reflects elements of recent communitarian approaches to justice[7]. Although superheroes continue to enjoy violence and jump at an opportunity for a good fight, the retributive justice they administer is toned down to a necessary minimum. In *Ben Ten*, *Superhero Squad*, and all other modern superhero narratives, villains are captured or incapacitated rather than killed[8].

Unlike the Darwinist survival track, the defensive war track remains a much more widely adaptable cross-generic pattern in YA speculative fiction. It can be found in works of science fiction and fantasy alike—for example, Orson Scott Card's Ender's Game series (1985–99) and Brian Jacques Redwall series (1986–2011)—as well as in comics, movies and games, such as the Transformers or Avengers franchises. This track accommodates plots involving large-scale conflicts but only those that envision no mid-way solution and project one side's survival as entailing the other side's annihilation. Many blockbuster films like *Independence Day* (1996), *Transformers* (2007), *The Avengers* (2012), *Ender's Game* (2013), and *Pacific Rim* (2013) are structured on the defensive war track and invoke the myth of pure evil. The aggressors are aliens, whose motive is a total conquest informed by a desire to inflict chaos and suffering on innocent victims. All such aggressors must die. This is also a norm in animated films, from *A Bug's Life* (1998) to *Monsters vs. Aliens* (2009). Narratives that problematize the sharp distinction between good and evil, question violence, or the effectiveness of retributive justice—like *The Battle for Terra* (2007), or *Avatar, the Last Airbender* TV series (2005–8), and graphic novels (2013–14), which I will later discuss for their use of the global justice script—often start with the retributive justice script but then discard it in favor of the restorative and/or global justice solutions.

The most recent track of the retributive justice script—and one predominant in contemporary YA speculative fiction—is the rebellion track. This structure borrows from the Darwinist survival track the focus on an individual's survival but couples it with the moral desert of standing up for a greater communal good that informs the defensive war track. Indebted to the ideology of civil rights struggles and to the post-Vietnam awareness of the ambiguous character of military conflicts, this track accommodates the basic quest structure of most YA narratives. In the rebellion track, a humble individual, often a child, becomes a victim of abuse or injustice that is part of the oppressive system, in which he or she lives. Propelled by circumstances, the protagonist becomes a rebel-fugitive, either coalescing forces against oppression or becoming implicated in the preexisting resistance and crucial for its eventual triumph. Each trajectory leads to a coup or revolution. This track found its paradigmatic example in the Star Wars saga, with its division into the rebels versus the empire. In recent YA fiction it informs, for example, Lyra's adventures in Philip Pullman's *His Dark Material Trilogy* (1995–2000), Kitty's rebellion against the tyranny of the magicians in Jonathan Stroud's *The Bartimeaus Trilogy* (2003–6), Katsa's joining the secret Council that seeks to right the abuses of power in *Graceling* (2008), Eragon's flight from Galbatorix and his joining of the Varden rebels in the Inheritance Cycle (2003–11), Katniss' rebellion against the Capitol in Suzanne Collins' *The Hunger Games* trilogy (2008–10), and righteous rebellions in countless other novels. The retributive righting of wrongs executed in this script is also subject of numerous action-adventure role-playing computer games, like *Assassin's Creed*, *Skyrim*, *The Witcher*, and others.

Given how many different actualizations of the retributive justice script there are, I now want to take a closer look at representative examples that illustrate each track's potential. A look at revenge of the dead motif in Nancy Farmer's *The Islands of the Blessed* (2009) will suggest why even this form of the Darwinist survival track is seen as problematic in modern fiction. The continuing appeal and limitations of the defensive war track will be discussed on the basis of Brian Jacques' *Redwall* (1986). Philip Pullman's *The Golden Compass* (1995) will highlight the relevance of the rebellion track for challenging political oppression.

Like its cognate motif, the curse, the revenge of the dead is a literary device that goes back to pre-modern literatures and rests on the script that recognizes retribution as indispensable for meeting the claim for justice, even when held by the dead. Yet, when the imperative for retribution is frustrated by the limited possibility for action the dead have, the result is that their revenge is often lashed out against the innocent who happen to be available. Such acts, of course, invalidate the claim for justice the dead may have and lock them, as well as their victims, in ever-expanding circles of revenge. In its assumption that the dead will not be able to "move on" until their claim for justice is met—and in pitting the often innocent protagonist against vengeful spirits who must be vanquished if the protagonist is to survive—the revenge of the dead motif is a mirror blend of the Darwinist survival track. It both validates and questions retribution as a way to meet one's claim for justice.

This revenge-justice dynamic is conspicuous in the dead mermaid subplot framing *The Islands of the Blessed*, the third volume of Nancy Farmer's mythopoeic fantasy series The Saxon Saga (2004–9). *The Islands* recounts Jack's and his mentor the Bard's attempts to appease the *draugr*: an undead spirit who seeks revenge for the wrong it suffered. The *draugr* was once a beautiful mermaid who fell in love with a young monk performing solitary penances on a storm-lashed island. Called to the island by the sound of a magic bell that Father Severus had discovered in a cave, the mermaid fell in love with him. All winter the mermaid kept him alive by bringing supplies, and all these months Father Severus assumed this to be the work of angels. When the mermaid eventually showed herself, Father Severus rejected her as a soulless beast; eventually, he burned the scaled fishtail the mermaid could put on or off and made her his slave. Within months of living on land, she turned into a mute beast. Father Severus abandoned her when he was picked up by a ship. The mermaid, now a sea hag, swam after him but drowned and became a *draugr*.

As the novel opens, the *draugr* arrives in England, thirsting for revenge. The Bard recognizes that the demon "has a genuine grievance [and] has earned the right to seek justice" (Farmer, *Islands* 86). By having already killed innocent victims, the *draugr* has forfeited the right to Father Severus' life. Yet, as the Bard declares, justice demands that Father Severus pay for what he did and that the *draugr's* claim be met. The fact that the *draugr* will not be satisfied with anything less than Father Severus' death, and that Father Severus considers his action merely a "bad judgment … [with] no harm done" (185), brings about a deadlock. Seeing Father Severus as guilty yet not deserving death, the Bard chooses to resolve the dispute by laying down his own life in exchange for Severus'. In a sacrificial plot-twist, he enters the grave with the *draugr* and allows himself to be entombed with her.

The Bard's willing assumption of responsibility for someone else's mistake is a recognition of the fact that the *draugr* is legitimately "owed life for life" (372); this act, the *draugr* subplot as well as the episode with a freeing of the vengeful spirits of sacrificial victims all acknowledge the retributive principle of revenge as a valid claim for justice. At the same time, as is made clear to the *draugr* and to these ghost victims, even righteous revenge does not quench one's pain or undo the hurt. The only way to put these behind is to let go of the past; to replace the focus on grief with a longing for the good things that went before or may likely await in the future. "You will find … joy again," the Bard tells the *draugr*, "but only if you let go of this world" (371); "Now your long vigil is over," he likewise tells the spirits. "You must go into the west, there to be restored and in time to return with the sun" (308). Thus, even though the revenge of the dead motif is based on the recognition that murder calls for retributive violence, narratives such as Farmer's *The Islands* problematize revenge as a satisfying solution. While acknowledging a certain unavoidability of retributive violence, they recognize that it tends to trap the living and the dead in an unending round of destruction.

The same paradox of continuing appeal mixed with the awareness of inherent limitations can be found in works that actualize the defensive war track—for example, in Brian Jacques' *Redwall*. This opening novel of what eventually became a 21-volume animal-epic fantasy, The Redwall Series (1986–2011), recounts adventures of anthropomorphized mice and other woodland creatures in a quasi-medieval world clustered around Redwall Abbey. All Redwall books employ the retributive justice script, and the majority are actualizations of the defensive war track.

Published almost exactly 100 years after *The House of the Wolfings*, *Redwall* has all elements of this track as expressed by Morris. The plot recounts how a peace-loving community is forced to defend its way of life against a diabolical enemy and how the slaying of this enemy ensures the return of peace. The novel's first ten chapters delineate the sides of the conflict and its character. The chapters that introduce Redwall characters idealize a peace-loving community of mice, who long ago "took a solemn vow never to harm another living creature, unless it was an enemy that sought to harm our Order by violence" (Jacques 5). Mossflower Woods are then invaded by the rat Cluny and his horde of wild rats. The chapters that introduce the invaders make sure that there is not one redeeming quality about them. The one-eyed Cluny the Scourge is "an evil rat with ragged fur and curved, jagged teeth," the most savage bilge rat imaginable, and "the biggest, fiercest, most evil-looking rat that ever slunk out of a nightmare" (19). Looking "like the Devil himself" (20), Cluny lives up to these descriptions in his words and actions. Constantly squabbling among themselves, Cluny's horde are no better. The group consists of "huge, ragged rodents, bigger than any [Matthias] had ever seen [whose] heavy tattooed arms waved a variety of weapons—pikes, knives, spears and long rusty cutlasses" (19). Following this introduction, the horde shows up at the gates of Redwall and Cluny spells out his terms: unconditional surrender, all property, and all lives at his mercy. The enraged Abbot rejects these demands and swears resistance to the last creature.

With the conflict thus drawn, the rest of the book concerns the proper response to the threat Cluny brings. It also splits into two subplots that are eventually merged by the end. The first involves young Matthias' quest to recover the long lost sword of Martin the Warrior. This quest first takes Matthias to the realm of the sparrow society, called the Sparra, then to the communist country of shrew guerilla comrades, then to an aristocratic realm of a squire cat and captain owl, and eventually pits him against the ancient evil of Mossflower Wood, the adder Assmoreus, whom Matthias kills. The other subplot involves a series of Cluny's moves to conquer the Abbey. From stealing the relic tapestry of Martin the Warrior, through impressing local weasels, stoats and ferrets into his army, to the assaults on the Abbey, Cluny uses it all. When Redwall falls, it falls to betrayal. Threatening to kill his entire family, Cluny forces a dormouse to open a side door of Redwall and let his horde in. With the defenders in chains, the next morning Cluny stages a mock trial to deliver his judgments. "There is only one law," Cluny announces, "my word! There is none to stop me, not badgers or hares or otters or mice. I will kill you all. Kill, kill, kill" (324). This spectacle is interrupted by the appearance of Matthias and the attack of shrews and Sparra braves who also release the Redwall captives. Matthias kills Cluny, and the horde are given no quarter. When the battle ends, "[n]owhere was there one of Cluny's infamous horde left alive" (329). The final chapter of the book, set one year later, describes the return of peace and prosperity. As the chronicler says, "the grass is green, the sky is blue, and the honey sweeter than ever before" (333).

The plot of *Redwall* highlights all four signature markers of the defensive war track: the innocence of the attacked, the viciousness of the aggressors, the inescapability of a justified retribution, and the effectiveness of this measure to ensure the return of peace. This clear-cut solution is also its weakness, though, for it requires absolutizing evil. This move, in turn, entails stereotyping entire groups/species into inherently good-natured—like mice, badgers, hares, moles, squirrels, and hedgehogs—and those irredeemably evil: rats, weasels, stoats, ferrets, rooks, magpies, crows, adders, and foxes. As exemplified in *Redwall*, the defensive war track demands moral absolutism that is highly problematic today. And it is exactly this moral absolutism that makes Jacques' novels, in one critic's words, "internally troubling." Besides negative stereotyping, it may lead to "terribly misguided, although ... likely messages for the reader to take away from this book: that 'right action is identical with the position held by myself and my friends,' or worse, that 'right action is the slaughter of one's enemies'" (Rostanowski 92).

These limitations are largely missing in the third track of the retributive justice script, the rebellion track. This seemingly least objectionable type of retributive justice script in modern YA speculative fiction recounts how a harm done to a humble individual becomes a pebble that brings down an oppressive regime. This track is the dominant cognitive model for violent resistance to state-sanctioned injustice. In the rebellion track, ordinary individuals become instrumental in bringing about a coup or revolution that does away with tyranny. One of the best recent examples of this track is the adventures of Lyra in *The Golden Compass*. The novel tells the story of an 11-year-old girl on a quest to free her abducted friend Roger, a quest that turns out to be the beginning of a much larger sequence. Although she is unaware of it, Lyra lives in a world in which the Church's power over every aspect of life is absolute and a major rebellion is in the making. The injustice that moves Lyra to action is the disappearance of the kitchen boy Roger, apparently taken by the mysterious Gobblers. The Gobblers, an acronym for the General Oblation Board, is a semi-private initiative with murky ties to the government, whose objective is to eliminate the human capacity for sin—or independent thinking—by means of a medical procedure. In order to harvest humans for experiments, the Gobblers kidnap pre-puberty street children, many of them a minority called gyptians, and send them to their facility in the North.

When Roger disappears, Lyra is determined to rescue him. Seeking a way to travel to the North, she joins a group of Gobbler-hunting gyptians. At the gathering of clans she learns that the gyptians, worst hit by losing children, are sending out a rescue party. They decide to rescue all children they will find—gyptian and not—but they also discuss the question of revenge. A woman whose child is missing asks Lord Faa to "take powerful revenge" (Pullman, *Golden* 137). To this the gyptian king replies by assuring her and the assembly that once their primary goal, saving the children, is achieved, those to blame will suffer. "When the time comes to punish," he declares,

> we shall strike such a blow as'll make their hearts faint and fearful. We shall strike the strength out of 'em. We shall leave them ruined and wasted, broken and shattered, torn in a thousand pieces and scattered to the four winds. Don't you worry that John Faa's heart is too soft to strike a blow when the time comes. And the time will come under judgment. Not under passion. (138)

When the expedition is sent out, the gyptians secure the friendly neutrality of the Lapland witch clans, win the services of a Texan aeronaut Lee Scoresby, and finally the help of an exiled armored bear, Iorek Byrnison. With such assistance, the group assaults Bolvangar and frees the captive children.

Although the novel continues to develop other plotlines, the destruction of Bolvangar and the freeing of captive children end the Trouble of injustice in *The Golden Compass*. Lyra's actions in this volume establish her importance to Lord Asriel's rebellion and she will be instrumental in ensuring its success. In this sense, the entire trilogy's as well as *The Golden Compass*' story arcs are actualizations of the rebellion track. Unlike the defensive war track, the rebellion track does not project complete innocence on the victims nor does it absolutize evil. In *The Golden Compass* specifically, the leader of the rebellion, Lord Asriel, is a power-hungry murderer. On the other side of the spectrum, descriptions of the Gobblers as a power-seeking political organization, of the Bolvangar medical personnel as just doing their jobs, and of the fierce Tartar guards as honorable mercenaries all acknowledge the banality of evil and motives that lead to it. Like the defensive war track, the rebellion track recognizes the need for retributive violence to deal with certain injustices. Some guards and other personnel must be killed if the children are to be freed; the usurper king of armored bears must be killed if Iorek is to reclaim his throne; many will also die on both sides in the course of the rebellion. For all that, there is in the rebellion track, at least as actualized in *The Golden Compass*, a recognition of the danger of retributive violence getting out of hand. Combined with awareness that responsibility for injustice is often spread systemically on a supra-individual level, this accounts for the minimum violence principle pronounced in this track. There is no "kill them all" imperative as in the Darwinist survival and defensive war tracks. In the rebellion track, John Faa does not allow thirst for vengeance to cloud his judgment and Lyra makes Iorek promise not to take vengeance on the people of Trollesund after she reveals where they had hid his armor (197). What matters most is ending the injustice, by use of force if necessary, with punishment being a different consideration. Animated films that employ the rebellion track—including *Robots* (2005), *Rango* (2011), *The Nut Job* (2014), and many others—are focused on maverick protagonists whose actions help expose and end the Trouble of the oppressive status quo. Due to its focus on the-individual-versus-the-unjust-system, and due to its hopeful affirmation that even large-scale systemic injustice can be brought down by grassroots efforts, the rebellion track is widely prevalent in modern YA speculative fiction. Like Lyra, other protagonists such as Katniss Everdeen and Harry Potter lead the way to a more just world, thus affirming the high relevance of this script for challenging various forms of large-scale political oppression.

The examples I mentioned are actualizations of the retributive justice script where the script is central to the story and structures the main plot. In each narrative the reader is offered clues as to the material and social premises of the world described that allow making connections between the *draugr's* thirst for revenge and the harm she had suffered, Cluny's invasion and the Redwall animals' determination to defend their freedom, and finally between Roger's abduction and the many other forms of oppression that permeate Lyra's world. In each case, many things are unclear and implicit, but that is exactly what stimulates the reader to fill in the cognitive gap between what harm or injustice is the source of Trouble, and how to best set it right. Although actions are not made right by virtue of being done by the protagonist, it is the protagonists—Matthias, Jack, and the Bard, Lyra—whose goals and motivations the reader is invited to evaluate in the first place. Different configurations of motivation are possible, and each helps to particularize components and roles in the script.

This is also true of narratives where the retributive justice script functions as a subplot, or where it is problematized by the ambivalent distribution of roles. In non-Western speculative fiction—from *Princess Mononoke* (1997), *Spirited Away* (2001), or *Howl's Moving Castle* (2004), to Nahoko Uehashi's *Moribito: Guardian of the Spirit* (1996) or Noriko Ogiwara's *Dragon Sword and Wind Child* (2007)—most characters have legitimate reasons for acting the way they do and the rightness of their actions is a function of other factors, especially duty or responsibility. In *Moribito*, for example, the Mikado—ruler of New Yogo—is bound to kill his own son, whose spirit possession may discredit the royal family. At the same time, the queen is bound to try to save the prince. The bodyguard Balsa, who accepts responsibility for the boy's safety, fends off the equally honorable and duty-bound royal hunters sent out to assassinate the prince. Where the right is depends on where one stands, and each narrative offers a different particularization of the retributive justice script.

For all its popularity, the retributive justice script also has its limitations. In projecting the necessity of retribution, this script encourages violent solutions and suggests that the rebels/victims are morally bound to win. It ignores the fact that in many circumstances—especially in conflicts within one community or one ecosystem—retribution is less feasible than forgiveness and that revolutions often tend to change the power dynamics, with the oppressed now becoming the oppressors, rather than redressing an injustice they sought to right. The retributive justice script also ignores the fact that moral claims are not guarantees of military victory. The longer the violence continues, the less transparent are moral claims advanced by each side and the more devastating the consequences. This awareness informs a handful of novels set in worlds where the retributive justice script is assumed as a default and where the protagonists' solution is not an armed rebellion but a conscientious refusal to participate in the oppressive order. In Lois Lowry's post-apocalyptic dystopia *The Giver* (1993) the sameness is a free choice of the community and the price they have willingly decided to pay for eliminating possible conflicts in the future. In Ursula K. Le Guin's feminist fantasy *Gifts* (2004), the principles of retribution are woven so deeply into the cultural fabric of the Uplander society that any attempt to eradicate them would likely tear that society apart. In both cases the protagonists realize that the most feasible form of resistance to the injustice in their societies is to move away—what amounts to enacting the freedom track of the social justice script. In probing the frontiers of the retributive justice, such narratives offer an alternative to the shortcomings of the retributive justice script—especially its violence for violence imperative. Another possibility, of course, is to change the script altogether and replace it with a solution in which harm is redressed without recourse to violence. The most powerful such alternative is the restorative justice script, the subject of the next chapter.

NOTES

1. For discussion of this trend and its manifestations, see, for example, Norbert Elias' *The Civilizing Process: Sociogenetic and Psychogenetic Investigations* (1939/2000), Julius Ruff's *Violence in Early Modern Europe 1500–1800* (2001), William Ury's *Must We Fight? From the Battlefield to the Schoolyard: A New Perspective on Violent Conflict and Its Resolution* (2002), Matt Ridley's *The Rational Optimist: How Prosperity Evolves* (2011), and Steven Pinker's *The Better Angels of Our Nature: Why Violence Has Declined* (2011). Although

the scope of each book is different, each of them adds unique evidence that supports the global tendency of the decline of violence.

2. The *Monty Python's Life of Brian* scene about "What have the bloody Romans ever done for us?"—with the painfully honest Judean patriots naming a long list of improvements the bloody imperialist Romans had indeed brought them—is a facetious but eye-opening recognition that the various historical *paxes* did reduce endemic violence among the conquered. For statistics, see Pinker, *Angels* 47–128.
3. The assumption that a tit-for-tat response to harm is less effective than a translation of harm into punishment is grounded in another assumption, namely that an impartial, emotionally, and ideologically neutral response to harm is more effective than a partisan, emotionally, and ideologically engaged one. Each of these two assumptions has been questioned, including whether or not it is possible or desirable for the judges of the law to be fully impartial. The supposedly discredited tit-for-tat response has enjoyed a renewed appreciation since 1971, when Robert Trivers demonstrated it as foundational to his theory about the evolution of reciprocal altruism, and when a few years later it was shown as a highly effective strategy in the evolutionary game theory. More recently, in the work of McCullough and others, the tit-for-tat of revenge has been reevaluated as a universally human psychological endowment. This research refutes the three most common approaches to revenge in Western thought—ones that see it as immoral, irrational, or indicative of psychological dysfunction. Instead, revenge is a human adaptation like its alternative, forgiveness; the choice of each strategy depends on computations of expected value. When the costs of revenge are too high relative to its expected deterrence benefits, an organism will likely choose forgiveness.
4. The idea of impartiality of judgment and punishment executed by the state has been denounced as masking highly non-egalitarian power relations. This argument harkens back to Michael Foucault's endlessly provocative *Discipline and Punish* (1977), which suggests that the post-Beccarian rise of the prison is just one of the institutions of control in a vast network of mutually-reinforcing mechanisms of the modern panoptic society. "[A]lthough the universal juridicism of modern society seems to fix limits on the exercise of power," Foucault declares, "its universally widespread panopticism enables it to operate, on the underside of the law, a machinery that is both immense and minute, which supports, reinforces, multiplies the asymmetry of power, and undermines the limits that are traced around the law" (223). The idea of translatability of harm based on neutral and objective adequation has been criticized too, as based on the false assumption that justice is a self-evident concept. This position has been voiced by literary scholars such as Wei Chi Dimock in her *Residues of Justice* (1996) but also by all communitarian philosophers who advocate the *Open Justice* position, where justice is a conditional, situational, and non-absolute concept. Communitarianists, finally, have also criticized the idea of an impartial and value-natural response to harm. Sandel's *Justice* (2009) is a representative voice for this trend in its argument that value-neutral judgments disregard specific gender, race, age, and other situatedness—along with their corresponding beliefs and intentions—that create unavoidably unequal obligations among the citizens of a just society.
5. For current and nuanced discussions about the meaning of retribution and its controversial position in legal and political philosophy, see the collections

edited by Tonry and White. For proposals that reconcile retributive, restorative, and communitarian ideas, see Anthony Duff's *Punishment, Communication, and Community* (2001) and his "Responsibility, Restoration, and Retribution" (2011).

6. From the mid-1860s to his death in 1896 Morris worked as a co-translator with Eirikr Magnusson on a number of Icelandic sagas and epics. Their most remarkable translation was *The Tale of Beowulf* (1895).
7. In fact, the only obstacle for superheroes teaming up in the same story are exclusive movie rights to different characters held by the three rival studios: Disney, Sony Pictures, and Twentieth Century Fox. See Fritz, D1–D2.
8. This tendency is ubiquitous. For details, see collections such as Tom Morris and Matt Morris, *Superheroes and Philosophy: Truth, Justice, and the Socratic Way* (2005), Terrence R. Wandtke, *The Amazing Transforming Superhero! Essays on the Revision of Characters in Comic Books, Film and Television* (2007), Jonas Prida, *Conan Meets the Academy: Multidisciplinary Essays on the Enduring Barbarian* (2012) as well as such studies as Terrence R. Wandtke's *The Meaning of Superhero Comic Books* and Jeffrey K. Johnson's *Super-History: Comic Book Superheroes and American Society, 1938 to the Present* (both 2012).

WORKS CITED

Barash, David P. *Natural Selections: Selfish Altruists, Honest Liars, and Other Realities of Evolution*. New York: Bellevue Literary Press, 2008.

Baron-Cohen, Simon. *The Science of Evil: On Empathy and the Origins of Cruelty*. New York: Basic Books, 2011.

Bloch, Marc. *Feudal Society: The Growth and Ties of Dependence*. New York: Routledge 1962.

Bowers, Rick. *Superman Versus the Ku Klux Klan: The True Story of How the Iconic Superhero Battled the Men of Hate*. Washington, DC: National Geographic, 2012.

Capeheart, Loretta and Dragan Milovanovic. *Social Justice: Theories, Issues, and Movements*. Piscataway, NJ: Rutgers UP, 2007.

Cragg, Wesley. *The Practice of Punishment: Towards a Theory of Restorative Justice*. London: Routledge, 1992.

De Camp, Sprague L. *Literary Swordsmen and Sorcerers: The Makers of Heroic Fantasy*. Sauk City, WI: Arkham House, 1976.

Dimock, Wei Chi. *Residues of Justice: Literature, Law, Philosophy*. Berkeley and Los Angeles: U. of California P., 1996.

Duff, Anthony. *Punishment, Communication, and Community*. New York: Oxford UP 2001.

Duff, Anthony. "Responsibility, Restoration, and Retribution." *Retributivism Has a Past. Has It a Future?* Ed. Michael Tonry. New York: Oxford UP, 2011: 3–24.

Ehrenreich, Barbara. *Blood Rites: Origins and History of the Passions of War*. New York: Henry Holt and Company, 1997.

Eng, Steve. "Barbarian Bard: The Poetry of Robert E. Howard." *The Dark Barbarian: The Writings of Robert E Howard, a Critical Anthology*. Ed. Don Herron. Westport, CT: Greenwood Press, 1984: 23–64.

Farmer, Nancy. *The Sea of Trolls*. New York: Atheneum Books, 2004.
Farmer, Nancy. *The Islands of the Blessed*. New York: Atheneum Books, 2009.
Finn, Mark. *Blood and Thunder: The Life and Art of Robert E. Howard*. Austin, TX: MonkeyBrain Books, 2006.
Foucault, Michel. *Discipline and Punish: The Birth of the Prison*. Transl. by Alan Sheridan. New York: Pantheon Books, 1977.
Fritz, Ben. "Why Can't These Superheroes Be Friends?" *The Wall Street Journal*. May 2, 2014: D1–D2.
Gerber, Monica M. and Jonathan Jackson. "Retribution as Revenge and Retribution as Just Deserts." *Social Justice Research* 26 (2013): 61–80.
Gottschall, Jonathan. *The Rape of Troy: Evolution, Violence, and the World of Homer*. New York: Cambridge UP, 2008.
Howard, Robert E. *The Coming of Conan the Cimmerian*. Ed. Patrice Louinet. New York: Random House, 2003.
Jacques, Brian. *Redwall*. New York: Ace Books, 1998.
Johnson, Jeffrey K. *Super-History: Comic Book Superheroes and American Society, 1938 to the Present*. Jefferson, NC: McFarland, 2012.
Knight, George. "Robert E. Howard: Hard-Boiled Heroic Fantasist." *The Dark Barbarian: The Writings of Robert E Howard: A Critical Anthology*. Ed. Don Herron. Westport, CT: Greenwood Press, 1984: 117–134.
Lawrence, John Shelton and Robert Jewett. *The American Monomyth*. Garden City, NY: Anchor Press/Doubleday, 1977.
Lawrence, John Shelton and Robert Jewett. *The Myth of the American Superhero*. Grand Rapids, MI: Eerdmans Publishing, 2002.
Leiber, Fritz. "Howard's Fantasy." *The Dark Barbarian: The Writings of Robert E Howard, a Critical Anthology*. Ed. Don Herron. Westport, CT: Greenwood Press, 1984. 3–15.
McCullough, Michael E. *Beyond Revenge: The Evolution of the Forgiveness Instinct*. San Francisco: Jossey-Bass, 2008.
Morris, Tom and Matt Morris. Eds. *Superheroes and Philosophy: Truth, Justice, and the Socratic Way*. Peru, IL: Open Court, 2005.
Morris, William. *The House of the Wolfings*. London: Reeves and Turner, 1889. http://morrisedition.lib.uiowa.edu/Images/Wolfings/. May 6, 2014.
Panksepp, Jaak and Lucy Biven. *The Archeology of Mind: Neuroevolutionary Origins of Human Emotions*. New York: W.W. Norton, 2012.
Pinker, Steven. *The Better Angels of Our Nature: Why Violence Has Declined*. New York: Viking, 2011.
Prida, Jonas. Ed. *Conan Meets the Academy: Multidisciplinary Essays on the Enduring Barbarian*. Jefferson, NC: McFarland, 2012.
Pullman, Philip. *The Golden Compass*. New York: Alfred A. Knopf, 1996.
Rostanowski, Cynthia C. "The Monastic Life and the Warrior's Quest: The Middle Ages from the Viewpoint of Animals in Brian Jacques's Redwall Novels." *The Lion and the Unicorn* 27.1 (2003): 83–97.
Silver, Carole. "Myth and Ritual in the Last Romances of William Morris." *Studies in the Late Romances of William Morris*. Eds. Blue Calhoun, John Hollow, Norman Kelvin, Charlotte Oberg, and Carole Silver. New York: The William Morris Society, 1976: 115–39.
Turner, Mark and Gilles Fauconnier. *The Way We Think: Conceptual Blending and the Mind's Hidden Complexities*. New York: Basic Books, 2002.

Tonry, Michael. Ed. *Retributivism Has a Past. Has It a Future?* New York: Oxford UP, 2011.
Wandtke, Terrence R. Ed. *The Amazing Transforming Superhero! Essays on the Revision of Characters in Comic Books, Film and Television.* Jefferson, NC: McFarland, 2007.
Wandtke, Terrence R. *The Meaning of Superhero Comic Books.* Jefferson, NC: McFarland, 2012.
White, Mark D. *Retributivism: Essays on Theory and Policy.* New York: Oxford UP, 2011.

5 No Future without Forgiveness

Restorative Justice Scripts

On March 18, 1806, five days before their departure from Fort Clatsop, Lewis and Clark sent four of their men to steal a canoe from the hospitable Chinook whose assistance helped keep the expedition alive through the wet and wretched winter. This was the first time Lewis and Clark had stolen anything from the Indians. According to James Ronda, the act was not merely a violation of expedition policy. It was a betrayal of friendship the captains had enjoyed with Chief Coboway and an offense to a community, where the beautifully carved canoes were a sacred part of the Chinook culture.

Troubled by the dim realization of how the theft tarnished the expedition's honor, in his journal Lewis used the word "take" rather than "steal" and then whitewashed it as justified "in lieu of the six elk which they stole from us in the winter" (Rhonda 211). He did not mention that Coboway had apologized for the theft and, following traditional practice, had sent them a couple of dogs for meat. With proper restitution made and accepted, rights and wrongs had been balanced. Now, however, the explorers were desperate to head back home even if it meant compromising their principles.

The theft became a sore subject with the Chinook and had been passed over in the accounts of exploits of Jefferson's captains. Two centuries later, however, Clark's eighth-generation descendant Lotsie Clark Holton found herself working on the same project with the chairman of the Chinook Nation's tribal council Ray Gardner. Following their conversations, the Clark family came up with an idea to present the tribe with a 36-foot replica of the stolen canoe. As Lotsie's mother put it, "We talked about what happened 205 years ago, and we believed that things could be restored if something like this were done. I think everyone acknowledges that it was wrong, and we wanted to right a wrong. The family was very much behind it" ("Clark's descendants" np). The canoe was delivered to the Chinook in an official ceremony on September 24, 2011. The restoration was emotionally overwhelming for both sides. Clark's descendants acknowledged the past wrong and made it right through restitution and apology. The Chinook accepted the return of the canoe as the healing of past harm but also as recognition of their traditions and culture—something they are still denied by the US government.

The story of the purloined elk and canoes, as well as the account of how each wrong was settled, illustrates the basic mechanisms of restorative justice. Although the term is of modern coinage, restorative justice practices have been known to human societies for millennia. Just as in retributive justice, they stem from the perception of harm as a violation of specific people and relationships. However, unlike retributive justice, the goal of restorative justice is the healing of the victims, reintegration of the victims and offenders into the community, and restoration of the relations the harm damaged. Based on resolutions developed at the community and interpersonal level—usually by those directly impacted by the harm—restorative justice seeks ways to make things right through apology, forgiveness, and restitution.

Restorative justice has been present in the *Old* and *New Justice* worlds but was a nameless, non-judicial practice, seemingly relevant only for women, children, and trivial situations, until it was named and its power was revealed in the toppling of the Apartheid in South Africa. In this sense, restorative justice is an *Open Justice* concept but one that also embodies the *New Justice* imperative for equality. If poetic justice was the script for the powerless in relation to their superiors, and retributive justice is the script of the powerful in relation to the powerless, restorative justice has long been the script for peer, family, or community relationships.

Restorative justice is more than a nice sentiment. Based on apology, forgiveness, and restitution—cultural equivalents of reconciliation signals known in the mammal world: self-abasing displays, cooperative reintegration with the group, and compensation (McCullough, *Beyond* 160ff)—restorative justice is a manifestation of an evolutionary adaptation called the "forgiveness system," which I take to be the neuropsychological foundation of restorative justice. Shared by all primates and most mammals with the curious exception of domestic cats (120), the forgiveness system operates to counterbalance impulses for retribution. It enables restoring relationships torn by harm, resuming cooperation after conflict, and avoiding costly and unproductive cycles of revenge. The restorative justice script informs narratives, which envision as the goal of justice the healing of the victim and ending of the harm rather than punishing the offender. Dialogical in its procedures, it seeks provisional rather than final solutions and achieves them through negotiation and compromise. My proposal in this chapter is that the restorative justice script comes in three tracks determined by their primary narrative focus: on the victim, on the offender, and on the community. I suggest that the restorative justice script is a recent arrival in YA speculative fiction but with important antecedents in the fantasy works by C. S. Lewis and J. R. R. Tolkien.

RESTORATION, FORGIVENESS, AND THE RESTORATIVE JUSTICE SCRIPT

Restorative justice is both an old and a new concept. It is primordial as the basic mechanism people everywhere have used to deal with harm in their everyday functioning, especially that occurring in close relationships within a family or a local community, where the use of force would further antagonize the parties and deepen the rift between them. Any parent who has seen their child hurt another or break their toy, and who has then made the child apologize and compensated for the other child's loss—any such parent has enacted restorative justice. Any teacher who has noted a student's misbehavior and made them apologize to the hurt party has been inculcating restorative justice. Anyone who has recognized they had hurt someone and apologized for it to restore that particular relationship has enacted restorative justice. Unlike retributive justice, focused on punishing the offender, restorative justice aims to heal the community, victims and offenders alike. In restorative justice, crime is seen as a violation of people and relationships rather than a violation of the law. The offender's responsibility is seen as primarily toward specific person(s) and relationships rather than toward the law or the state. In this basic sense, restorative justice can be traced back to pre-state societies in which most harm was not considered harm against the state or the collective but as specifically inflicted on a victim and/or his or her family. All resolution techniques based on the principles of acknowledgement of the harm, apology, forgiveness, and reconciliation are forms of restorative justice. Based on negotiation and arbitration, these non-judicial techniques used to be the normative protocol for peer relations in the Western civilization throughout the *Old Justice* period (Zehr, "Retributive" 7)[1]. The rise of the modern state—whose primary means of expressing power is the *New Justice* criminal law—changed this in favor of state justice based on the retributive model. Yet, even today most people in their everyday interactions with others engage in negotiated, restitutive practices that are the essence of restorative justice.

For all that, restorative justice is also a new concept. In the first place, it is an *Open Justice* reaction to a totalizing world of excessive rationalism characteristic of the *New Justice* law. When Rawls argues that children should have equal opportunities, and notes that different families offer different opportunities, his conclusion is that the achievement of fair equality of opportunity would mean the elimination of the family (Brighouse 55)[2]. This and many similar syllogisms inherent in the *New Justice* paradigm, gesture toward its highest logical aspiration, where justice is a coherence of premises but has nothing to do with the world "out there." In terms used by McGilchrist, such form of rationality reflects the left-hemispheric tendency to prioritize the system of closed signs, where the truth does not correspond to anything outside of the system. Similarly, the *New Justice* ideal of the due process of law—where justice is most just when it is most mechanical (see Baron)—is a left-hemispheric product, unchecked by the right hemisphere's "bullshit detector" (McGilchrist 193). The rise of restorative justice is a critique of this misplaced rationalism. It changes focus from the system to actual people whose lives it affects.

This, I think, is the second factor behind the appeal of restorative justice. The ascendency of justice as a philosophical concept envisioned in the *New Justice* framework—and of the *New Justice* legal systems in the West—was tied to redefining the notions of harm and crime as offenses against the king, state, or law rather than against their actual victims. As law embraced the retributive justice model—emphasizing procedures driven by prosecution and defense, establishing the guilt or innocence of the accused, and determining the appropriate punishment—the victim became irrelevant in the process of making amends. Although this model was successful in some areas, it failed in others. First, by focusing largely on the offender, criminal law has disregarded the needs of victims—its goal being deterring future offenses rather than making victims whole. Second, by severing the link between personal responsibilities for harm in actual human contacts in favor of responsibility to impersonal law, it has little to offer in terms of healing human relationships and the communities the offense disrupted. Third, by assuming that the punishment of the guilty party makes things right, criminal law has led to ever-rising incarceration rates. By the late twentieth century, the result—especially in the US—has been crowded prisons, soaring recidivism, and startling statistics of incarceration, including the fact that "America incarcerates more of its children than any country in the world" (Young-Bruehl 2). As some experts argue, the emphasis on procedures and retribution has made the *New Justice* criminal law ineffective and undermining to the social peace it supposedly seeks to preserve (Stuntz 55)[3]. In countries that simply could not afford to imprison so many people, the retributive criminal justice system came close to a collapse. This was made evident first in South Africa (from the early 1980s), later in former Yugoslavia (1991–95), and Rwanda (1994–6). It was out of this climate that alternatives to retributive justice were sought. By the early 1990s, the search resulted in the development of a movement toward restorative justice.

The credit for introducing the term "restorative justice" belongs to two Americans: the psychologist Albert Eglash and the legal scholar Howard Zehr. Working with juvenile delinquents in the 1950s, Eglash developed the concept of creative restitution[4], which he saw as more effective than anything the criminal justice system could offer. In 1977 Eglash then coined the term "restorative justice" (91) but admitted to being concerned "primarily with offenders," not victims (99). It took the legal scholar Howard Zehr, however, to integrate restorative justice and the victim perspective into criminological thinking. Zehr's "Retributive Justice, Restorative Justice" (1985) brought the first formulation of the term in its current understanding as an alternative to retributive justice. Restorative justice then took off in earnest in 1990, the year that saw the publication of the first collection on restorative justice: Burt Galoway and Joe Hudson's *Criminal Justice, Restitution and Reconciliation* and of Zehr's groundbreaking *Changing Lenses*. The fact that neither of these studies used the phrase "restorative justice" points to how new the term still was; both, however, were pioneering works in the restorative justice field.

Since 1990 restorative justice has been field-tested all over the world, proving especially effective in dealing with situations where violence and conflict within the same nation or community have seemed endemic[5]. The work of the Truth and Reconciliation Commission that operated in South Africa between 1995 and 1998 was groundbreaking in setting a model for healing nations that had been traumatized by civil wars, genocide, and gross human rights violations. The success of the TRC—described in Desmond Tutu's *No Future Without Forgiveness* (1999) from which I have borrowed the title of this chapter—helped establish a paradigm that was later successfully applied for reconciling the warring groups in Northern Ireland, the Balkans, Rwanda, and elsewhere. What all these applications have demonstrated is that restorative justice unfolds as a pattern. It starts with the acknowledgment of the harm, one that "exposes the awfulness, the abuse, the pain, the degradation, the truth" (Tutu 270). The telling of the story—the ability of being heard and getting at the truth, rather than punishment, apology, or restitution—has profound cathartic effects on the victim but also on the offender and all parties involved. It enables empathetic identification, when pain is shared and thus somehow released. The process then makes parties confront the meaning of apology and forgiveness, empowering them to embrace these emotions in place of hurt or guilt. The end result is that the parties are free from the roles in which past events had cast them and can move on.

When described in cognitive terms, restorative justice is based on four behavioral-psychosocial phenomena: apology and forgiveness, empathy and cooperation. Apology and forgiveness are related; although forgiveness can be offered even in the absence of apology, studies show that forgiveness is offered more easily and is more likely to come in its highest stages when following apologies, acknowledgement of responsibility, and/or public confession about the harm (Mullet and Girard 112–13). Forgiveness is also a cognate of empathy and altruism. All three are made possible by the capacity for empathetic identification mediated by the right hemisphere (McGilchrist 57) and developed in the course of our evolution as a mode of human reasoning in social contexts[6]. Empathy—which Baron-Cohen refers to as the Empathizing SyStem, or TESS—is the lynchpin of our Theory of Mind, "the real jewel in the crown [that] carries with it the adaptive benefit of ensuring that organisms feel a drive to help each other" (Baron-Cohen, "Empathizing" 473). Finally, forgiveness is an important precondition for any social cooperation. As studied in the field of game theory, any strategy of cooperation that has a claim to evolutionary stability in a world full of "noise"—misunderstandings and misinterpretations of other people's intentions or actions—is a strategy that includes multiple-layer forgiveness protocols that make possible cooperation even after the most prolonged competition[7]. These mathematical calculations have been confirmed in studies of primate behavior, especially these conducted by the Dutch primatologist Frans de Waal. Apology, forgiveness, and restitution as engaged in by humans are cultural equivalents of reconciliation behaviors in the mammal world: self-abasing displays, cooperative reintegration with the group, and compensation (McCullough, *Beyond* 160ff). Also the median Conciliatory Tendency value for primates and young humans was discovered to be about the same (120). All these studies demonstrate that conflicts among primates "lead most commonly to friendly contact, not interminable cycles of revenge or alienation" (119).

The statistical dominance of reconciliation has also been demonstrated by studies on the developmental, cognitive, and neuropsychological aspects of forgiveness in humans. A gradational phenomenon with at least six main types (Enright, Santos, and Al-Mabuk 96)[8], forgiveness has been defined as a cross-cultural developmental phenomenon, with strong correlation between the respondent's age and a forgiveness stage she embraces: the older one is, the more likely one is to forgive (Mullet and Girard 113–20). The fact that reactions related to forgiveness and revenge are processed by the same neural hardware—especially the seeking system—implies that both are context-sensitive "*conditional* adaptations," with forgiveness being a switch on revenge (McCullough, *Beyond* 132). In other words, although we are equipotentially hardwired for violence, competition, and revenge as well as for cooperation, forgiveness, and empathy, the latter statistically predominate (see McCullough, *Beyond*; Baron-Cohen, *Evil*; Pinker, *Angels*). This refutes the still ubiquitous Darwinist-Malthusian assumptions that violence and competition are the norm whereas forgiveness and cooperation are less frequent or more conditional. It also indicates that revenge—a reaction that is more Old Brain-based and thus "cheaper" in terms of emotional energy investment than forgiveness—is a bad alternative to the more complicated forgiveness, even though forgiveness needs sustaining over the entire process. Yes, it may be "simpler" to quarrel with someone and break the relationship than to reconcile differences and deal with the same or similar problems in the future. And yet, who is the winner in any feud? What justice is achieved if parents punish their teenage child while they can but then the young adult leaves home and never speaks to them again? The retributive option is a losing strategy for human relationships, whereas the more effort-demanding forgiveness and empathy sustain relationships and help them grow. That is why forgiveness, as a switch on revenge, is so highly effective for maintaining relationships within small networks such as those between friends and family. In such networks agents statistically forgive "up to 80% of other network members' defections" (McCullough, Kurzban, and Tabak 232). The complexity of forgiveness in its interaction with other factors makes it not only more complicated than revenge but accounts for why recent studies examine it not in isolation but as the "forgiveness system" (230)[9]. The forgiveness system is a psychological endowment that provides us with tools to resolve disputes peacefully and restore relationships torn by harm; restorative justice builds on that foundation.

If the affective basis for retributive justice is revenge, forgiveness is the foundation for restorative justice. Given that forgiveness has been described in cognitive terms as a schema, or script (Mullet and Girard 125–6)—a species-universal neurocognitive process that unfolds within a relatively prescribed four-step pattern: "(1) recognition of the injury to the self; (2) commitment to forgive; (3) cognitive and affective affinity; and (4) behavioral action" (Newberg et al. 101)—it is no surprise that the processes of restorative justice follow an analogous path. If the end result of forgiveness is a new understanding of the relationship between the self and the world, one "that includes the offending individual, the offense, and the resolution of that offense" (104), that is also the goal of restorative justice. Integrating apology, forgiveness, and restitution, restorative justice is a tertiary-process affect (Panksepp and Biven 147) arising out of the biologically evolved forgiveness system, which itself is an encephalized manifestation of our reptilian self-preservation and mammalian cooperation instincts. Restorative justice is an open-ended combinatorial engine for generating possibilities that offer victims and offenders alike a sense of closure, that help us resume cooperation after conflict, and enable coexistence despite past harm. In short, restorative justice emerges from the cognitive faculties that make us smart and socially capable.

The various models of restorative justice are usually grouped into three categories: circles, conferences, and victim-offender mediations. In all of them crime is understood as a "violation of people and relationships [that] creates obligations to make things right" (Zehr, *Changing* 181). The process of making things right is carried out through a series of discussion meetings involving the victim, the offender, and the community that lead to a situational resolution that helps heal both sides, ends the conflict, and seeks to repair the damage. The process usually requires mediation by a third party, whose role is to help the victim feel safe enough to talk about their suffering and to help the offender acknowledge the harm done, understand, and apologize for it. Punishment may or may not be projected as a necessary component of restorative justice, but it is never its central focus. As an extended causal sequence, with its specific goals, character roles, and constituent slots, the process of restorative justice operates as a script.

The restorative justice script is a cognitive program that envisions as the goal of justice the healing of the victim and ending of the harm. This script prompts agents away from the punitive action-protocols of the retributive justice script and cues the search for non-violent solutions that promote repair and reconciliation. Unlike the inherently monological retributive justice script, it is dialogical both in its procedures and goal. It invites the victim, the offender, and the community to express their different perspectives, and it achieves its goal through negotiation rather than force. Unlike the retributive justice script, it does not offer one final solution, but a series of strategies based on compromise. These strategies, however, enable such reintegration of the victim, the offender, and the community that they are capable of peaceful coexistence and cooperation in the future.

As defined by Zehr, restorative justice is a distillate of three ingredients: truth or acknowledgment of the harm, apology-forgiveness, and reconciliation or reparation. Accordingly, I suggest that the restorative justice script can be described in terms of three signature phenomena: recognition of the harm as a disregard for the needs of the hurt party; the restorative justice encounter—composed of five general elements: "*meeting* (the offender and victim meet face to face); *narrative* (both present their particular story); *emotion* (each expresses anger, fear, sorrow); *understanding* (each begins to empathize with the other); and *agreement* (some kind of resolution is attained)" (Capeheart and Milovanovic 56); and the resolution which—through the exchange of apology and forgiveness, restitution and generosity—leads to personal, behavioral, and social-structural change that puts all sides on the path toward healing and reintegration.

THE RESTORATIVE JUSTICE SCRIPT IN MODERN YA SPECULATIVE FICTION

Where and when the first attempts to use the restorative justice script in speculative fiction were made is hard to say, but some early groping toward it can be detected in the work of the two founders of the mythopoeic fantasy genre: C. S. Lewis and J. R. R. Tolkien. Both committed Christians, Lewis and Tolkien structured their fantasy worlds largely on the transcendentalist justice script, projecting the achievement of perfect justice in the heavenly beyond: the ultimate West and the eternal Narnia. At the same time, they acknowledged that some imperfect justice also has a place in this world. Both were WWI veterans who knew first-hand the insane destructiveness of conflicts informed by retributive logic. Their Christian understanding of justice was also tempered by concepts of apology and forgiveness. Drawing on these two backgrounds, each stumbled toward some version of the restorative justice script in their fantasy fiction.

Lewis seemed to have envisioned restorative justice as a norm in personal relationships but only a slight possibility in communal and international relations. The Chronicles of Narnia abound in examples of violations of people and relationships resolved by apology and forgiveness and followed by reconciliation. Edmund's betrayal in *The Lion, The Witch and The Wardrobe* (1950) ends with his conference with Aslan and then the boy's apology to his brothers and sisters. Following their meeting with the Professor, the siblings also apologize to Lucy for questioning her Narnian experience. Similar episodes are found in other volumes too[10], but in each case Lewis sets it as a rule that forgiveness can be offered only after receiving a sincere apology from the offender. This textual practice contradicted his belief expressed in "On Forgiveness" (1947), where Lewis claims that Christians are expected to forgive other people's sins, "however spiteful, however mean, however often they are repeated" (178), and even forgive the inexcusable, "because God has forgiven the inexcusable in you" (182)[11]. On the other hand, Lewis' insistence on apology was in line with his conviction that it is critical for the offender to realize *why* what they did was wrong and that some punishment is always necessary[12].

Lewis also believed in righteous war. "The doctrine that war is always a greater evil," he argued, "seems to imply a materialist ethic, a belief that death and pain are the greatest evils. But I do not think they are. I think that a suppression of a higher religion by a lower, or even a higher secular culture by a lower, a much greater evil" (Lewis, "Why" 77). Thus, in the two cases when The Chronicles approach the restorative justice script for resolving conflicts, Lewis projects it as an alternative only *after* the retributive justice script has been used first. In *Prince Caspian*—whose 14 out of 15 chapters actualize the rebellion track of the retributive justice script—the victorious Narnians must deal with the large population of the defeated Telmarines. The dilemma is solved by Aslan who gives the Telmarines a choice of staying in Narnia—thus accepting the new status quo—or leaving the country. When the Telmarines pass through the door to an unnamed island in the South Seas, they can never return. The problem of the Telmarines is solved by their expulsion. Where *Prince Caspian* deals with a civil war, *The Horse and His Boy* is a tale of a military aggression. In the second part of the novel—which in its final chapters employs the defensive war track of the retributive justice script—Shasta's journey becomes a quest for Archenland's and Narnia's freedom, as he carries the last-minute warning about an impending attack. Because he succeeds, the Calormene cavalry is defeated. The victors must then decide what to do with Prince Rabadash and the survivors. Rabadash is offered freedom if he apologizes and promises never to invade again. He is unrepentant, though, offensive, and threatens vengeance (207). As a result, Aslan turns him into an ass and declares that Rabadash will be able to regain his human shape only in Tashbaan. Adding that "[j]ustice shall be mixed with mercy," Aslan warns Rabadash that if he ever moves more than ten miles from his capital, he will again and permanently become an ass (211). Following this treatment, Rabadash returns to his country to become "the most peaceable Tisroc [ruler] Calormen had ever known" (212).

In both novels Lewis used elements of restorative justice but did not envisage the restorative justice script. First, he saw restorative justice as mercy, effective only under the threat of retribution. Second, his solution for communal and international conflicts was separation rather than integration. Third, the resolution was monological, dictated by one side, rather than dialogically negotiated. Fourth, because it was offered by divine Aslan, justice was projected as absolute and final rather than, as is the case in restorative justice, provisional and based on some compromise.

On the last three of those points Tolkien came much closer to restorative justice, both in *The Hobbit* and later in *The Lord of the Rings*. As can be glimpsed from his only reflection on the principles of justice—a 1956 note on W. H. Auden's review of *The Return of the King*—Tolkien realized that in real life causes are ambiguous and situations in which "*right* is from the beginning wholly on one side" are rare. At the same time, he believed in objective justice, where the rightness or goodness of certain causes is not established by any party's claims but is dependent "on values and beliefs above and independent of the particular conflict" (242). The rightness of the cause, he was careful to stress, does not justify that its supporters use morally contemptible means, nor do individual good actions by those on the wrong side justify their cause. It does imply, however, that "[t]he aggressors are themselves primarily to blame for the evil deeds that proceed from their original violation of justice" and have no moral right to expect from their victims anything but retribution (243)[13].

This affirmation of retributive justice did not mean Tolkien saw it as necessary. In *The Hobbit* (1937), for example, the conflict among the dwarfs, Wood-elves, and the Men of the Lake is resolved through a restorative justice negotiation carried out by Bilbo. When the three sides are locked in a seemingly unresolvable stalemate, Bilbo delivers the Arkenstone to Bard and the Elvenking. He knows that Thorin's desire for the gem may be the only way to force him to consider the demands of the other side. Thus armed, Bard indeed succeeds in negotiating a settlement, but both sides are embittered. The Elvenking's hope "for something that will bring reconciliation" (338) is realized through the Battle of Five Armies, when men, elves, and dwarfs unite against goblins and wolves. Chastised by the experience, the dying Thorin apologizes to Bilbo and acknowledges the courage it took on the hobbit's part to seek a peaceful solution. Recognizing Thorin's right to Arkenstone, Bard returns it unconditionally by laying it on Thorin's breast. Also, the Elvenking, in a gesture of apology for having imprisoning the company, returns Thorin's sword taken from him in captivity by placing it on his tomb. The new dwarf king, Dain, in turn, offers one-fourteenth of the treasure to Bard, who then generously shares it with the Master of Lake-town and with the Elvenking. The resolution means compromising everyone's demands but heals each party's resentment and a sense of having been victimized. While the negotiation is extorted, the resolution meets the basic requirements for restorative justice. It heals past harms of all sides; it restores good relationships between dwarfs, humans, and elves; it is achieved through dialogue and compromise; and it involves the exchange of apology and forgiveness, restitution, and generosity.

Restorative justice encounters can also be found in *The Lord of the Rings*, with characters such as Boromir and the army of the dead serving as especially poignant illustrations of the power of apology. The most extended articulation of the restorative justice script, however, is to be found in the last two chapters of *The Return of the King*, which offer one of the earliest actualizations of the restorative justice script in YA speculative fiction.

When Frodo and his companions return home, they discover that the Shire has become a police state. At some point in the past, a greedy hobbit Lotho Pimple had became a local robber baron and invited big men to become the Shire's police. This soon led to a coup, with Pimple taking over, before he, in turn, was imprisoned by the mysterious Sharkey whom the ruffians now call Boss. In their first encounter, the four hobbits stand up to the ruffians, announce that Gondor has a king, and declare that the times of tyranny are over. They let the ruffians go but warn them not to trouble the village again. Realizing that Lotho is now a hostage of the ruffians, Frodo decides they must rescue him. He insists that "there is to be no slaying of hobbits, not even if they have gone over to the other side. ... And nobody is to be killed at all, if it can be helped" (310). Frodo's command sounds hard, but the hobbits recognize that chasing out the ruffians at the cost of dividing the Shire will not bring lasting peace. That same night they raise the Shire to community action. When another group of the ruffians arrives, the hobbits again demand that they peacefully leave the Shire. The leader ignores the advice, attacks Merry, and is shot; the other ruffians surrender. The next day, over a hundred ruffians attack Bywater. Again, the battle begins with Merry's warning and a call to surrender. When they ignore it, 70 ruffians and 19 hobbits die in the fight. "Frodo had been in the battle," the narrator recounts, "but he had not drawn sword, and his chief part had been to prevent the hobbits in their wrath at their losses, from slaying those of their enemies who threw down their weapons" (321). The prisoners will eventually be escorted to the borders of the Shire and released.

All these sequences cue the reader about the motivational-affective components of the restorative justice script, suggesting answers to how it can be applied and what it involves. These cues are reiterated with particular force in the episode when Frodo discovers Saruman, aka Sharkey, in Bag End. Saruman exults in having been able to ruin the Shire to the extent that will be hard to undo and gloats about the pleasure of "set[ting] it against my injuries" (324). Despite other hobbits' angry murmuring to kill him, Frodo rejects retribution. "I will not have him slain," he declares. "It is useless to meet revenge with revenge: it will heal nothing" (325). Frodo refuses to have Saruman killed even when the wizard stabs him, his knife breaking on Frodo's mailcoat; the protagonist also offers healing and reconciliation to Wormtongue, Saruman's wretched slave.

Concluding Tolkien's vision of restorative justice is a series of restoration acts, as the hobbit prisoners are freed, the last remnants of the ruffians are shown off to the borders, ugly industrial plants are pulled down, and the countryside is restored to its former shape. The old family feud is resolved. Lotho's mother, Lobelia, hobbles out of prison leaning on Frodo's arm, won over by his kindness and offer of Bag End, which she returns to him. When she dies the next year, Frodo learns that she had left him all of her money to use for helping hobbits "made homeless by the troubles" (328). The end of the feud and the scouring of the Shire are thus accomplished on the principles of restorative justice.

The episodes in which Tolkien uses the restorative justice script are so different from the bulk of his story that they have been passed over in silence by critics, were edited out from film adaptations, and are still considered odd exceptions, not only in Tolkien's oeuvre but in speculative fiction as a whole. Over the past few decades, however, the restorative justice script has slowly been taking shape in public consciousness as an effective action-protocol for addressing a variety of harms. The arrival of *Open Justice* in the 1980s has deepened our understanding and appreciation of restorative justice across the globe. Although this script is not yet normative in modern YA speculative fiction, it is not uncommon.

The restorative justice script is characterized by three signature phenomena: acknowledgment of the harm as affecting specific people, the restorative justice encounter that responds equally but distinctly to the needs of the victim and the offender, and resolution of the harm that involves reconciliation and/or reparation. Inasmuch as the search for restorative justice involves multiple parties pursuing multiple goals—and because it is dialogical, decentralized, and played out in specific circumstances—this script functions in three tracks determined by the primary narrative focus: on the victim, on the offender, and on the community. These tracks are the three dominant facets of restorative justice actualized in YA speculative fiction, and I want to compare them by looking at Ursula K. Le Guin's *Voices* (2006), Terry Pratchett's *Nation* (2008), and Jeanne DuPrau's *The People of Sparks* (2004). The most interesting question, as always, is not whether it is possible to see the restorative justice as a script but what happens when we do.

Set in the universe of the Western Shore trilogy, Le Guin's *Voices* is a story of a conquered city-state Ansul on the brink of a revolt following a seventeen-year long occupation by the Alds. It tells how—through the mediation of a wandering poet-couple Orrec and Gry—the victims and the offenders are able to negotiate an almost impossible rapprochement. Focalization through the Ansulian protagonist Memer, a girl born out of rape and thus also half-Ald, cues the reader to privilege the victim's perspective and makes *Voices* a model actualization of the victim track of the restorative justice script. Speaking for herself, her mother, her family, and her people, Memer has to deal with the legacy of hurts they have incurred as a result of the conquest of their city. The novel's central question for the Ansulians is thus how to be free of the Alds.

The answer to this question, in a retributive justice script, would be a power-based solution such as revolution. In a restorative justice script, however, the answer is not a glamorous victory but a healing compromise. In *Voices* Le Guin develops these two solutions side by side, thus comparing their efficiency and challenging the reader to do the same. On the one hand, there is Desac, the leader of the Ansulian opposition, whose arguments for an uprising and a retributive solution have strong Old Brain appeal to the characters and the reader alike. On the other hand, there is a small group who favor a less glamorous but feasible negotiation. Initially the rebellion takes the upper hand, channeling the community's anger at years of oppression. Yet, the rebellion fails. Desac and his conspirators die. Leaderless, the angry citizens take over the city except the barracks and the Council House where strong Ald forces barricade themselves. Nothing is won. Simultaneously with the rebellion track, however, *Voices* develops the restorative justice script centered on Orrec, his wife Gry, Memer, the old Ansulian leader the Waylord, and the Ald governor the Gand. Among themselves, these five become the restorative justice circle that slowly grows throughout the novel, rechanneling the anger of the rebellion, and eventually encompassing the citizens and the Alds alike.

The restorative justice script in *Voices* highlights the role of a mediator in the character of Orrec. Brought to the city at the invitation of the Ald governor, Orrec also gives poetry performances to the enslaved population. Because Ald religion forbids housing unbelievers, Orrec and Gry find accommodation in the household of Memer and the Waylord. Orrec becomes the first person ever to talk as an equal to both sides. He sees that the Alds realize the harm they had caused and are open to an honorable solution. He also sees that the Ansulians are desperate to leave their victimhood behind. Orrec's visit thus triggers an extended restorative justice encounter. In the course of this encounter, each side tells its version of the story, begins to understand the other, and works toward a resolution.

The first imperative of restorative justice is that those who suffered must be able to tell their stories. This happens in Chapters 1–4, especially after the arrival of Orrec, when the Ansulians, for the first time since the conquest, are offered a sympathetic audience and a safe community to have their pain acknowledged. Later chapters also abound in stories about harm; their telling is shown to transform Ansulians from being locked in victimhood to envisioning healing and a vengeance-free future. The Alds too are given a chance to speak. In his interaction with the Gand, Orrec learns that only a small Ald faction supports the politics of religious persecution. Throughout this crucial period, each side has to deal with its own retributive justice faction: Desac's revolutionaries on the one hand and Prince Iddor's fanatical priests on the other. When in Chapter 12 the rapprochement seems to have failed because of Desac's unsuccessful rebellion and Prince Iddor's coup, Orrec, the restorative justice mediator, again steps in to save both sides from escalating violence. Addressing the Ald officers and the citizens of Ansul, Orrec announces the restorative peace proposal:

> We have set free the master, we have set free the slave. Let the men of Asudar know that they have no slaves, let the people of Ansul know they have no masters. Let the Alds keep peace and Ansul will keep peace with them. Let them sue for alliance and we will grant them alliance. (258)

Following this announcement, the restorative justice circle grows to encompass the citizens and soldiers alike. The Gand commands arms-down, following which "not one soldier raised his sword even when shoved, struck or pushed aside by exultant, vengeful civilians" (260). The citizens of Ansul, in turn, form themselves into a barrier between the crowd and the soldiers, "preventing random insults and forays against the Alds by young men looking for a fight" (283). These behaviors are not spontaneous; to each side they are an enactment of a new justice script triggered through Orrec's mediation, fleshed out in each side's narrative expression, and taking shape as piecemeal solutions before their own eyes. As the Gand sends a messenger to his king, requesting home rule for Ansul, both sides have to wait. These additional days give each side more time to tell stories and express emotions associated with the conflict and its consequences.

One effect of this sharing is that understanding between the two sides becomes possible. Memer, who has always been ashamed of being an Ald-looking "siege brat" (102), discovers that when she lets go of her prejudices, she can no longer hate the Alds. The Gand is tired of ignorance about the people of Ansul and eager to know their epics and poetry. The Waylord too is ashamed "that our people and theirs know so little of one another, after so many years" (112–3). Both sides see the wisdom of negotiations and realize how much they can benefit from being allies. Already from chapter three *Voices* thus presents a series of meetings, each being a stage in the ongoing restorative rapprochement. First the mediators meet with the victims, then with the offenders, then rank-and-file victims meet with rank-and-file offenders, and eventually the leaders communicate directly too. The end result is an agreement that embodies all principles of a successful restorative justice resolution: apology, behavioral change, restitution, and generosity. The Alds' behavioral change is both personal and political, allowing reintegration of former victims and offenders into a new society. Ansul is relieved from tribute and becomes a protectorate. The government will be shared between the Gand, now Prince-Legate, and the people, whose representatives will be the democratically elected Waylords. As Memer comments on those developments:

> I had to listen to … what [everyone was saying] about it, all day long, before I was able to understand that in fact Asudar was offering us our freedom—at a price—and that my people saw it clearly and truly as a victory.
>
> Maybe they could see it so clearly because it had a price on it, in money and trade agreements, matters my people understand.
>
> Maybe I had so much trouble seeing it because nobody died bravely for it. No heroes fighting on Mount Sul. No more fiery speeches in the square. Only two middle-aged men, both crippled, sending messages across the city, cautious and wary, working out an agreement. (313)

Accompanying this process is a number of restoration acts, as the community is remaking itself and the city is "coming back to life" (295). While the process will obviously continue beyond what is described, *Voices* suggests that restoration heals communities in a way that no retribution ever could. Although Memer admits to being "disappointed by this ignominious, uncertain outcome" (317), she and the citizens of Ansul realize that hate rooted in having been victimized cannot heal the harms suffered. Nor can it become a basis for the future. This basis can only be created through the flexible and situational processes of reconciliation characteristic of restorative justice.

In *Voices* it is the victims who initiate the restorative justice encounter. It is also possible, however, that the first step is taken by the offender. This is the case in the offender track, which structures narratives of apology, personal healing, and reconciliation with the community focalized from the perspective of the offender. Cole's story in Ben Michaelson's *Touching Spirit Bear* (2001) or Surly's in the animated movie *The Nut Job* (2014) are actualizations of this script where the offender is clear-cut, but I want to use an example where the offender is also a victim of circumstances, forced to make a tragic choice. This is the story of Daphne's committing murder in Terry Pratchett's *Nation*. With its many themes, *Nation* is a bundle of justice scripts, all of which develop from an unusual colonial encounter[14]. A white girl, the daughter of a governor of a distant archipelago, is shipwrecked by a tsunami on an island where the same tsunami leaves only one survivor. Thrown together, a Polynesian youth and a Victorian-era girl build a new world, the Nation, that accommodates straggling refugees from nearby islands.

Written in the aftermath of the 2004 South Asian tsunami, *Nation* is set in an alternate world sufficiently unfamiliar to the reader to demand significant cognitive effort. At the same time, it employs recognizable concepts: native vs. European civilization, colonialism, cultural clash, natural disasters, and so forth. The framing plot in Chapters 1 and 15 concerns the efforts of the British Crown to locate the next two living heirs to the throne: the local governor and his daughter who was at sea when the wave hit. The main plot in Chapters 2–14 recounts the story of two teenagers: Mau and Ermintrude, who reinvents herself as Daphne. The fact that Daphne uses a different name is a powerful cue to the reader—although not to Mau—that she is a different person from the one she would have been in her prim and hierarchical world of colonial dominance. On the island, she and Mau are equals. The restorative justice script in its offender track is a subplot developed in Chapter 11 but a subplot that turns out to be the catalyst for the global justice arrangement at the novel's conclusion when the Nation becomes an exterritorial outpost of the Royal Society: not a colony, but an independent territory under the British Empire's protection.

The Trouble begins when it turns out that Daphne was not the only survivor from her ship. Two mutineers from *Sweet Judy*, Foxlip and Polegrave—who had been abandoned at sea—made it to an island of a cannibal tribe and taken control over it. Now they arrive on the island to terrorize the Nation. Foxlip shoots the priest Ataba, and Daphne is taken hostage. Realizing that the pirates will kill other islanders, Daphne offers them local beer that she knows is poison in its raw form. As brewed on the island, the beer needs a human spitting into and singing to give it time to become nourishing. As she offers it to the men, Daphne spits into her bowl, sings the curing song, and tells the men to do the same. Foxlip, who ignores her advice, tumbles back dead; Polegrave runs away. Defending herself and others, Daphne has killed a man.

Arguing with herself about what she should have done, Daphne asks Mau and the Nation for a trial. What follows is the Nation's first court in which "everyone sat around in a circle, children too. ... No one was more important than anyone else" (263). Because Foxlip is dead and because his death saved the community, the focus is on healing Daphne. In the restorative justice encounter, Mau acts as a mediator, Daphne as the offender, and the community as judges and victims. The circle proceeds through questions, asked by Mau, and stories told by Daphne in response. As "[t]he whole population of an island sat in a circle around Daphne, trying to be helpful" (265), Mau first asks Daphne about the men. The girl recounts the story of mutiny on *Sweet Judy* that involved Foxlip, Polegrave, and the First Mate Cox and how the three were left adrift in a small boat just before the tsunami arrived.

As the awareness of the mutineers' murderous nature dawns on the islanders, a wave of whispers travels around the circle. Speaking to and for all gathered, Mau then reiterates the facts of the case and asks the circle: "The ghost girl thinks she killed a man. Did she? You must decide" (275). The people start talking at once, but it soon becomes clear that "something was happening. Little conversations got bigger, and then were picked up and rolled from tongue to tongue around the circle" (275). Watching this jumble of words and emotions, Daphne becomes aware that she is seeing "pure democracy [where p]eople don't just get a vote. They have a say" (276). When the roar settles down, the translator Pilu announces that a man who would kill another man, a priest, and a dolphin for pleasure was no man, but "a demon haunting the shell of a man." Thus, "the ghost girl could not have killed him, because he was already dead." Following this pronouncement, Mau asks if this is what everyone thinks. The circle responds with "a roar of agreement" (276).

This verdict puts a sense of closure on the episode by evaluating it in the context of the needs of the community and of the offender/victim Daphne. Offered the support of the Nation, Daphne is empowered to deal with having been victimized and having killed a man. She confesses to a wrongdoing, acknowledges the harm, and admits responsibility. The expression of guilt, in turn, makes her eligible for forgiveness, which she receives from the community. While the standard constellation of roles in this script is garbled—Daphne and Foxlip are both offenders and victims at the same time—the restorative justice script is shown to work well even in such complex situations, as long as it involves the entire community.

Obviously, had Foxlip survived, been unrepentant, or refused to come to the table, the restorative justice script would be impossible. This contingency is stressed in the episodes that follow, which remind the reader and the characters alike that restorative justice is not always a viable option. When the cannibal raiders land on the island, Mau is forced to fight and eventually kill Cox. However, as the reader understands from Mau's explanation to Daphne, Mau's motivation for the duel is to avoid large-scale bloodshed. The imperative to avoid violence as a conflict-resolution strategy runs strong through *Nation* and adds to its message about the validity of solutions based on dialogue.

The need to be open to dialogue is also stressed by the fact that much of the novel is about "noise"—the misreading of intentions in human communication. For example, when Daphne meets Mau for the first time, she fires a pistol at him, but the result is only a fountain of sparks from the lock. Mau, in turn, joyfully "accepts" the gift of the pistol as a fire lighter. Signal misreading is soon repeated when Daphne leaves Mau a drawing, intended as an invitation for him to come to the wreck at noon. Mau reads it as "[w]hen the sun is just above the last tree …, you must throw a spear at the big wrecked canoe" (55). Frozen in the cabin by having just seen a spear fly through it, Daphne then reasons out that this must be a signal meaning that Mau comes in peace. "It's just like shaking hands," she thinks, congratulating herself. "Well, I'm glad that's one little mystery solved" (64). In recounting these and other instances of "noise"—compounded, for Daphne and the islanders, by cultural, technological, and language differences—*Nation* repeatedly stresses how easy it is to get things wrong and how openness to cooperation is a saving power in the noise-ridden human interaction.

This, too, is one of the themes of Jeanne DuPrau's *People of Sparks*—a dystopian SF novel set in a post-apocalyptic United States. Unlike the two previous books, this novel is an example of the conflict of needs, in which both sides are right. This second book in the City of Ember tetralogy recounts how, after the technical death of their underground city, the people of Ember come to the surface and wind up in the village of Sparks. Led by the teen protagonists Doon and Lina, the Emberites request food and shelter. The village leaders welcome the newcomers but realize that the resources accumulated by 322 people of Sparks will not be enough to accommodate 417 refugees or to last both groups through the winter. The conflict that gradually builds is thus one about legitimate needs: the people of Sparks are unwilling to risk starvation by depleting their resources. The people of Ember, in turn, need food and assistance until they can stand on their own. Framed by each group's struggle to assert their needs in consideration of those of the other group, *The People of Sparks* develops the community track of the restorative justice script. The negotiation takes the form of two restorative justice encounters, the first one encompassing Chapters 3–26 and the other one spanning Chapters 27–28. In both encounters several characters—Mary Waters, Dr. Hester, and Kenny on the Sparks side, and Lina, Doon, and Mrs. Murdo on the Ember side—act as mediators between their communities.

The first encounter is unsuccessful because the leaders of Sparks, concerned with their own community's survival, are not honest with the Emberites. They agree to accommodate the refugees for six months, but none of them dares to say out loud that even after six months, in winter, the Emberites will not be able to start a new town or grow their own food. By failing to inform the refugees of this basic fact, the leaders of Sparks lay the ground for the future conflict, when the Emberites realize that leaving Sparks would mean certain death. The leaders' lack of courage not only fails to solve the problem but also sets the tone to the relationship between the two groups.

This first incomplete resolution creates a situation of a limited integration, which favors the buildup of tension. When neither side trusts the other, cultural differences become a source of prejudice. For the Emberites the living standards of the people of Sparks—extremely proud of their material culture—are primitive when compared to the comforts of the technologically advanced life in Ember. They are grateful for their hosts' hospitality, but their stories of life in Ember are seen as lies or complaints against what they are offered. The people of Sparks, in turn, resent the Emberites as helpless in the most irritating way. Emberites have never seen buildings, animals, trees, fruit, or most of the tools and food used in Sparks. They are afraid of fire and not used to hard physical work. They do not know natural phenomena like seasons or temperature changes. They have not heard of the Disaster—the nuclear holocaust that almost wiped out the human race. In all those ways, the Emberites seem "unbelievably dumb" (126) to their reluctant hosts.

Adding to these problems is the attempt by the people of Sparks to save on food offered to their temporary guests. As the Emberites grow tired of being grateful and feeling starved at the same time—and as the people of Sparks grow convinced that the Emberites are too large a drain on their resources—the ground is created for individuals to use personal grudges to turn the two communities against each other. On the Sparks side, this ignominious role is played by the 9-year-old boy Torren, who at one point smashes two crates of tomatoes and then blames the act on the Emberite Doon. With the word of one boy against the other's, each community believes their own and resentments flourish. On the Ember side, the conflict monger turns out to be the charismatic teenager Tick who incites animosity by writing offensive slogans and plotting shenanigans against the Emberites, while making it seem like the work of the Sparks people. He convinces the Emberites that their hosts hate them, organizes a militia, and leads the armed Emberites against the people of Sparks.

The episode in Chapter 26 that recounts a violent confrontation between Tick and Ben Barlow marks the failure of the solution proposed in the first restorative justice encounter. The Emberites know that they are unable to leave and are convinced by Tick that their only option is to fight. As tempers flare up, the Sparks' leaders roll out a machine gun against the Emberites. The first discharge goes in the air, but the gun explodes catching the pine tree standing next to the town hall on fire. In the confusion that follows, Doon saves Torren, who happened to be sitting up in the tree. Lina, for her part, is the first of the Emberites to join the people of Sparks as they desperately fight the fire spreading to the entire city. After Doon joins the fire line too, other Emberites follow and the town is saved.

The outcome of this confrontation for both sides is the realization that violence will only bring more damage and that they need to be honest with one another. Accordingly, Chapter 28 offers a restorative justice encounter when the two groups meet, apologize to each other, acknowledge each other's needs, and declare their willingness to create a future together. Mary, the leader of Sparks, announces that the Emberites will not be asked to leave and apologizes for trying to make them do so. The meeting also addresses past resentments that led to the escalation of conflict. Doon announces what he just learned—namely, that the slogans against the Emberites had been written by one of their own. Mary, in turn, declares that the smashing of the tomatoes had not been done by Doon but by Torren who now admitted to his lie. As the two groups reconcile, Mary declares: "We will refuse to be each other's enemies. We will renounce violence which is so easy to start and so hard to control. We will build a place where we can all live in peace ... From now on, we are *all* the people of Sparks" (324). In the new spirit, new projects are undertaken and the community will profit from this cooperation. As recounted in the sequel, *The Diamond of Darkhold* (2008), they undertake a salvage expedition to Ember where they retrieve extra food and electricity-generating solar-power crystals that will reintroduce electricity into their world.

Now, in all three tracks perhaps the biggest cognitive challenge—for the characters and the reader as well—is to know when forgiveness should be given and why. At any point during our interaction with others, we risk that forgiving may be interpreted as a sign of weakness and make us subject to further exploitation. This is a complex challenge in life as in stories. Thus, in *Voices* neither side of the conflict first offers forgiveness as this may be seen as a sign of weakness. Instead, forgiveness is proposed to both sides by the mediator Orrec. In *Nation*, forgiveness is asked by the victim/offender who would otherwise be unable to deal with the burden of her guilt. In *Sparks*, finally, it is first acted out in a situation crucial for the community's future and only later verbalized, with each side apologizing to the other. In all these examples the restorative justice script ties the resulting slot "forgiveness" to a preceding slot "trust," which crucial connection I now want to illustrate on the example of the animated film *Battle for Terra* (2007).

The film tells a story of a human invasion of a peaceful alien planet and represents it through the eyes of the Terran protagonist Mala. While this may suggest the defensive war track structuring, the fact that (1) neither party is presented as evil, (2) most screen time is devoted to Mala's interaction with the human protagonist James Stanton, and (3) the resolution of Trouble is not the extermination of one group but their peaceful coexistence makes this film an actualization of the restorative justice script in its community track. The first 17 minutes of the film introduce the invasion as Trouble, or, in restorative justice terms: harm that affects specific people. For Mala, of course, the personal harm is that the invaders capture her father. The next 50 minutes of the movie, from Mala's first meeting with James to his sacrificial death, amounts to being an extended restorative encounter, where James and Mala learn to understand the other's community, acknowledge each community's needs and rights, and work toward a solution. The last six minutes are a fast-forward to the future after that solution—reconciliation and forgiveness—has been achieved.

Because the story is told through a film medium, the opportunities it offers for embedded mind-reading are different than in a textual narrative and come only through external focalization: what characters see or say. We have no access to their thoughts, but this is sufficiently compensated for by their bodily movements and facial expressions: after all, 90 percent of human emotional communication is nonverbal and most of it is processed through reading the face (McGilchrist 257). The opening scene introduces Mala and Senn flying their gliders, but the sequence serves several purposes. First, it is an introduction to their world and its natural beauty. Second, it gives us hints about its dangerous places: the forbidden area, the wind tunnel. Third, and most importantly, it gives us plenty of hints about Mala's and Senn's personalities and their friendship. Mala is curious, likes risk and adventure. Senn is cautious, considerate, yet steadfast. When Mala gets caught in the wind tunnel, Senn risks his life and pulls her out. Although nothing is stated explicitly, this sequence illustrates friendship and trust between Mala and Senn.

However, when Mala rescues one of the human invaders and harbors Lieutenant Stanton in her house, Senn is scared out of his wits. He has just been attacked by Stanton who wants to keep his presence hidden, but Mala makes Stanton let Senn go. Intimidated and confused, Senn whispers "you betrayed us" and runs away to get the guards. Yet, a moment later, when they arrive at her house, Senn lets Mala and Stanton slip away. Why does he do it? He sees the man's shadow and Mala as she leaves the room. Unlike in a textual narrative, where this scene would be described as "before slipping out of the room, she paused and gave him a long, wistful look," Senn must figure out the meaning of Mala's facial expression. We then understand that he does not alert the guards because he infers that Mala is *not* betraying her people. The audience, of course, know better: we have seen how Mala's father had been taken, how Mala crashed Stanton's ship, rescued him, and finally made a deal with him that if she helps him repair the ship, he will take her to her father. The audience knows Mala's motivations that are hidden from Senn, yet Senn can bridge this knowledge gap because of his trust in Mala. This is an important cognitive lesson about friendship in a noise-laden world of social relationships: trust allows us to keep the relationship alive even when our friends' actions seem puzzling. When they meet at the end of the movie, Senn apologizes saying, "I'm sorry I doubted you," and Mala is happy to forgive him.

Mala's considerate behavior also becomes a factor in changing Stanton's attitude toward Terrans. Although he is the aggressor, Mala rescues him; Stanton also relates to Mala's determination to free her father and is astounded by the beauty of the Terran world celebrated in the Ceremony of Life. The positioning of Mala as a young female and James as a young male may be part of the rapprochement too. Mala's growing empathy toward Stanton stems partly from the fact that she pulled him out from the crash. When the robot Giddy explains about wars that destroyed the Earth, she understands why humans are attacking Terra: they are looking for a world to live in. She then shares with Stanton the Terran history, how they too fought wars that almost wiped them out but at the last moment realized the foolishness of it and remade themselves into a non-violent civilization living in harmony with nature. Even before Mala and Stanton discover a hidden Terran military complex—with the record of past wars and some aircraft that is now being readied to fend off the invasion—they and the audience realize that the two sides have more in common than they thought. A bond of trust is formed when Stanton and Mala—whom he used to call "it"—switch to using their first names.

From this moment on, James and Mala develop a rapport based on an empathetic connection. James smuggles Mala to the human spaceship. Although Mala's father dies, James rescues Mala by ordering Giddy to take her back to her people. There Giddy explains to the Terran leaders how the humans want to "terraform" the planet. As we and the characters understand, this would mean death to all Terran life forms, since their atmosphere contains less oxygen than is necessary for humans. The attack is imminent, though. General Hammer exacts a coup and locks away the council, whose members are against the extermination of the Terrans. It is true that the spaceship is falling apart and humans must find another planet soon if they are to survive. However, Hammer's answer to this predicament is based on the "either us or them" logic that does not allow alternatives. James, in turn, believes there has to be another way. When the conflict and each side's motivations have been laid clear, the audience is called upon to evaluate available options. Who is the victim and who is the aggressor? As human beings, we may feel that human survival is a priority that legitimizes General Hammer's extermination plan. But having come to see Mala and her people as a wonderful species, we suspect that the two groups can reach an agreement. This latter inference is foregrounded by our earlier evaluation of James' and Mala's actions as a workable, peaceful cooperation.

As the invasion starts and a gigantic terraformer is planted on Terra, both sides are fighting for survival. Besides necessity, Hammer invokes the death of two men who perished in the scuffle with Mala's father, and so he projects the retributive justice script as the action protocol for the invasion. The outgunned Terrans fight back, but their strength weakens as the terraformer pumps more oxygen into the atmosphere. The tragic dilemma of the situation is made clear when James' younger brother Steward tries to shoot down Senn's glider and finds himself chased by Mala. James saves him, but seeing that Mala is piloting the enemy glider, he does not take her down. As he looks at the destruction around him and sees a beautiful Terran whale he accidentally shot go down, James understands that retributive violence is not the right answer. He positions his aircraft between Mala's and Steward's, says "I'm sorry" to his brother, and then rams into the terraformer, ending the invasion and human hopes for conquering Terra. In the context of the story so far, Mala, Steward, and the audience are cued to understand James' act as the alternative to Hammer's "us or them" logic, similar to Bilbo's offer of the Arkenstone that forces the two parties to meet and negotiate. The last scenes of the film confirm that James was right. Mala and Senn again fly their gliders and circle around a huge air dome, with the human settlement in it. In the middle of the settlement is a huge statue of James Stanton: a hero who made peaceful coexistence of the two societies possible.

What is the young audience taking from these and similar stories? While they are not fully aware of their own cognitive processes, their minds register the importance and quiet heroism of forgiveness and the scripted path for restorative justice. Although each actualization of the restorative justice script recounted in this chapter is different, they all share the focus on seeing harm as affecting specific people, and all offer a vision of justice as the healing of the victim and the offender alike. They all involve restorative justice encounters, in which different needs of each side are taken into consideration, and they all conclude with a peaceful resolution of the conflict that involves reconciliation, integration and/or reparation. In each narrative the restorative justice solution is projected as an alternative to the retributive one and involves multiple parties and multiple goals. In each it is built on a dialogue. For Lewis, Tolkien, and Le Guin, one can argue that a narrative quest for restorative solutions derives from the authors' spiritual beliefs, Protestantism, Catholicism, or Taoism; however, the imperatives of restorative justice hold just as strong within the framework of the secular humanism that informs the work of Pratchett, DuPrau, and Tsirbas' *Battle for Terra*. The restorative justice script thus accommodates but does not require spiritual or religious underwriting. Instead, it offers viable alternatives to the destructiveness of violence as a strategy of conflict resolution. Although different, all narratives examined in this chapter acknowledge how easy it is to get caught in a retributive mindset and all narratives stress the courage it takes to break the circle of violence in favor of restorative solutions. In this sense, and perhaps most importantly, novels and stories structured on the restorative justice script advance a new understanding of justice and ways to achieve it: an understanding that sees harm as a disregard for the needs of another person and justice as a process commencing with the recognition of those needs.

NOTES

1. Restorative justice practices, albeit often involving some form of punishment, have also been the norm in most indigenous societies. For an interesting example of "The Great Law of Peace" which bound five North American tribes from the twelfth through the seventeenth centuries, see Mann 357–366.
2. At bottom, this criticism of Rawls and of *New Justice* boils down to whether or not there is an independent argument for fair equality of opportunity. In his last book Rawls himself registered his uncertainty as to whether fair equality of opportunity should as absolutized as it had been in his work. See footnote 44 to section 50 in *Justice as Fairness, A Restatement.*
3. For other criticism of the system's failures see Michael Tonry's *Retributivism Has a Past. Has It a Future?* (2011) and Mark White's *Retributivism: Essays on Theory and Policy* (2011).
4. Eglash's creative restitution—in which "an offender, under appropriate supervision, is helped to find some way to make amends to those he has hurt by his offense, and to 'walk a second mile' by helping other offenders" (qtd. in Mirsky 1)—has been foundational to the restorative justice movement.
5. One study of its efficacy over the 25-year period found that restorative justice programs are significantly more effective in "improving victim and/or offender satisfaction, increasing offender compliance with restitution, and decreasing the recidivism of offenders" than traditional criminal justice responses (Latimer, Dowden, and Muise 138). The restorative turn in the entire justice system has been noted even by scholars not specifically concerned with restorative justice; Stuntz, for example, observes that over the past decade or so "[t]hose who enforce the law and those most tempted to break it [have] turned ... toward each other—toward relationship rather than enmity" (310–11) and sees hope in the growing recognition that "[b]oth sides are us"—that "[d]emocracy and justice alike depend on getting that most basic principle of human relations right" (312). Another powerful testimony about the shift toward the extra-judicial restorative justice strategies, this time in crime prevention, can be found in David M. Kennedy's *Don't Shoot: One Man, A Street Fellowship, and the End of Violence in Inner City America* (2011). Although Kennedy does not use the term "restorative justice," his approach is nothing but restorative. He treats crime as primarily a violation of people and relationships; he views of prevention as aimed at restoring the health of the community; he relies on the community—including the offenders' families, neighbors, community leaders and local law enforcement; and he addresses urban violence from the perspective of the people affected, the victims and offenders alike. Kennedy's accounts of how some of America's most dangerous neighborhoods were rid of their supposedly intractable violence and drug pushing through the application of remarkably effective strategies based on cooperation rather than persecution reads like a fairy tale but is true. As of 2015, Kennedy is deemed one of the most innovative criminologists in the nation and his methods have been successfully applied in over seventy US cities to date.

6. This, at least, is the argument of Peter Singer's *The Expanding Circle* (1981): the first book on the implications of evolutionary psychology for the study of ethics. Singer argues that ethics is a kind of biological adaptation that builds "on a more limited, biologically based forms of altruism" (149) and stems from the human capacity for reason. Since the capacity for reason is an open-ended one, with ever expanding and unpredictable consequences, so ethics and empathy have been the eponymous expanding circle. Over the centuries, ethics and empathy have expanded beyond kin, tribe, community, nation, gender and cultural difference, even species—a change "so tremendous that it is only just beginning to be accepted on the level of ethical reasoning and is still a long way from acceptance on the level of practice" (113). Singer's belief that the expansion of the moral sphere has happened because of the autonomy of reasoning is shared by Pinker in *Better Angels*, who attributes ethical advancement largely to the increase of human intelligence, especially in the segment of abstract reasoning amply documented over the twentieth century in the studies of the so-called Flynn Effect (650).
7. The concern of game theory is strategies of action that self-interested organisms employ when interacting, competitively, with other organisms who are equally self-interested and whose future behavior cannot be predicted. Ever since its infancy in the 1940s, game theory has been in search of its holy grail, the Nash Equilibrium, which is the most stable strategy in evolutionary terms. Based on analyses of the famous Prisoner Dilemma, game theory in the 1960s held that self-interested actors will always defect rather than cooperate. However, the Prisoner Dilemma describes a one-shot interaction with a random partner, which model fails to capture the reality of social organisms' cooperation, which happens in iterated games—open-ended sequences of interactions—and with a limited set of partners. When these insights were applied by Robert Axelrod to a series of experiments in the 1970s, the most evolutionarily stable strategy was tit-for-tat: a strategy where organisms cooperate as a default starting reaction, and then copy the other player's previous move, responding to defection with defection and to cooperation with cooperation. The strength of tit-for-tat turned out to be not its raw power, but "its ability to encourage win-win behavior" (McCullough, *Beyond* 96). The key to long-term evolutionary success was tit-for-tat's capacity to forgive those who express remorse and are willing to cooperate again. More recent work since the mid-1990s has refined these findings by including in computations the big problem for tit-for-tat called noise. Noise is a term describing mistakes in cooperative strategies, when one side defects when it really meant to cooperate or the other side interprets cooperative behavior as defection. Since noise is a significant problem in real-life cooperation, computer simulations that include the possibility of unintentional harm or defection have led researchers to propose three candidates for the most evolutionarily stable strategy: generous tit-for-tat that "grants forgiveness *unconditionally* ... about one third of the time" (101), contrite tit-for-tat that, if it has defected without justification or through noise, "allows its partner to defect without itself retaliating in return" (102), and firm-but-fair that makes its next decision based on its own and partner's recent choices. Although none of these seem to be the final answer, all game theory research to date indicates that always-defecting strategies or even moderately forgiving tit-for-tat are not evolutionarily stable.

8. These types are distinguishable by the agent's different capacity for forgiveness—each of them corresponding to a specific stage of moral development and reasoning about justice as described by Kohlberg. In the lowest stages of *Revengeful Forgiveness* and *Restitutional Forgiveness*, forgiveness can only occur after the wrongdoer has been subjected to revenge or appropriate punishment; in the highest ones, *Forgiveness as Social Harmony* and *Forgiveness as Love*, forgiveness is offered because it restores peaceful relations in society and promotes a true sense of love (see Enright, Santos, and Al-Mabuk 96ff).
9. The forgiveness system—which McCullough, Kurzban, and Tabak posit as a possibly plural structure of interconnected systems that produce motivational changes—serves two functions. First, it restrains the revenge impulse that, if acted upon in relation to the kin, would reduce the avenger's "inclusive fitness"—the pool of one's genetic relatives (McCullough, *Beyond* 98), and if acted upon in relation to nonrelatives would terminate cooperation needed to thrive in large social groups. Second, it enables the restoration of "potentially valuable relationships" in the aftermath of interpersonal harms (McCullough, Kurzban, and Tabak 230).
10. This is the case with the horse Bree who apologizes and is forgiven for his pride in *The Horse and His Boy* (1954), with Eustace who apologizes and is forgiven for his spiteful former self in *The Voyage of the Dawn Treader* (1952), with Digory who apologizes and is forgiven for his inadvertent bringing Jadis to Narnia in *The Magician's Nephew* (1955), and even with the donkey Puzzle who apologizes and is forgiven his part in the false-Aslan scheme in *The Last Battle* (1956).
11. Like Tutu later, Lewis stressed that forgiveness is not forgetting or excusing. Instead, it means "[l]ooking steadily at the sin, the sin that is left over without any excuse, after all allowances were made, and seeing it in all its horror, dirt, meanness, and malice, and nevertheless being wholly reconciled to the man who has done it. That, and only that, is forgiveness. ..." (Lewis, "Forgiveness" 181).
12. Lewis was a strong believer in punishment. In "The Humanitarian Theory of Punishment" (1949) he maintained that punishment and justice are a continuum linked by the concept of responsibility; in "Why I Am Not a Pacifist" (1949) he stated that punishment is an important social practice "as an expression of the moral importance of certain crimes" (76). Throughout, he vigorously defended desert. Thus, instances of forgiveness and reconciliation in The Chronicles are always preceded by some form of punishment, like Edmund languishing in the White Witch's dungeon or Aravis being slashed on the back by the lion's claws.
13. This sentiment is widely shared even today. For example, Doris Wolf's "The Suffering of the Perpetrators" (2014)—an analysis of two YA novels set in Germany during and shortly after WWII—highlights the problematic nature of representing the suffering of young Germans in the hands of the Allies. Although it is true that many German youths and civilians did suffer, Wolf concludes that attempts to represent their painful wartime experiences is always done at the expense of guilt and amounts to "a refusal to provide crucial historical context" (67). "[A]cknowledging how Germans were victimised during the war," she says, "minimises the evils they perpetrated throughout the Third Reich and the suffering of their victims" (66).
14. As the community of Nation knits itself, one important theme is social justice: what customs and power relations from the world before the disaster should be embraced by the reconstituted society. Although social justice themes are raised, they do not entail a script. The latter part of the novel, however, offers

an actualization of the global justice script. In this part, Pratchett reimagines the myth of the Atlantis by making Nation its last surviving island and reimagines the history of British colonization by making the two sides arrive at a just and mutually beneficial relationship.

WORKS CITED

Baron-Cohen, Simon. "An Empathizing System: A Revision of the 1994 Model of the Mindreading System." *Origins of the Social Mind: Evolutionary Psychology and Child Development*. Eds. Bruce J. Ellis and David F. Bjorklund. New York: Guilford, 2005: 468–492.

Baron, Jane B. "Storytelling and Legal Legitimacy." *Un-Disciplining Literature: Literature, Law, and Culture*. Eds. Kostas Myrciades and Linda Myrciades. New York: Peter Lang, 1999: 13–27.

Battle for Terra. Dir. Aristomenis Tsirbas. SnootToons and MeniThings Productions, 2007.

Brighouse, Harry. *Justice*. Malden, MA: Polity Press, 2004.

Capeheart, Loretta and Dragan Milovanovic. *Social Justice: Theories, Issues, and Movements*. New Brunswick, NJ: Rutgers University Press, 2007.

"Clark descendants give canoe replica to tribe." Associated Press. *AZCentral*. Sep 23, 2011. http://www.azcentral.com/news/articles/2011/09/23/20110923lewis-clark-stolen-canoe.html. June 29, 2014.

DuPrau, Jeanne. *The People of Sparks*. New York: Random House, 2004.

Eglash, Albert. "Beyond Restitution: Creative Restitution." *Restitution in Criminal Justice: A Critical Assessment of Sanctions*. Eds. Joe Hudson and Burt Galaway. Lexington, MA: D.C. Heath and Company, 1977: 91–99.

Enright, Robert D., Maria J. D. Santos, and Radhi Al-Mabuk. "The Adolescent as Forgiver." *Journal of Adolescence* 12 (1989): 95–110.

Kennedy, David M. *Don't Shoot: One Man, A Street Fellowship, and the End of Violence in Inner City America*. New York: Bloomsbury, 2011.

Latimer, Jeff, Craig Dowden, and Danielle Muise. "The Effectiveness of Restorative Justice Practices: A Meta-Analysis." *The Prison Journal* 85 (2005): 127–144.

Le Guin, Ursula K. *Voices*. Orlando, FL: Harcourt, 2006.

Lewis, Clive, Staples. "Why I Am Not a Pacifist." *The Weight of Glory*. Ed. Walter Hooper. San Francisco: HarperCollins, 2001: 64–90.

Lewis, Clive, Staples. "On Forgiveness." *The Weight of Glory*. Ed. Walter Hooper. San Francisco: HarperCollins, 2001: 177–183.

Lewis, Clive, Staples. "The Humanitarian Theory of Punishment." *God in the Dock: Essays on Theology and Ethics*. Ed. Walter Hooper. Grand Rapids, MI: William B. Eerdmans, 1996: 287–300.

Lewis, Clive, Staples. *The Horse and His Boy*. New York: Collier Books, 1970.

Lincoln, Abraham. "Second Inaugural Address 1856." http://www.let.rug.nl/usa/presidents/abraham-lincoln/second-inaugural-address-1865.php. July 20, 2014.

Mann, Charles C. 1491: *New Revelations of the Americas Before Columbus*. New York: Alfred A. Knopf, 2005.

McCullough, Michael E. *Beyond Revenge: The Evolution of the Forgiveness Instinct*. San Francisco: Jossey-Bass, 2008.

McCullough, Michael, Robert Kurzban, and Benjamin A. Tabak. "Evolved Mechanisms for Revenge and Forgiveness." *Human Aggression and Violence: Causes,*

Manifestations, and Consequences. *Eds*. Philip R. Shaver and Mario Mikulincer. Washington, DC: American Psychological Association, 2011: 221–239.

McGilchrist, Iain. *The Master and His Emissary: The Divided Brain and the Making of the Western World*. London: Yale UP, 2009.

Mikaelsen, Ben. *Touching Spirit Bear*. New York: Harper Collins, 2001.

Mirsky, Laura. "Albert Eglash and Creative Restitution: A Precursor to Restorative Practices." *Restorative Practices EForum*, Dec. 3, 2003: 1–4. http://www.iirp.edu/iirpWebsites/web/uploads/article_pdfs/eglash.pdf. July 23, 2014.

Mullet, Étienne and Michèle Girard. "Developmental and Cognitive Points of View on Forgiveness." *Forgiveness: Theory, Research, and Practice*. Eds. Michael E. McCullough, Kenneth I. Pargament, and Carl E. Thoresen. New York: Guilford Press, 2000: 111–132.

Newberg, Andrew B., Eugene G. d'Aquilli, Stephanie K. Newberg, and Verushka deMarici. "The Neuropsychological Correlates of Forgiveness." *Forgiveness: Theory, Research, and Practice*. Eds. Michael E. McCullough, Kenneth I. Pargament, and Carl E. Thoresen. New York: Guilford Press, 2000: 91–110.

Panksepp, Jaak and Lucy Biven. *The Archeology of Mind: Neuroevolutionary Origins of Human Emotions*. New York: W.W. Norton, 2012.

Pinker, Steven. *The Better Angels of Our Nature: Why Violence Has Declined*. New York: Viking, 2011.

Pratchett, Terry. *Nation*. New York: Harper Collins, 2008.

Rawls, John. *Justice as Fairness: A Restatement*. Cambridge, MA: Harvard UP, 2001.

Rhonda, James P. *Lewis and Clark Among the Indians*. Lincoln, NE and London: The U. of Nebraska P., 2002.

Singer, Peter. *The Expanding Circle: Ethics, Evolution, and Moral Progress*. 2nd ed. Princeton, MA: Princeton UP, 2011.

Stuntz, William J. *The Collapse of American Criminal Justice*. Cambridge, MA: Harvard UP, 2012.

Tolkien, J. R. R. "Notes on W. H. Auden's review of *The Return of the King*." 1956. *The Letters of J. R. R. Tolkien*. Ed. Humphrey Carpenter. New York: Houghton Mifflin Company, 2000: 238–244.

Tolkien, J. R. R. *The Hobbit: Or There and Back Again*. 1937. *The Annotated Hobbit*. Rev. ed. Annotated by Douglas A. Anderson. London: HarperCollins, 2003.

Tolkien, J. R. R. *The Return of the King*. New York: Ballantine Books, 1994.

Tonry, Michael. Ed. *Retributivism Has a Past. Has It a Future?* New York: Oxford UP, 2011.

Tutu, Desmond Mpilo. *No Future Without Forgiveness*. New York: Doubleday, 1999.

White, Mark D. *Retributivism: Essays on Theory and Policy*. New York: Oxford UP, 2011.

Wolf, Doris. "The Suffering of the Perpetrators: The Ethics of Traumatic German Historicity in Karen Bass's Young Adult World War II Novels." *International Research in Children's Literature* 7.1 (2014): 64–77.

Young-Bruehl, Elisabeth. *Childism: Confronting Prejudice Against Children*. New Haven, CT: Yale UP, 2012.

Zehr, Howard J. *Changing Lenses: A New Focus for Crime and Justice*. Scottsdale, PA: Herald Press, 1990.

Zehr, Howard J. "Retributive Justice, Restorative Justice." *New Perspectives on Crime and Justice* #4. Akron, PA: Mennonite Central Committee Office of Criminal Justice, 1985: 1–16.

6 Humans Are Animals Too
Environmental Justice Scripts

The previous chapters have established retributive and restorative justice as two basic modes of reacting to harm hardwired into our cognitive architecture. The first of these modes prioritizes punishment, while the other seeks healing. The ability to seek vengeance and the ability to forgive are complex, tertiary-process operations processed by the New Brain, yet they draw on the primary- and secondary-process Old Brain affective systems: the fear/rage circuits and the anger affect on the one hand; care and the forgiveness system on the other. Since the evolution of the reptilian-brain potential for retribution preceded the development of our mammalian-brain potential for forgiveness, revenge and forgiveness are developmentally staggered. Young and preadolescent children find it easier to punish than to forgive because their world is dominated by the right hemisphere[1]. Inasmuch as forgiveness requires more complex processing, the ability to forgive and cooperate must be initially taught, but it becomes second nature once the left hemisphere catches up in middle childhood, around the age of eight (Crago 91). Revenge and forgiveness are thus asymmetrical on the New Brain level of interhemispheric competition and remain so throughout the entire human lifespan. Because the transfer of information from the left to the right hemisphere is always slower than the other way around (McGilchrist 218), the revenge impulse is biologically quicker to emerge than the forgiveness one.

The other four types of justice—poetic, environmental, social, and global—are cultural adaptations that draw on these two basic evolutionary endowments. Just as poetic justice was a cultural adaptation for the *Old Justice* feudal world of hierarchical and inescapable oppression, so too environmental, social, and global justice are cultural adaptations that draw on *New* and *Open Justice* paradigms to challenge various forms of injustice, including non-criminalized oppression in an increasingly interconnected world. Although these types of justice are late twentieth-century adaptations, each has important antecedents in the history of Western thought and literature.

For the emergence of environmental justice, perhaps the principal challenge has been overcoming the legacy of thinking in terms of the absolute divide between culture and nature. The sharp distinction between the human and the natural world was one of the key assumptions within the

New Justice paradigm and a conceptual carryover of *Old Justice*. When in the late nineteenth century one of the most celebrated moral voices of the period claimed that "we … have no duties toward the brute creation," that "there is no relation of justice between them and us," but instead "into our hands they are absolutely delivered [so that w]e may use them, we may destroy them at our pleasure … for our own ends, for our own benefit or satisfaction" (Newman 79–80), he was expressing the mindset that was perfectly acceptable at that time but became indefensible and morally suspect within a century. Beatified in 2010 by Pope Benedict XVI, Cardinal John Henry Newman who said these words *was* a person of deep ethical commitments. However, like most of his contemporaries, he believed in an unbreachable abyss between people and God. If "no justice … could exist between them and Him" (187), the same absolute divide separated people from animals and the natural world.

That Newman's position finds few supporters today is testimony to the fact that the connection between environmentalism and justice is historically located. As I argue in this chapter, environmental justice is a modern application of the principles of distributive justice in the environmentally conscious world of the twenty-first century. In its narrower, anthropocentric form it recognizes the rights of humans—but only humans—to environmentally sound living conditions and challenges environmental racism: exposing certain peoples to unequal and discriminatory levels of environmental hazards. In its broader, biospheric form, however, environmental justice encompasses concerns for the rights of non-human animals and the natural environment writ large. On this level, it is used to protest environmental speciesism, which assumes human superiority, disregards the well-being of the biosphere, and leads to environmental destabilization.

As a motivational and behavioral response to environmental inequality, environmental justice operates as a script. This distinctly modern script recognizes that humans are animals too and extends the idea of environmental rights to non-human animals and the natural world. Its goal is the protection and/or restoration of environmental well-being for all parties involved. Drawing on restorative or retributive solutions, this script hones our capacity for biospheric cooperation and seeks to reduce the negative impact of environmental consequences resulting from human operations. Its cognitive function is to provide us with tools to identify and remedy the ongoing forms of environmental injustice as well as to prevent future activities that are environmentally unjust.

ENVIRONMENTALISM, NATURE, AND THE ENVIRONMENTAL JUSTICE SCRIPT

The rise of the environmental justice movement in the 1990s has a complex history and resulted from a convergence of several distinct processes. With no pretense to any comprehensiveness, I want to start by recounting a rivulet of the process—cited in Mark Neuzil and William Kovarik's *Mass Media and Environmental Conflict* (1996)—that helps situate environmental justice as an *Open Justice* phenomenon, albeit with origins harking back to the late nienteenth-century *New Justice* developments.

In 1852 a group of gold-seeking prospectors stumbled across a grove of giant trees in northern California's present-day Calaveras County. Word of the colossal trees spread fast, and one of the first who traveled to see them was a businessman named George Gale. Gale's admiration of the sequoias prompted him to call the largest tree he discovered "the Mother of the Forest." At 325 feet high, flawlessly symmetrical, and 92 feet in circumference at the ground, the tree was so magnificent that Gale decided to bring it down for display in a traveling show (56–7). The five men he employed worked tirelessly for 25 days, but the sequoia was so well-balanced that it remained upright even after it had been sawed completely through. Then, after Gale had given up, a night storm blew across the valley and felled the behemoth that none of the axmen's wedges could topple. The immense sequoia buried itself 12 feet deep into the muck of a creek bed. There were no witnesses to its fall, but the sound of the crash carried 15 miles to a mining camp and eventually reached Gale too. His team returned to the site in 1854 and stripped the bark in sections that could be reassembled for a display. The two-feet-thick bark cylinder was 50 feet high, 30 feet in diameter, and 90 feet in circumference. The rest of the tree—later estimated to have been 2,520 years old—was left to rot. It stored so much water in its system that it kept its leaves alive for five more years (57).

Although the Mother of the Forest operation was a success, the exhibit was not. As the show traveled through San Francisco, the Atlantic States, and finally London, Gale's vaudevillian stunt was uniformly condemned. The audiences "either thought the bark to be a fake, or, more surprisingly, were hostile to the killing of what was billed the largest tree in the world" (58). Commentators bemoaned the barbarity of exploitation the show represented and took solace in the thought that the entrepreneurs were financial losers. More importantly, they called for laws that would protect American natural wonders. As Neuzil and Kovarik contend, the Mother of the Forest show triggered one of America's first green crusades. It helped usher in the emergence of environmentalism and led to the establishment of the Yosemite Valley as the first state park in 1864. It bolstered the forces that led to the creation of the national park system, starting with Yellowstone in 1872 (60–70). In a larger perspective, the event was a catalyst for the conservationist and preservationist movements of the nineteenth century (Brenner), which in turn paved the way for the emergence of ecological thinking in the early twentieth century (Crane). This new ecological thinking then absorbed elements of rights movements of the 1960s and new sciences of ecology and systems theory in the 1970s (Jamison and Eyerman), which in turn helped redefine environmentalism as based on biospheric thinking and central to the ongoing Sustainability Revolution that had been gaining momentum since the 1980s (Rifkin, Edwards, Sze, and London). One blossom of this organic, accretive expansion has been the environmental justice movement. By the mid-1990s, this movement embedded environmental justice issues into the discourse and policies of environmental protection and led to a recognition that justice concerns lie at the heart of scientifically viable environmentalism.

Although few people today remember the Mother of the Forest controversy, millions enjoy its legacy in the form of the national parks system. Even more people embrace its conceptual offspring in the form of environmental awareness. While the leap from environmentalism to environmental justice occurred only within the *Open Justice* framework in the 1990s, the environmental awareness of the 1860s and then the modern environmentalism that burst onto the stage in the 1960s both emerged as part of the *New Justice* discourse. The connection between *New Justice* and environmentalism is not absolute, but at least in the Euro-American context it seems that most environmental issues were invisible in the *Old Justice* paradigm. While premodern societies experienced the ills of air and water pollution, deforestation, soil erosion, and the like, few of them had practical means to deal with these challenges. As made clear in Paul and Anne Ehrlich's *One with Nineveh* (2004), even once-thriving civilizations have been known to perish due to unsustainable practices that weakened their natural resource base. *Old Justice* societies not only took the natural world for granted but lacked what Max Weber has called an ideological model—a combination of ideas, values, and ideologies—to address environmental issues.

The Industrial Revolution, which overlapped with the emergence of *New Justice* thinking, launched the unstoppable acceleration of human impact on the planet—an imprint so unrelenting that it threatens to undermine the Earth's ability to support life (see Ehrlich, *Dominant* and *Tightrope*; Capra, *Connections*; Lovelock, *Revenge* and *Vanishing*). Its side effect was to force us—too slowly it seems—to confront the exponential surge and myriad forms of environmental degradation. In Western Europe and the US by the mid-nineteenth century, horrifying living conditions in mushrooming urban areas brought the realization that pure drinking water, sanitation, and other environmental issues must be regulated for the benefit of the entire society[2]. There was a rising consciousness that nature outside of the cities needed protection too. European Romantics and American Transcendentalists claimed that nature was essential to our physical and spiritual well-being. Within the conceptual framework that arose between the publication of Emerson's *Nature* (1835) and Thoreau's *Walden* (1854), exploits such as the felling of the Mother of the Forest were doomed to public censure. Transcendentalism was also largely responsible for the creation of what became the two main poles of environmentalism in America: conservation and preservation (Brenner 1255). The first seeking to regulate human use of nature and the latter striving to preserve the natural environment in its pristine form, the two movements were like Johnny Appleseed and Paul Bunyan pitted against each other. They shared an overlap of interest but had unequal political leverage[3]. Closely aligned with the *New Justice* imperatives for rational adequation, computability, and usefulness, conservation appealed to progressive reformers led by Gifford Pinchot and President Theodore Roosevelt who then "imposed a conservationist model on American environmentalism that would last until years following World War II" (Crane 1262). Throughout this period, even the most vocal preservationists, such as the "Wilderness Prophet" John Muir, were excluded from the conservationist-dominated political decision processes that tended to accommodate industry in ways that ignored legitimate preservationist goals.

Without denying its genuine achievements, progressive conservationism had a serious blind spot. Its foundational premise was the absolute divide, illustrated in the Newman quote, between humanity and nature. This "myth of Civilization," as Ursula Le Guin calls it (Le Guin, *Buffalo* 11), was foundational to the Enlightenment distinction between rationality and savagery, mind and matter, culture and nature. It saw the environment as a resource: a collection of objects valuable only inasmuch as they can be made useful. This stress on utility and the emphasis on the parts rather than the whole exemplifies an unchecked left-hemispheric perception (McGilchrist 55, 401) that McGilchrist identifies as the root of Western civilization's current problems—a perception that *Open Justice* theorists have denounced as *New Justice*'s greatest failure. Within this mechanistic paradigm, conservationism embraced "a colonial vision of nature" (Rifkin 600) that failed to capture the embeddedness of phenomena. In political practice, conservationism was tainted with an elitist myopia that tended to conserve natural resources for industry and upper-class citizens but disregarded the environmental costs for low-income citizens and oppressed ethnic groups—a realization that in the early 1980s led to the emergence of the environmental justice movement. In scientific practice, conservationism created such considerable gaps in understanding that as recently as 1998 the president of the American Society for Environmental History John Opie declared America as "still mostly an unknown in environmental terms" (x). Its environment, he affirmed, has not been understood holistically.

The awareness of the relational embeddedness of natural phenomena began to emerge in the first decades of the twentieth century with the introduction of the term "biosphere"[4]. In the 1950s, before the scientific community began the holistic turn, the American environmental movement became more preservationist in outlook. The concept of nature as encompassing rather than existing outside of society was first presented to a wider public in Fairfield Osborn's *Our Plundered Planet* (1948). The call for and the concept of "land ethic" burst onto the stage with Aldo Leopold's *A Sand County Almanac* (1949). The idea of ecology and ecosystemic connections, finally, was first popularized with Eugene Odum's *Fundamentals of Ecology* (1953). However, it took Rachel Carson's *The Silent Spring* (1962) to drive home the awareness of how fragile ecosystems are and how modern technology is capable of destroying them[5].

The Silent Spring—published at the height of the civil rights movement—cemented the link between "holistic" environmentalism and the consciousness of expanded rights. In a way, environmentalism has always been about rights. However, it was only the Rights Revolutions of the second part of the twentieth century that expanded the concept of rights beyond the species line and the human vs. nature distinction. The idea of rights applied to the environment led to initiatives such as Earth Day, first celebrated in 1970, which forced environmental protection onto the national political agenda and paved the way for the creation of the US Environmental Protection Agency in December of that year. Earth Day united diverse environmental groups across the US but also inspired, by the 1980s, the emergence of two new forms of environmentalism: animal rights and the environmental justice movement. The former, although not without history[6], took off in earnest after the publication of Peter Singer's *Animal Liberation* (1975) that sparked debate about animal cognition and provided the animal rights cause with "a rational ideology that sought to stigmatize speciesism as a form of discrimination, like sexism or racism" (Unti 1313). It was also in 1975 that young Dutch biologist Frans de Waal began a six-year project on the world's largest captive colony of chimpanzees at the Arnhem Zoo. The results of de Waal's study—published as *Chimpanzee Politics* (1982)—inspired the now thriving field of primate cognition and provided empirical support for Singer's attack on speciesism[7]. There was only a small step from speciesism as a form of discrimination to speciesism as a form of environmental injustice; the two perceptions soon blended within one larger "Rights" framework.

The environmental justice movement (EJM) evolved upon a somewhat bumpier road. Unlike animal rights or earlier forms of environmentalism, it had no clear precedents and emerged as a conceptual blend only within the *Open Justice* framework that redefined justice as a plural concept operating differently in different spheres of life. An expression of this awareness, the EJM unfolded as a grassroots reaction of low-income and non-white communities to a disproportionate concentration of pollution in their neighborhoods[8]. Although there was no direct causal connection between the two, the first EJM protests—such as the Warren County, NC sit-ins of 1982—occurred soon after the publication of the first work of the *Open Justice* tradition, MacIntrye's *After Virtue*, which argued that in any social arrangements parties are always unequal and the search for justice must be situational and community-focused. The sit-ins were a practical expression of this sentiment. The "discovery" of environmental racism in the 1980s[9] inspired the emergence of the "Principles of Environmental Justice" platform in 1991, the creation of EPA's Office of Environmental Justice in 1992, and then the government's recognition of environmental justice in President Clinton's Executive Order 12898 of 1994. The order—requiring all federal agencies to take environmental justice consequences into account in their decision making—firmly embedded issues of environmental justice into the discourse and policies of environmental protection. Since then environmental justice "has undergone spectacular growth and diffusion" (Sze and London 1331), crossing the lines of color, social class, gender, and nationality, and expanding its focus to ever-new environmental challenges.

Environmental justice is first and foremost a politics of protest. Exposing and resisting "social power disparities associated with the environment" (1332), it seeks the achievement of environmental equality. It condemns "environmental inequality (or environmental injustice), which refers to a situation in which a specific social group is disproportionately affected by environmental hazards" (Brulle and Pellow 3.2). In its narrower form, environmental justice means that "all people and communities are entitled to equal protection of environmental and public health laws and regulations"—a distributive justice principle seeking to protect human beings against environmental hazards (3.2). In a larger sense, environmental justice encompasses concerns for the rights of non-human animals and the natural environment, recognizing the fact that "exploitation of the environment and exploitation of human populations are linked" (3.6). Within its narrower, anthropocentric focus, environmental justice opposes environmental racism. Its goal is the improvement of environmental health for the human population. In its broader, biospheric focus, environmental justice challenges environmental speciesism. It is concerned with the well being of humans but also with the welfare of non-human animals and the entire biosphere.

When seen from a cognitive perspective, environmental justice is a conceptual blend in which elements of the two domains—justice and nature—are integrated to create emergent structures. Represented by concepts such as environmental racism, speciesism, biospheric awareness, or unsustainable practices, these and other emergent structures within the superblend of environmental justice are processed by our cognitive system in terms of "exemplars," or prototypical instances (Hogan, *Cognitive* 47). Thus, environmental racism is best grasped through specific examples—or stories—about, for instance, dumping toxic waste in a poor neighborhood, as is the case in Nancy Farmer's *The House of the Scorpion* (2002). Likewise, environmental speciesism is best understood through a story told from the perspective of the victims of a speciesist arrangement, as is the case in the animated films *Chicken Run* (2000) or *Free Birds* (2013). Each exemplar is either a version of a standard case I refer to as a script, or it establishes a script if a clear precedent is not available. The latter happens, for example, in *Mr. Peabody and Sherman* (2014), which differs from standard narratives about anthropomorphized animals by reversing the roles in the adoption script: the dog, Mr. Peabody, adopts Sherman the boy, thus rewriting the power dynamic between the human and the animal. In other words, the film goes beyond using the standard blend of animal + human = anthropomorphized animal, by couching it within another, unusual blend of an animal as a human parent.

Whether filmic or textual, all narratives that challenge environmental racism or speciesism employ causal schematization or a script structure that cues participants about how to identify environmental injustice, casts them in roles relative to that injustice, and frames relevant activities and goals they seek to achieve. Responding to environmental injustice always involves taking a moral stand and is informed by what Jeremy Rifkin has called "biosphere awareness" (600)[10] and John Stephens refers to as "ecoconsciousness" (Stephens, "Impartiality" 207). Since ethical behaviors and attitudes informing them, as Stephens notes, are modeled in modern YA fiction "by plot and narrative rather than explicated by narrators" (206), the environmental justice script is a type of plot structure and a motivational protocol that the reader reconstructs from the characters' beliefs and actions. If retributive and restorative justice scripts draw, respectively, on the human capacity for revenge/violence and forgiveness/cooperation, the environmental justice script draws on our newly discovered capacity for biospheric cooperation. It is a cognitive programming characterized by two signature phenomena: recognition of the environmental harm or threat, and then its reparation or elimination in order to ensure restoration of the environmental well-being. The definition of "well-being" in this case spans the spectrum from the minimum requirements for survival to the ideal of long-term sustainable growth.

Although this structure may sound rather obvious, I want to stress that the environmental justice script is a very recent phenomenon. It emerged from the confluence of social activism, rights movements, the holistic, biosphere awareness, and the sustainability imperative[11], all of which are expressions of *Open Justice* thinking. This script began taking shape in the public consciousness in the mid-1980s and surfaced in young people's literature about a decade later. Stephens, for example, distinguishes earlier strands of environmental YA literature from "a recent ... sub-genre that appeared after the mid-1990s in which ethical issues are negotiated by constructing a parallel between ecoconsciousness and cathectic [i.e., close and personal] relationships"—narratives that create a mirror-network of "threatened or damaged nature" and "threatened or damaged [human] lives" (207). This distinction, I believe, highlights the rift between environmentalism in young people's literature *before* and *after* the rise of the environmental justice script. Although different stands, or wisps, of environmentalism in young people's fiction can be traced as far back as the nineteenth century, the environmental justice script radically changed the picture. "[I]n the first half of the 1990s," it played a fundamental role in the emergence of the analytical discourse of *ecopoiesis* or *ecocriticism* (Bradford at al. 79). In its cognitive substrate this script differs from anything that came before it in at least three ways. First, it considers environmental derailing the greatest threat to the survival of the human race. Second, it is predicated on the realization that humans are animals too. Third, it extends the idea of environmental rights to non-human animals and the natural world. All of these assumptions are revolutionary, and all spread with lightning speed in less than two decades. In the early 1990s the idea of "extending the concepts of justice and rights to non-human moral objects [was] a highly controversial matter" (Nevers, Gebhard, and Billmann-Mahecha 173). By the mid 2000s the perception that "justice is both conceptually and historically at the heart of environmentalism" had become the norm (Jamieson 86). It took the environmental justice script to bring this realization home.

THE ENVIRONMENTAL JUSTICE SCRIPT IN MODERN YA SPECULATIVE FICTION

Although it is a recent arrival, the environmental justice script draws on two historical roots: the literary trope of anthropomorphism and the esthetic-philosophical legacy of Romantic antimodernism. The former ensures this script a couching within a broad literary tradition that stretches back to Aesopica and over centuries has reminded readers of the many connections they share with animals. The latter invests the environmental script with the moral authority of a tradition that attributes environmental degradation and human spiritual impoverishment to the mechanistic Enlightenment outlook. Informed by these two traditions, the environmental justice script is actualized in modern narratives that dismantle the culture-over-nature dualism and related hierarchical binaries embedded in the *New Justice* paradigm, like male-over-female or reason-over-emotion.

Many of these narratives involve anthropomorphized animals, even though traditional talking animal stories were not structured on the environmental justice script. In classical fables, for example, talking animals reflected the subservience of animals, women, and children; they spoke in "the language of the subordinate ... associated with the serving classes from earliest antiquity" (Lerer 36). When the fable transmogrified into Christian allegory, talking animals became representations of moral qualities or aspects of the human character—a pattern dominant in the nineteenth century and very much alive today, for example, in the distinct personalities of the Penguins of DreamWorks' *Madagascar*. Most turn-of-the-century classics, such as *The Jungle Book* (1894), *The Tale of Peter Rabbit* (1902), or *The Wind in the Willows* (1908) were essentially fables that used animals to inculcate moral lessons. This dynamic is exemplified in Kipling's praise of friendship and other virtues that sustain us even in the most trying circumstances, in Potter's cautionary tales for little children about the dangers outside the home, and in Grahame's celebration of the foibles and dangers of gentlemanly life in his pastoral vision of countryside Britain. All three authors subscribed to a rigid human-animal divide, even though each also incorporated elements of the emerging outlook—implied in Darwin's *On the Origin of Species* (1859)—that humans are animals of a sort. While each author depicted animals as three-dimensional creatures with personalities, emotions, moral commitments, and equal-to-human intelligence, these characters were stand-ins for humans rather than animals as a non-human life form.

The turn of the century saw first speculative fiction where animals' plight in the wild was increasingly seen to reflect the human predicament after the Industrial Revolution. None of these texts, however, came close to actualizing the environmental justice script. In Kipling's tale "The White Seal," for example, a young white-furred seal named Kotick realizes that his people are subject to genocide, so he sets out to search for a new home where they will be safe. With this story, Kipling was probably the first to offer a preservationist perspective on real animals killed by humans in the real world[12], but he nevertheless saw the absolute power of humans over animals as inevitable. The tension between preservationism and the culturally entrenched imperative for extermination of "dangerous species," in turn, was first made central in Kenneth Grahame's *The Reluctant Dragon* (1898). Even though the dragon is not a vicious beast but a pacifist bohemian poet, he and St. George have to fake a joust so that traditional roles are satisfied and a monster-human dichotomy is upheld. It takes this false display of human domination for the dragon to be granted "popularity and a sure footing in society" (40). Although Kipling and Grahame included elements that would later be central to the environmental justice script, they, like Newman, perceived justice between humans and animals as an impossibility. In Kipling's tale, the thing that saves the white seal is religious taboo, whereas in Grahame's the saving factor is human fancy. There is no biosphere awareness, and human efforts to save the animals are conditional favors that lack environmental imperatives. Because they were informed by similar assumptions and set in a thoroughly anthropocentric world, classical works of toy animal fantasy—Margery Williams' *Velveteen Rabbit* (1922), A. A. Milne's *Winnie the Pooh* (1926), or Russell Hoban's *The Mouse and His Child* (1967)—are not actualizations of the environmental justice script.

While the themes of the degradation of animals and the environment grew stronger in young people's fiction, until roughly the 1970s these works did not yet offer a script to reverse those trends. The shift of sensibility to animal suffering, for example, was prominent in Hugh Lofting's *The Story of Doctor Doolittle* (1920). Moved by his experience in the trenches "to envision situations in which the unfair balance of power between humans and animals might be shifted" (Elick 324), Lofting wrote stories suggesting that animals have reason and should be accorded certain rights. Although he thus groped toward the environmental justice script, Lofting was hampered by his views of native peoples and the White Man's Burden. The trope of interchanging human and animal perspectives, in turn, was first extensively used in T. H. White's *The Sword in the Stone* (1938). The fact that young Wart's education consists of being magically turned into various animals and participating in the sentience of trees and stones makes White's book probably the earliest narrative example to gesture toward biosphere awareness. At the same time, White's narrative retains an anthropocentric focus and its descriptions of animal life project unquestionable human supremacy. Other components of what later crystallized into the environmental justice script also inform Dr. Seuss's recently revived *The Lorax* (1971). Conservationist in its representation of a treeless landscape and in its poignant message "UNLESS someone like you // cares a whole awful lot, // nothing is going to get better. It's not" (59), *The Lorax* is a symptomatic work that accurately portrays the process of environmental degradation but is vague about its reversal. Although for the repentant Once-ler "Truffula Trees are what everyone needs" (Seuss 61), this is exactly what he had insisted about thneeds. While he claims that since the Lorax's disappearance "I've sat here and worried // and worried away" (58), Once-ler does not move a finger to restore the countryside. Also, the Lorax himself is presented as powerless to stop the destruction. The story's trajectory from a green paradise to a smog-covered desert expresses the fear of ecological degradation and eventually places the responsibility in the hands of the reader. However, it does not offer a sustainable alternative to the unchecked business expansion Once-ler's enterprise represents. Like Grahame's *Reluctant Dragon*, Seuss's *The Lorax* exemplifies pre-sustainability environmentalism.

When the environmental justice script finally crystallized in the 1990s, it drew on the sense of environmental threat voiced by Seuss, on the biospheric awareness hinted at by White, and on the idea of rights extended to encompass the natural world embraced by Lofting. In its challenge to the anthropocentric assumptions at the root of the false divide between humans and nature, this script is essentially ecofeminist insofar as it structures narratives that "develop a sustainable ethic of care that can encompass … non-humans and the environment" (Curry, *Environmental* 1) and seeks to provide "restorative responses to disembodiment and dislocation in human-earth relationships" (129). The environmental justice script recognizes that humans are animals and that animals are persons too—the latter realization with history going back to the 1960s[13] both being central to the recent category of "ecopoietic" fiction (Bradford et al. 82).

The two signature phenomena of the environmental justice script include recognition of environmental harm or injustice and then actions taken toward its resolution in the spirit of biospheric awareness. Depending on the narrative focus, there can be many different—often hybridic—actualizations of this script. In modern YA speculative fiction this script functions in one of three dominant tracks. The endangered species and preservationist tracks frame stories focalized through human protagonists, in which the challenge is to ensure the survival of an endangered species or an ecosystem. The sustainability track, in turn, accommodates stories of anthropomorphized animals who seek sustainable arrangements with humans within a larger biospheric framework[14].

The endangered species track can be found actualized in novels and movies where the plot revolves around human encounters with creatures whose lives are in jeopardy and whose survival is projected as an environmental imperative, even though human protagonists may not understand the creatures' role in the ecosystem. Those species may be real, as in the *Madagascar* and *Ice Age* franchises, but in YA speculative fiction they usually tend to be imaginary, like in *Shrek* or *Monsters Inc.* movies. Among imaginary creatures, the most frequently featured are dragons, whose animal otherness makes them statistically even more popular than vampires and werewolves. Featured as endangered species in YA fiction and film—R. L. LaFevers' *The Wyverns' Treasure* (2010), Troy Howell's *The Dragon of Cripple Creek* (2011), the *How to Train Your Dragon* movies, and countless others—contemporary tales about dragons involve human protagonists' efforts to understand these creatures in the context of their own biological niche and protect their survival at all cost. Thusly conceived dragons are also central to Susan Fletcher's *Dragon Chronicles* (1989–2010), whose most recent, stand-alone companion I want to examine as an extended actualization of the endangered species track.

Ancient, Strange, and Lovely (2010) features a young protagonist on a quest to save what she recognizes as a cryptid: a species thought extinct. The story is set in the Northwest and Alaska in a post-ecocatastrophe world of a not-so-distant future. Mutant and deformed babies are born to humans and animals alike, biodiversity disappears, and poaching becomes increasingly profitable as rich people collect animal artifacts the way they used to collect art. Rare animals are disappearing from the zoos everywhere as black-market syndicates sell them to collectors and speculators who stockpile any rarities. Biopirates and cryptid hunters collect genetic resources wherever they can find them. This context presents a cognitive challenge to the reader, who must figure out the importance of saving a single animal in a world of a ruthless black-market trade in rare and endangered animals (82).

Ancient tells a story of 15-year-old Bryn whose biochemist mother had disappeared in Alaska the previous fall. Her father is gone too, searching for his wife. Bryn finds herself taking care of a mysterious critter hatched from an egg her mother had shipped from a research trip before her disappearance. After the critter is hatched, it comes to the attention of local speculators. Fearing for the lizard's safety, and not knowing it to be a dragon, Bryn and her friend Sasha flee for Anchorage, where they hope to entrust the critter to her mother's friend, Dr. Mungo Jones. Chased by speculators, the young fugitives make it to Alaska, but the dracling is fading. As Dr. Jones had found traces of an adult dragon's presence, Bryn takes the baby to its mother, but is followed by the poachers who mortally wound the old dragon. As Bryn saves the dracling, the dying mother saves Bryn by flying them to a small island where the girl's parents happen to be too. Before she disappears over the sea to die, the old dragon implores Bryn to take care of her baby. "It wasn't words, but I caught her meaning. She wanted … a promise to tend, take care. To protect her baby from the world. A promise that took in a whole of his lifespan, hundreds of years. I promised—and I meant it—though I had no idea how I would keep it" (306). The last chapter, set one year later, outlines a possible arrangement in which that promise may be kept: Bryn's family members establish secret lives in remote Oregon. They raise the dragon and continue Bryn's mother's groundbreaking research on dragon-related microbes—the whole enterprise supported by a small circle of friends who provide funds and other assistance.

Because the novel uses first-person narration through two focalizes—Bryn and Josh—it offers two parallel plots, each actualizing the endangered species track from a different angle. The first person focalization gives us access to Bryn's and Josh's minds, enabling empathetic identification with their emotions and understanding of their motivations, as exemplified in Bryn's promise above. The larger context of the story, however, invites us to use embedded mind-reading of these and other characters' thoughts and cues us to figure out certain connections even before the characters make them. Perhaps the most obvious one is that we are expected to know that the critter is a dragon, even before Bryn herself realizes that. Josh's plot takes only four of the novel's 46 chapters and is a tale of a boy whose hunting guide father has become a poacher of fossils. His father makes Josh part of his search for the dracling. From the moment they discover fossilized eggs, Josh is cast into two contrary roles: an accomplice in his father's search yet also a conscientious objector in the name of environmental justice. How Josh will respond to this ethical conflict is a cognitive/affective challenge for him, but it creates an ethical dilemma for the reader as well. Thus, when we witness Josh's thoughts that "It was wrong. Cap [his father] wouldn't admit it, but it was. Not just illegal. Wrong" (72), we are cued by the text to embrace the same position: that the fossilized eggs should be in the museum, studied by scientists, rather than locked in a private collection. These intuitions are confirmed when the group begins tracking Bryn and when Josh understands that "Cap didn't intend to save the animal. ... He wanted to collect it. Preferably alive, but dead if necessary" (284). Acting on this realization, Josh helps Bryn escape twice and then fights his own brother to stop him from shooting the dragon.

The set of imperatives that informs Josh's actions is even more pronounced in Bryn's plot, which highlights interconnectedness of life and biospheric awareness essential to the environmental justice script in at least four ways. Here, again, the protagonist is cast into a specific role that embodies a biospheric motivational scheme and makes it recognizable, if not typical. First, the women in Bryn's family are gifted with the ability of "kenning" with animals. Thought-communicating—a cognitive metonymy for the deep organic bond between the human and animal world—enables Bryn to connect with the critter and "converse" in mental pictures with its mother. Second, the dragons retain their irreducible animal Otherness and are not anthropomorphized. Instead, both the dracling and its mother are represented through the lens of personalism, a view that recognizes the autonomy of non-human animals, allowing Bryn and the reader to relate to dragon feelings through their subjective emotional knowledge of fear, loss, and the like. Third, interconnectedness and biosphere awareness are foregrounded in the environmental details of the story's setting. The births of deformed human babies, as well as the emergence of insect and animal swarms that ravage fields and forests are all linked to environmental toxins, especially endocrine disruptors that derail hormonal systems and cause chromosome damage (233). Bryn's mother's research on toxin abatement leads her to Alaska where she finds the microbes that neutralize endocrine disruptors. These microbes are then discovered to be the same as the ones on the dracling's droppings, shell, and saliva. Fourth, while the protagonists are not aware that the dragons' existence may hold the genetic key to cleaning the planet (314), the dragons and humans alike are biocentrically conceived as elements of the same ecosystem. The saving of the dragon is presented as an environmental imperative *per se* rather than a merely utilitarian choice—a position that highlights the tragic irresponsibility of the dragon's pursuers. Although the role of dragons in the environment is not initially understood, they turn out to offer an important benefit for the survival of humanity.

In developing those concerns and themes, *Ancient* offers an actualization of the endangered species track that not only identifies the danger to the dragonkind as a form of environmental threat but also exemplifies an activist formula to avert it. The script starts with Bryn's realization of her own responsibility toward the creature but soon spreads to involve other people: Bryn's sister, her mother's assistant Taj, her friend Sasha, Dr. Jones, the pilot Samantha Mills, and even random strangers, such as a boy named Anderson Brown. Anderson's description and picture of the dracling in his blog entry, in turn, sends the scientific community buzzing and creates a large international fan base for the endangered "lizard." As a result, a crowd of activists blocks the poachers at Anchorage airport, helping Bryn escape and reach Dr. Jones. By the end of the novel, those concerted actions of environmentally conscious individuals have ensured the dracling's survival and are projected as a viable foundation for its future. As Bryn wonders about how the dracling will survive without its mother and how to keep him safe, she concludes, "I'd have to find a way" (307). Bryn's determination reflects deep connectedness to the natural world and its mystery and a sense of moral obligation for humans to act from a perspective of biospheric awareness. It affirms that in an unbalanced world environmental justice requires protecting *all* endangered species as part of a larger effort to safeguard the integrity of the entire ecosystem.

That same concept informs the preservationist track of the environmental justice script, in which the threat of a species' extinction is linked to the destruction of its natural habitat. In this track, the focus is on preserving multiple, interrelated levels of ecosystemic wilderness rather than its one element. The rationale for such protection is based on environmental criteria rather than on concerns about human interests, even though the focalization is usually through human characters.

In its basic form the preservationist track involves an outsider character who arrives in a wilderness setting in which the protagonist assists indigenous preservationist forces in forestalling the imminent destruction of the local ecosystem. The environmental threat is either denied or explained away by the requirements of technological progress; figuring out what the threat is thus requires substantial cognitive effort on the part of the characters and the readers alike. Because forces of global capitalism show no regard for the environmental consequences of their actions, protagonists in the preservationist track expose these vested interests and thus prevent the ongoing ecosystemic destruction. In all these ways, the preservationist track is best described as scripting a set of formulas and attitudes that Kevin DeLuca has identified as "wilderness environmentalism": one that requires putting the non-human and the ecosystem first (28). Although "wilderness" is a rhetorical construct, novels and movies based on this track—for example, *Up* (2009) or *Rio* (2011)—acknowledge this relativity only to a degree. They simultaneously position wilderness as an *a priori* reality that makes the human life possible. This double ontological character of wilderness is nowhere more clearly reflected than in stories set in the paradigmatic wilderness of the planet: the Amazon.

One of the best recent examples that successfully challenges anthropocentrism and actualizes the preservationist track is Isabel Allende's *City of the Beasts* (2002). The novel is the first in the fantasy trilogy chronicling the adventures of 15-year-old Alexander Cold who accompanies his journalist grandmother Kate on expeditions to remote corners of the globe: the Amazon in *City*, the Himalayas in *Kingdom of the Golden Dragon* (2004), and the forests of Kenya in *Forest of the Pygmies* (2005). Wherever they go, Alexander and his companions encounter and defend local cultures, species, and environments threatened with extinction or enslavement by greedy business magnates. In *City*, Alexander's California life is disrupted by his mother's cancer, following which he is sent to live with his eccentric grandmother. As she had already signed on for the *International Geographic* expedition to the Amazon, Alexander finds himself accompanying Kate on the trip to capture the gigantic humanoid creature that had been reportedly killing people there.

Led by the famous anthropologist Professor Ludovic Leblanc, the expedition arrives at the heart of Indian territory, where they are equipped for further travel by the wealthiest local entrepreneur, Mauro Carías. At Carías' insistence, the travelers are given military protection and are joined by doctor Omayra Torres, an employee of the National Health Service, who comes along to vaccinate the Indians before they make first contact with foreigners. The expedition fails to capture the Beast, but Alexander and his new friend Nadia befriend the elusive tribe of the People of the Mist. They discover that the Beast is one of thirteen surviving prehistoric creatures that have coexisted with the tribe for millennia. Alexander and Nadia convince the People of the Mist to get vaccinated, but at the very last moment they learn that the vaccination is Carías and Torres' ploy to inject the entire tribe with a deadly virus—an echo of an actual extermination scheme deployed in

1968 (Bradford et al. 98). They prevent the genocide, expose the criminals, save the tribe, and rescue the beasts as well. Their actions also preserve a swath of the Amazon.

The *City* has all four signature markers of the preservationist track. Alexander is an outsider, who becomes a preservationist quester ready to risk his life to ensure the survival of the local ecosystem. The environmental injustice to be prevented is Carías' land grab plan, which had already wiped out several tribes and—if carried through—would have exterminated the People of the Mist and the beasts, thus opening the newly "uninhabited" territories to rapacious mining and irreparable environmental damage. The wilderness to protect is the Amazonian rainforest, which is treated not as the setting for human actions but almost as a character: an autonomous ecosystem that supports a variety of human, animal, and plant life and requires active preservationist efforts for its own sake. Lastly, the novel's conclusion brings a successful resolution along environmental justice lines. Carías' genocidal scheme is thwarted. Plans are also made to establish a foundation that will protect the region inhabited by the People of the Mist and the beasts. Local preservationists such as Padre Valdomero, the guide César Santos, and his daughter Nadia will oversee the foundation's work in the Amazon; the renowned journalist Kate and the world-famous Professor Lablanc will lead the foundation's work internationally.

While the primary focus is on Alexander's adventures and his quest for self-identity, the novel "dismantles the anthropocentric orientation to the natural world" (Bradford et al. 98) and builds a case for the ecological value of the Amazon wilderness as a place of alternative modes of thinking and feeling. While Allende occasionally succumbs to romanticizing the native peoples and their relationship with the environment, she never idealizes the wilderness itself. The Amazon is neither sublime nor ideal but is an *a priori* condition that grounds the human. Wilderness becomes an effective trope that enables Allende to call attention to the problems of the Amazon and also to the structural-global challenges of preserving the natural world in a modern capitalist economy. The fact that it was Carías who helped organize the expedition only to use it as a cover for his genocidal plan suggests how easy it is to appropriate the ethos of scientific research in the service of murky political and economic interests. The corruption of Captain Ariosto, who puts himself and his soldiers at the service of Carías rather than the law—and the blind infatuation of Dr. Torres, who becomes Carías' *de facto* henchwoman—brings home the grim reality of how protection of the native people often looks in practice. Despite these dangers, Allende stresses that there is no alternative to preservation of wilderness. The indigenous people have no alternative for "[t]hose Indians who had been incorporated into civilization had become beggars, [losing] their dignity as warriors, and their lands" (326). The world at large has no alternative either, for large swaths of wilderness such as the Amazon rainforest are crucial to preserving ecosystems and maintaining biodiversity. In a world where mass extinction of species is a major environmental threat, wilderness is a strategic resource for the entire planet.

Another feature of the preservationist script in *City* is how many species and relationships are jeopardized by threats to the integrity of the Amazon. For example, Alexander and the reader are increasingly made aware of the fact that "this vast area, the last paradise on the planet, was being systematically destroyed by the greed of entrepreneurs and adventurers" (48). The real villains are those who traffic weapons or natural resources and those who pollute and exploit the region. Some of these things are stated explicitly by characters, but others need to be deduced because things are not always what they seem. The seemingly gentle Carías is a mastermind behind Indian genocide, and the supposedly murderous Beast turns out to be a peaceful creature. Due to their extreme longevity, infallible memory, human-like intelligence, and the ability to use language, the beasts have been the Indians' memory banks or "gods" who remember the tribe's history and can preserve it verbatim. The valley where they live is "an ecological archive where species that had vanished from the rest of the earth still survived" (267): a place where Alexander finds herbs and water that may help cure his mother's cancer. Despite the "uses" of the wild, underlying the preservationist script in *City* is recognition of the autonomy of the beasts and of the entire non-human world. The Amazonian wilderness is acknowledged as a realm that humans did not create and that has reasons for its own existence separate from human designs or concerns.

In scripting ways to preserve the ecosystem of the Eye of the World, the last chapters of the novel also grapple with the question of how preservation can be accomplished. As they discuss plans for the foundation, Father Valdomero, Kate, and Professor Lablanc decide "[t]hey would go to churches, political parties, international organizations, governments … They would knock on every door until they found the necessary funds" (401–2). The adults do not yet realize that a large portion of those funds may come from three large diamond eggs that were given to Nadia. "They belong to the People of the Mist," she tells Alexander, as she hands him the gems. "[T]hese eggs can save the Indians and the rainforest where they have always lived" (403). With the initial endowment in place and with the determination of expedition members, the foundation is likely to succeed. Given that the complete effacement of the human is impossible (Bradford et al. 99), the preservationist effort is focalized through a human group. Yet, it rests on recognition of exploitation of the environment as linked with exploitation of indigenous populations, especially in the Amazon. Because the Indians do not distinguish between human and animal (267) and because their land-use practices support the survival of the Amazonian wilderness, protection offered to the tribe secures protection of the entire ecosystem, which in turn amounts to achieving environmental justice in its broad, biospheric application.

The endangered species and preservationist tracks share an overlap that is also their limitation. Both frame stories focalized through human protagonists—usually misfits or outsiders who discover their place in society through developing biosphere awareness and acting on it—and so both are inescapably homocentric. When Bryn wonders what to do with the baby dragon or when Alexander and Nadia mediate between the People of the Mist and the expedition, the narrative's cognitive advantage is that it exemplifies specific ethical dilemmas and invites the reader to reconstruct and evaluate the characters' choices. At the same time, the perception of the natural environment as framed by such focalization, privileges the human, in fact, an adolescent perspective, which may not be adequate to grasp the complexity of the environmental challenge or may likewise look for simplistic solutions.

This limitation is unavoidable but may be masked in narratives in which the human protagonist magically becomes part of the animal community and learns about animal life. In *The Ant Bully* (2006), for example, Lucas is "resized" to live among the ants, whom he had tormented earlier. Because Lucas is the victim of a bully in his own world, the adventure not only transforms him into a defender of ants against the local exterminator but also teaches him lessons about ant cooperation, which he then applies in order to stand up to the bully and to break up his gang. The cognitive operation that the audience is expected to execute is to link two input spaces—bullying as a form of injustice in the human world and human abuse of animals as a form of bullying—to create a new blend in which "bullying" or abusing animals is a form of injustice. The bully's phrase that Lucas initially repeats to the ants, stating that he can do anything to them because he is big and they are small, is rejected. Instead, Lucas embraces an activist preservationist position, challenging the exterminator as well as the bully. Another strategy to mediate the homocentric focus is to use human *and* animal first-person participant narrators. This is well executed in the film *Rio* (2011), which features two parallel plots involving human and animal protagonists. The double romance plot between the macaws Blu and Jewel on the one hand and Blue's caretaker Linda and the ornithologist Túlio on the other reveals the endangered species track through double focalization, very much like in *Ancient*. The survival of macaws is presented as an act of restoration that counteracts environmental injustice in the form of wildlife smuggling.

In an increasing number of narratives structured on the environmental justice script, it is animals that stand up for their own rights. Most of these films and novels rely on young people's spontaneous empathetic identification with animals or "basic epistemological anthropocentrism" (Nevers et al. 178). The mere use of animal protagonists does not mean that the narrative will privilege the environmental justice script. For example, in the *Kung Fu Panda* and *Ice Age* movies, animal characters are stand-ins for people and the Trouble reflects conflicts typical for human society. As a result, these films are largely structured on social, restorative, and/or retributive justice scripts. In other films, focalization through animal characters is meant to highlight the plight of animals in our world but does not position humans and animals as partners within the same biospheric framework. This is the case in *Chicken Run* (2000), *Happy Feet* (2006), *Over the Hedge* (2006), the *Madagascar* movies (2005–12), and *Free Birds* (2013). In other films that involve environmental themes—such as *Battle for Terra* (2007) or *Horton Hears a Who* (2008)—the focus may be on two different societies seeking coexistence, and this overriding theme falls under the global justice script. However, when animals become the story's protagonists and speak for the oppressed non-humans in a human-dominated world, such narratives are likely to employ the sustainability track of the environmental justice script.

In this third dominant track of the environmental justice script the challenge is not to preserve a particular species or ecosystem but to redefine power relations between the disfranchised and the dominant species within the environment. The sustainability track erases the line between the human and the animal; it achieves its effect through cognitive estrangement resulting from inviting the reader/audience to identify with creatures that are otherwise culturally constructed as no more than objects for human use. Unlike pre-sustainability stories that merely raise questions about human vs. non-human relations—as in "The White Seal," whether it is fair for people to kill seals or, as in *Charlotte's Web*, whether it is fair to kill pigs—stories structured on the sustainability track propose answers that seek to amend species' inequality by turning difference into mutual advantage. Thus, the friendship between different animal protagonists in the *Madagascar* and *Ice Age* movies are all extended, speculative arguments for the benefit of species diversity. Yet, these movies fall short of actualizing the sustainability track in that they project no change in animal-human power relations. The sustainability track, by contrast, requires communication between the animal and the human; it involves transforming their unequal relationship into a symbiotic one. A model filmic actualization of this track is *How to Train Your Dragon* (2010), where the initial hostility between dragons and humans is transformed into an exciting symbiosis. Incidentally, the film has been much more successful than the book on which it was based because it presents dragons and humans as equals and offers a resolution whereby the two species can coexist to mutual benefit. This is not the case in the novel, where dragons are either threats to be killed or pets to be tamed in ways that affirm human dominance. This speciesism is missing from the film. The conclusion of the book, for example, pits two gigantic dragons against each other, their mutual deaths presenting a positive resolution for the humans. In the conclusion of the movie, by contrast, the oppressed dragons *and* people confront the dragon-tyrant, whose death sets *both* groups free.

In narrative fiction the sustainability track structures the plot of Terry Pratchett's *The Amazing Maurice and His Educated Rodents* (2001). A satirical rewriting of "The Pied Piper of Hamelin" for the Discworld universe, *The Amazing Maurice* is narrated from the perspective of an anthropomorphized cat named Maurice and talking rats who call themselves the Clan. Accompanied by a pipe-playing boy named Keith, the group travels around Discworld, tricking towns into paying them to get rid of "rat infestations."

The scam ends when the party arrives in Bad Blintz, a town on the edge of starvation. The inhabitants are already being leeched by two rat-catchers, who not only steal the townspeople's food but have also released a Rat King—a blind, spidery creature that has developed collective consciousness, hypnotically controls legions of ordinary rats, and plots revenge on humankind. In a tightly packed plot the Clan, Maurice, Keith, and the mayor's daughter Malicia end the double evil of rat-catchers and the Rat King. The Clan strikes a deal with the townspeople, making Bad Blintz the first Discworld town in which rats and humans establish peaceful, sustainable coexistence.

If the goal of the sustainability track is the achievement of ecological sustainability characterized by viable forms of animal-human coexistence, the first cognitive challenge for the reader is to identify what stands in the way of that goal. The culprit, of course, is speciesism: a radical and hierarchical separation between humans and animals on the one hand and between various animal species on the other. The opening chapters cue the reader to identify speciesism in at least two ways. First, after the protagonists, who have human-like mental capacities, are introduced as a group of talking rats and a talking cat, the reader must confront her own speciesist reservations about treating animals, especially rats, as persons. This dynamic is developed throughout the novel, all of which is focalized through animal participant narrators, with the expectation that the reader will develop an empathetic identification with the Clan, their choices, and their goals. Second, speciesism is gradually revealed as the status quo in Discworld, where sociopolitical structures place humans at odds with other species. This denial of a systemic view of life as a larger whole in which human and nonhuman living systems are in continual interaction has severe consequences. Speciesism condemns the humans of Bad Blintz to an ongoing struggle over resources with the town's local rat population. It produces an extremely dangerous environmental backlash, represented by the Rat King: a Frankenstein-like rat monster accidentally created by the rat catchers. Speciesism also condemns Maurice and the Clan to itinerant stealing from humans, as it is the only sociopolitical survival practice available to a hunted minority in this particular ecosystem. In the constant war between humans and animals, both sides are losers and sustain collateral damage. All these aspects invite the reader, who is seeing this world through animal eyes, to identify speciesism as the root of environmental injustice.

According to Fritjof Capra, there are two general steps to achieve ecological sustainability: namely, ecoliteracy and ecodesign. Ecoliteracy, or "the understanding of the principles of organization that ecosystems have evolved to sustain the web of life" (232–33), implies grasping such principles as networks, cycles, partnership, diversity, and dynamic balance. Ecodesign, in turn, is the application of this biospheric knowledge "to the fundamental redesign[ing] of our technologies and social institutions, so as to bridge the current gap between human design and the ecologically sustainable systems of nature" (233). Inasmuch as reaching out across the species line—or overcoming speciesism—is the defining marker of the sustainability track, these two steps is how it unfolds. As an actualization of this track, *The Amazing Maurice* chronicles the protagonists' education in ecoliteracy that, in the last chapter, culminates in a successful form of ecodesign. The whole process includes specific narrative illustrations of the five principles of ecoliteracy: networks, "all living systems communicate with one another and share resources across their boundaries"; cycles, "an ecosystem generates no net waste, one species' waste being another species' food"; partnership, "the exchanges of energy and resources in an ecosystem are sustained by pervasive cooperation"; diversity, "ecosystems achieve stability and resilience through the richness and complexity of their ecological webs"; and dynamic balance, "an ecosystem is a flexible, ever fluctuating network" (231).

The achievement of environmental justice is not easy. At the novel's outset, even the animal protagonists seem to embrace speciesist mentality, believing that their relationship is asymmetrical. Although they work together, the individualist Maurice patronizes the Clan's collectivism as a sign of gullibility and lower intelligence. The Clan, in turn, distrusts Maurice as selfish, irresponsible, and lacking morals. Each side holds to negative representations of biological differences. This is also true of humans in Bad Blintz who would never dream of treating the animals as equals. When Malicia first talks to Keith, for example, she says that "[t]he only good rat is a *dead* rat" (Pratchett, *Maurice* 57). In time, however, she not only overcomes her speciesist prejudices but helps her father the Mayor do the same. As the novel develops, first animals and then humans learn to reach out across the species line in search of mutually beneficial, sustainable arrangements. The reader witnesses Maurice mature morally and sees him develop a truly symbiotic, respectful relationship with the rats of the Clan. The cat and the Clan subsequently establish such a relationship with the human orphan Keith, and they eventually form respectful relationships with all the inhabitants of Bad Blintz. In a process strongly suggestive of Singer's "expanding circle" of empathy, the cat, the rats, and the humans grow to appreciate the benefits each has to offer the others. This amounts to more than just dismantling the anthropocentric perspective. The successive iteration of action sequences, dialogues, and behaviors provides a setting for characters to overcome their traditional prey-predator relationships and transform them into ecologically sustainable coexistence, cueing the reader to compare the consequences of speciesism in Discworld and in our reality. This cognitive operation of linking processes happening in the imagined and real world helps the reader conclude that, just as speciesism may be overcome in Discworld, so too it may be overcome in our world. Obviously, the novel does not include a set of instructions about how to eliminate speciesism in our world. Instead, its primary purpose—as is the case for all narratives that employ the environmental justice script—is to shape attitudes and create storied models showing how the biospheric mindset can translate into action.

As characters develop the awareness of others as "people," the reader is shown a number of episodes, when animals turn out to be more humane than humans. Examples include Maurice's sacrifice of one of his lives for the "prophet" rat Dangerous Beans (197) or Darktan's heroic generosity when he refuses to burn the barn with humans who tortured rats in the rat ring (158–9). In the novel's last chapter ecoliterate protagonists apply their knowledge to create an ecologically sustainable, inter-species community by negotiating a settlement: in return for feeding rats and guaranteeing them specific rights, the Clan will control vermin, assist with woodcarving, clock-making, and other duties, and also attract tourists. "If it was a story, and not real life," the narrator wryly comments, "then humans and rats would have shaken hands and gone on into a bright new future."

> But since it was real life, there had to be a contract. A war that had been going on since people first lived in houses could not end with just a happy smile. And there had to be a committee. There was much detail to be discussed. The town council was on it, and most of the senior rats, and Maurice marched up and down the table, joining in. (229)

The tenor of these negotiations suggest how hard it is to design an inter-species arrangement, but it also affirms that a transformation of human-animal power relations is possible and would result in creating a society that is both just and sustainable. In that arrangement, a sustainable settlement eliminates the trans-species inequality and exclusionism that lie at the root of environmental injustice.

Interestingly, these two tracks play a very minor role in the many textual and filmic narratives classified as postapocalyptic dystopia. Although many dystopias seem to develop environmental themes[15], these themes are usually static backdrops that allow very little space for environmental change. The bulk of postapocalyptic dystopias—including Philip Reeve's the Predator Cities quartet (2001–6), Susan Collins' the Hunger Games series (2006–10), Scott Westerfeld's the Uglies quartet (2005–11), and others—are set in worlds where environmental derailing has already occurred. The damage is either irreversible or, as in *Wall-E* (2008), will take so long to undo that the return of humans to the Earth amounts to a colonization of a new planet. In these worlds the Trouble is social, not environmental, injustice, and the focus is on social conflicts: the traction cities' predatory capitalism in Reeve, Capitol's colonial absolutism in Collins, ecophobic aesthetic standardization in Westerfeld, and the Autopilot's "no-return" directive in *Wall-E*. Although environmental themes offer an important context, with the rare exception of *Wall-E*, postapocalyptic dystopias are largely structured on retributive and social justice scripts that involve humans and offer very little hope for any environmental healing. This quiet resignation about the Earth's doomed fate also informs texts and films about human colonization of space, which often invoke the trope of "terraforming" other planets to make them habitable, as in *Battle for Terra*.

Humanity's destructive attitudes toward the environment are thus affirmed in James Cameron's *Avatar* (2009), which showcases the awe-inspiring biospheric unity of life but projects it as possible on Pandora rather than on Earth, where the environment had long been beyond repair. The film's opposition to environmental destruction is structured through the rebellion track of the retributive justice script and concludes with the expulsion of most humans as irredeemable species unable to grasp the meaning of biosphere awareness. The fate of the Earth is not a concern.

Although for the readers and viewers, the likelihood of hatching a dragon, encountering endangered Amazonian beasts, or sharing urban space with talking rats is rather slim, all tracks of the environmental justice script affirm that humans are not above but are part of an intricate web of life. If humanity is to survive, let alone flourish, individuals and societies alike must come to terms with the concept of species-partnership. This growing realization of the human capacity for biospheric cooperation is what informs the environmental justice script as a cognitive blueprint for survival. In all its tracks, this script communicates four key components. First, it makes it known that there is such a phenomenon as environmental injustice, understood as a complex history of political and economic interactions leading up to and continuing beyond the instance of perceived injustice: the hunt for the last dragon, the threatened destruction of a large ecosystem, and the "naturally" unequal power relations between humans and animals. Second, it communicates that environmental injustice results from environmental inequality, itself the outcome of the social dynamics of modern society, especially the functioning of the market economy, institutionalized racism, and speciesism. Third, the environmental justice script suggests that in order to save the planet's biodiversity, ecosystems, and long-term sustainability, environmental injustice must be addressed through redefining power relations among humans, animals, and the natural world within a biospheric framework. Fourth, it points out that the social change required to realize sustainable environmental justice is most likely to succeed if it is based on three principles: (1) The principle of participatory research, which democratizes scientific practices associated with analyses and resolutions of environmental threats by involving the community affected by these threats—as is Bryn's mother's scientific practice in *Ancient, Strange, and Lovely*; (2) The precautionary principle, which requires businesses and institutions to first prove that their projects are environmentally safe rather than presume so until data and research prove otherwise while the environment has already been adversely impacted—the principle violated by Carías in *City of the Beasts*; and (3) The principle of organizational support, which ensures that the achievement of environmental justice in any particular case is safeguarded through international policy changes or creation of organizations to monitor its continuing implementation—depicted as an informal NGO in *Ancient, Strange, and Lovely*, an international foundation in *City of the Beasts*, and an inter-species contract in *The Amazing Maurice*. Regardless of which track, the environmental justice script in YA speculative fiction usually involves the trope of biospheric accommodation—of rats, wolves, birds, but also dragons, vampires, and other species—which signals that humans and animals can coexist as equals on multiple levels.

In their 2004 *One With Nineveh*, Stanford biologists Paul and Anne Ehrlich drew a parallel between modern civilization and ancient Mesopotamia, the latter having collapsed due to the weakening of their resource base. The ancient Mesopotamians' unsustainable environmental practices, the Ehrlichs say, turned their once-fertile land into a sweltering desert. Within a few centuries, Mesopotamia committed ecological suicide, and that area of the planet has not recovered to this day (4–5). Nineveh's story is a cautionary example of human arrogance leading to ecological collapse. Unlike in the past, humanity's global domination today endangers not just one region but the entire planet. Unlike ancient Assyrians, we are aware of the symptoms of ecological decline. Whether or not we can still avert environmental catastrophe, the environmental justice script is the most powerful cognitive tool we have to steer clear of the future depicted in *Wall-E*.

NOTES

1. Another explanation is that evolution selects more strongly against cooperation and altruism in young children than it does in mature adults because the latter have already reproduced and can risk their lives without endangering their gene pool (Singer 56).
2. Some of the earliest reform proposals in the environmental area include Jeremy Bentham's *The Constitutional Code* (1830) and Edwin Chadwick's *The Sanitary Condition of the Labouring Population in Great Britain* (1842)—both largely preoccupied with the importance of water supply and sewer systems to ensure public health.
3. Yet another strand of environmentalism taking shape at this time was urban environmentalism that addressed the many problems of growing American cities. Rooted in the tradition of municipal reform, this strand is best exemplified by the work of Alice Hamilton, "the first great urban/industrial environmentalist" (Crane 1258), whose research on the relationship between disease and the human working-living environments pioneered a reform movement toward safer and cleaner workplaces.
4. The first to expand the notion of ecological relationships to the entire planet was the Russian scientist Vladimir Vernadsky who in 1911 introduced the term the biosphere and then, in his *Biospheria* (1926) "broke with the scientific orthodoxy of the day, arguing that geochemical and biological processes on Earth evolved together, each aiding the other" (Rifkin 596–7). The concept of the biosphere took about four decades to catch on. Vernadsky's conclusions were first confirmed by publications in the field of quantum physics in the mid-1930s—starting with Albert Einstein's *Essays in Science* and Niels Bohr's *Atomic Physics and the Description of Nature* (both 1934) and continuing in the post-WWII years with Robert Oppenheimer's *Science and the Common Understanding*

(1954) and Werner Heisenberg's *Physics and Philosophy* (1963)—all of which argued for a deeply holistic interrelatedness of time, space, matter, and energy. Vernadsky's idea of an interacting biosphere was vindicated only since the late 1970s, with the emergence of holistic theories that saw the Earth as a complex self-regulating, living organism. Among the first studies to demonstrate how various biological, chemical, and other forces work in a synergistic relationship within a unified system of life on Earth were the Gaia hypothesis formulated by the British chemist James Lovelock in his *Gaia: A New Look on Life on Earth* (1979), the theory of morphogenetic fields proposed by the British biologist Rupert Sheldrake in *A New Science of Life: The Hypothesis of Formative Causation* (1981), and Serial Endosymbiotic Theory put forth by the American biologist Lynn Margulis in her *Symbiosis in Cell Evolution* (1981).

5. While her book is often considered the single most important contribution to raising the American public's environmental awareness, Carson was part of a larger transformation of the 1960s and 1970s. In their *Seeds of the Sixties* Andrew Jamison and Ron Eyerman place her together with Fairfield Osborn and Lewis Mumford as the three key activist-authors who "planted some of the most important seeds of the many-branched tree we now call environmentalism" (67). A former banker with no illusions that business and industry have any other interest than making money, Osborn became a builder of important institutions that supported pioneering environmental research (76); a prolific public intellectual and a vigorous opponent of the emerging military-industrial complex, Mumford provided environmentalism with a sense of history and was "the strongest and most influential American voice that came to be raised against the environmental consequences of the postwar technological culture" (82); lastly, a marine biologist and a best-selling author of the documentary *The Sea Around Us* (1951), Carson commanded an audience large enough that she was able to fill the gap between science and society, translating scientific findings into evocative prose that could not only be understood but also appreciated.
6. The story of the animal rights is complex and could be traced back to Bentham's 1789 denouncement of unnecessary cruelty toward animals, to the emergence of animal law—from the first piece of animal protection legislation in the West, the British Cruel Treatment of Cattle Act of 1822—or to the forming of various animal protection organizations, the first in England in 1824 and in the US in 1866 (Unti 1311). All of these are examples of the *New Justice* Beccarian postulates as applied to animals.
7. Also in 1975—when sophomore Pinker tortured a lab rat to death, which he now recounts as the worst thing he had ever done (*Better* 454)—the incredible brutality of seal and whale hunts, to which Greenpeace activists were alerting the world, was a norm. Within a decade, though, the seal hunts were banned and what used to be standard practices in the treatment of lab animals had become either illegal or strictly regulated. This transformative thrust was informed by new sensibility to animal suffering but also by the new studies of animal cognition and consciousness.
8. The term environmental racism refers to a cultural practice whereby communities composed of low-income people and people of color are consistently exposed to higher levels of environmental risks—including exposures to air and water pollution, high levels of ambient noise, residential crowding, quality of housing, quality of local schools, and the work environment—than the predominantly

white, higher income communities. Environmental racism was confirmed in a number of influential reports—starting with the 1983 Government Accounting Office's report on *Sitting of Hazardous Waste Landfills* in eight southern states, and then the groundbreaking 1987 national study by the United Church of Christ, *Toxic Waste and Race in the United States*—which documented the unequal and discriminatory siting of toxic waste facilities across the nation.

9. The term environmental racism refers to a cultural practice whereby communities composed of low-income people and people of color are consistently exposed to higher levels of environmental risks—including exposures to air and water pollution, high levels of ambient noise, residential crowding, quality of housing, quality of local schools, and the work environment—than the predominantly white, higher-income communities. Environmental racism was confirmed in a number of influential reports—starting with the 1983 Government Accounting Office's report on *Sitting of Hazardous Waste Landfills* in eight southern states, and then the groundbreaking 1987 national study by the United Church of Christ, *Toxic Waste and Race in the United States*—which documented the unequal and discriminatory siting of toxic waste facilities across the nation.
10. Biosphere awareness involves the realization that we are part of an intricate network so that "[o]ur powers of intelligence and technology do not belong specifically to us but to all life" (Margulis and Sagan 22) and a recognition of "the continuous symbiotic relationship between every living creature and between living creatures and the geochemical processes," which symbiosis is the only thing that ensures "the survival of both the planetary organism and the individual species that live within its biospheric envelope" (Rifkin 598).
11. The concept of sustainability, meaning "development that meets the needs of the present without compromising the ability of future generations to meet their own needs" (Edwards 17), burst onto the scene in the early 1980s in response to the growing awareness of how humanity's use of natural resources threatens the long-term survival of the planet. Absorbing some earlier initiatives, notably the appropriate technology movement of the 1970s as well as pre-sustainability environmentalism of the 1960s and 1970s, sustainability spread like wildfire, leading to the creation of the UN's World Commission on Environment and Development in 1983, inspiring the first Earth Summit in 1992, generating new analytical tools—like the concept of the ecological footprint, since 1992—and helping forge the Earth Charter, which brought together a wide range of sustainability issues for worldwide acceptance in 2000. The idea of sustainability spawned an impressive body of studies, laws, and initiatives collectively referred to as "the Sustainability Revolution" (5). This revolution, in turn, is the most visible practical application of what Rifkin calls "the biosphere awareness" that has been on the rise since the early 1980s too.
12. This tradition of depicting real-world animals killed or enslaved by humans is represented in works such as E. B. White's *Charlotte's Web* (1952), Robert C. O'Brien's *Mrs Frisby and the Rats of NIMH* (1971), Kathi Appelt's *The Underneath* (2008), or recently Katherine Applegate's *The One and Only Ivan* (2013).
13. Animal personalism emerged first in science fiction—one of its first examples being Andre Norton's *Catseye* (1961)—and quickly generated related concepts, like psionic communication or "kything" as first featured in Madeleine L'Engle's *A Wrinkle in Time* (1962). By the 1980s it had become an important trope in all speculative fiction. In the meantime, the Dying Earth subgenre—starting from

Jack Vance's Dying Earth books (1950–84) and Brian Aldiss' *Hothouse* (1962) through Gene Wolfe's four-volume *The Book of the New Sun* (1980–83)—had prepared the ground for considerations of environmental justice by exploring the consequences of the planet's cosmic dusk.

14. It needs to be noted that the concept of environmental justice, as well as its accompanying script, emerged as reactions to practices and modes of thinking based on a sharp culture-nature dichotomy that is unique to the Western world. Many non-western cultures have never drawn such sharp lines between the human and natural worlds. As a result, much non-western and minority speculative fiction does not feature ecological terminology. Instead, human and ecological concerns are treated integrally. This "ecological hybridity" of human characters—who are at the same time part of nature and part of society—not only exposes "the artificiality of the stable dichotomies between self and other, human and non-human" (Curry, "Traitorousness" 43) but it also explains why non-western and Black Atlantic speculative fiction (see Thaler) tends to embed issues of environmental justice in narratives structured on the social justice script. This is the case in Pakistani-British Salman Rushdie's *Haroun and the Sea of Stories* (1991), Japanese Hayao Miyazaki's *Princess Mononoke* (1997) and Nahoko Uehashi's *Moribito: Guardian of the Spirit* novels and anime TV series (1996–ongoing), Aboriginal Australian Archie Weller's *Land of the Golden Clouds* (1999), Indian-American Chitra Bannerjee Devakaruni's the Brotherhood of the Conch trilogy (2005–9), Nigerian-American Nnedi Okorafor's West African Zahrah the Windseeker trilogy (2005–11), and even in novels by white "border crosser" authors such as Louise Erdrich or Ursula Le Guin.
15. For a theoretical discussion of how ecofeminist thought—which I see as an expression of *Open Justice*—sheds light on post-apocalyptic fiction, see Alice Curry's *Environmental Crisis in Young Adult Fiction* (2013).

WORKS CITED

Allende, Isabel. *City of the Beasts*. Trans. Margaret Sayers Pedan. New York: HarperCollins, 2002.

Battle for Terra. Dir. Aristomenis Tsirbas. SnootToons and MeniThings Productions, 2007.

Bradford, Clare, Kerry Mallan, John Stephens and Robyn McCallum. *New World Orders in Contemporary Children's Literature: Utopian Transformations*. Houndsmills, UK: Palgrave Macmillan, 2008.

Brenner, Aaron. "Environmental Movement: Introduction." *Encyclopedia of American Social Movements*. Vol. 4. Immanuel Ness. Ed. Armonk, NY: M. E. Sharpe Inc., 2004: 1255–57.

Brulle, Robert J. and David N. Pellow. "Environmental Justice: Human Health and Environmental Inequalities." *Annual Review of Public Health* 27 (2006): 3.1–3.22.

Capra, Fritjof. *The Hidden Connections: A Science for Sustainable Living*. New York: Anchor Books, 2004.

Chicken Run. Dir. Peter Lord and Nick Park. Aardman Animations, DreamWorks Animation, and Pathé Pictures, 2000.

Crago, Hugh. *Entranced by Story: Brain, Tale and Teller from Infancy to Old Age*. New York: Routledge, 2014.

Crane, Jeffrey. "Environmental Movement: Nineteenth and Twentieth Centuries." *Encyclopedia of American Social Movements*. Vol. 4. Immanuel Ness. Ed. Armonk, NY: M. E. Sharpe Inc., 2004: 1258–84.

Curry, Alice. *Environmental Crisis in Young Adult Fiction: A Poetics of Earth*. New York: Palgrave Macmillan, 2013.

Curry, Alice. "Traitorousness, Invisibility and Animism: An Ecocritical Reading of Nnedi Okorafor's West African Novels for Children." *International Research in Children's Literature* 7.1 (2014): 37–47.

DeLuca, Kevin. "A Wilderness Environmentalism Manifesto: Contesting the Infinite Self-Absorption of Humans." Ronald Sandler and Phaedra C. Pezzullo. Eds. *Environmental Justice and Environmentalism: The Social Justice Challenge to the Environmental Movement*. Cambridge, MA: MIT Press, 2007: 27–56.

Dr. Seuss. *The Lorax*. New York: Random House, 1971.

Edwards, Andres R. *The Sustainability Revolution: Portrait of a Paradigm Shift*. Gabriola Island, BC: New Society Publishers, 2005.

Ehrlich, Paul R. and Anne H. Ehrlich. *One With Nineveh: Politics, Consumption, and the Human Future*. Washington DC: Island Press, 2004.

Ehrlich, Paul R. and Anne H. Ehrlich. *The Dominant Animal: Human Evolution and the Environment*. Washington DC: Island Press, 2008.

Ehrlich, Paul R. and Robert E. Ornstein. *Humanity on a Tightrope: Thoughts on Empathy, Family, and Big Changes for a Viable Future*. Plymouth, UK: Rowman & Littlefield Publishers, 2010.

Elick, Catherine L. "Anxieties of an Animal Rights Activist: The Pressures of Modernity in Hugh Lofting's Doctor Doolittle Series." *ChLA Quarterly* 32.4 (2007): 323–39.Fletcher, Susan. *Ancient, Strange, and Lovely*. New York: Atheneum Books, 2010.

Free Birds. Dir. Jimmy Hayward. Relativity Media and Reel FX Creative Studios, 2013.

Grahame, Kenneth. *The Reluctant Dragon*. New York: Henry Holt and Company, 1988.

Happy Feet. Dir. George Miller. Kennedy Miller Productions and Animal Logic Films, 2006.

Hogan, Patrick Colm. *Cognitive Science, Literature, and the Arts: A Guide for Humanists*. New York: Routledge, 2003.

How to Train Your Dragon. Dir. Chris Sanders and Dean DeBois. DreamWorks Animation, 2010.

Jamieson, Dale. "Justice: The Heart of Environmentalism." *Environmental Justice and Environmentalism: The Social Justice Challenge to the Environmental Movement*. Ronald Sandler and Phaedra C. Pezzullo. Eds. Cambridge, Mass: The MIT Press, 2007: 85–101.

Jamison, Andrew and Ron Eyerman. *Seeds of the Sixties*. Berkeley: U. of California P., 1994.

Le Guin, Ursula K. *Buffalo Gals and Other Animal Presences*. Santa Barbara, CA: Capra Press, 1987.

Lerer, Seth. *Children's Literature: A Reader's History, from Aesop to Harry Potter*. Chicago: The U. of Chicago P., 2008.

Lovelock, James. *The Revenge of Gaia: A Final Warning*. Basic Books: New York, 2007.

Lovelock, James. *The Vanishing Face of Gaia: Earth's Climate Change and the Fate of Humanity*. Basic Books: New York, 2009.

Margulis, Lynn, and Dorion Sagan. *Microcosmos: Four Billion Years of Microbial Evolution*, New York: Summit Books, 1986.

McGilchrist, Iain. *The Master and His Emissary: The Divided Brain and the Making of the Western World*. London: Yale UP, 2009.

Mr. Peabody and Sherman. Dir. Rob Minkoff. DreamWorks Animation, Pacific data Images, and Bullwinkle Studios, 2014.

Neuzil, Mark and William Kovarik. *Mass Media and Environmental Conflict: America's Green Crusades*. Thousand Oaks, CA: Sage Publishing, 1996.

Newman, John Henry. *Sermons Preached on Various Occasions*. Notre dame, IN: U. of Notre Dame P., 2007. *Google Books*. Web. 31 January 2014.

Nevers, Patricia, Ulrich Gebhard, and Elfride Billmann-Mahecha. "patterns of Reasoning Exhibited by Children and Adolescents in Response to Moral Dilemmas Involving Plants, Animals and Ecosystems." *Journal of Moral Education* 26.2 (1997): 169–186.

Opie, John. *Nature's Nation: An Environmental History of the United States*. Belmont, CA: Wadsworth Publishing 1998.

Pinker, Steven. *The Better Angels of Our Nature: Why Violence Has Declined*. New York: Viking, 2011.

Pratchett, Terry. *The Amazing Maurice and His Educated Rodents*. New York: HarperCollins. 2001.

Rifkin, Jeremy. *The Empathic Civilization: The Race to Global Consciousness in a World in Crisis*. New York: Penguin Books, 2009.

Rio. Dir. Carlos Saldanha. Blue Sky Studios and Twentieth Century Fox Animation, 2011.

Singer, Peter. *The Expanding Circle: Ethics, Evolution, and Moral Progress*. 2nd ed. Princeton, MA: Princeton UP, 2011.

Stephens, John. "Impartiality and Attachment: Ethics and Ecopoeisis in Children's Narrative Texts." *International Research in Children's Literature* 3.2 (2010): 205–216.

Sze, Julie and Jonathan K. London. "Environmental Justice at the Crossroads." *Sociology Compass* 2/4 (2008): 1331–1354.

Thaler, Ingrid. *Black Atlantic Speculative Fictions: Octavia E. Butler, Jewelle Gomez, and Nalo Hopkinson*. New York: Routledge, 2010.

The Ant Bully. Dir. John A. Davis. Legendary Pictures, Playtone, DNA Productions, and Warner Bros. Animation, 2006.

Unti, Bernard. "Animal Rights Movement." *Encyclopedia of American Social Movements*. Vol. 4. Immanuel Ness. Ed. Armonk, NY: M. E. Sharpe Inc., 2004. 1309–16.

7 We *All* Have a Dream

Social Justice Scripts

Meeting the house-elf Winky transforms Hermione into a social justice activist. "It's slavery, that's what it is," she fumes indignantly, commenting on how Winky's master Barty Crouch had the elf save him a seat at the top of the stadium, even though Winky is terrified of heights (Rowling, *Goblet* 112). Later, when the pandemonium raised by Death Eaters culminates in the Dark Mark in the sky, unconscious Winky is found on the crime site, holding a wand in her hand. The Dark Mark, Voldemort's sign, is strictly prohibited magic; house-elves, like other non-human creatures, are banned from carrying or using wands. In a classic case of scapegoating, Winky is accused of both crimes. Although an impromptu investigation reveals that Winky did not cast the spell and only picked up Harry's lost wand to return it, she is punished with what for a house-elf is the equivalent of a death sentence: banishment from service. As disconsolate Winky throws herself at the feet of her master, he takes a step backward, "freeing himself from contact with the elf, whom he was surveying as though she was something filthy and rotten that was contaminating his overshined shoes" (124).

Having witnessed this individual case of elf mistreatment, Hermione soon realizes that she and other wizards are implicated in oppressing house-elves in general; her beloved Hogwarts, for example, turns out to be the largest slave-owning institution in Britain, with over a hundred house-elves working for its kitchens and maintenance. Incredulous, she asks Nearly-Headless Nick, "But they get *paid*? … They get *holidays*, don't they? And—and sick leave, and pensions and everything?" (161). When Nick bursts out laughing, Hermione pushes her plate away, refusing to eat what she declares to be the result of "[s]lave labor" (162). Hermione becomes a champion of elf rights. She sets up The Society for the Promotion of Elvish Welfare, whose immediate goal is to "secure house-elves' fair wages and working conditions" but eventually to achieve their equality in the wizarding world (162).

Sadly, none of these goals is realized. Because the underlying discourse of the series supports racial hierarchies (Ostry, "Accepting" 93), Rowling's otherwise sincere attempt at championing social justice fails. In this failure, the Harry Potter series illustrates how profoundly the ideal of social justice challenges assumptions of racial or other inequality unconsciously held even by the "good people." On another level, though, Rowling does a good job:

Hermione's arguments for social justice are strikingly similar to those that inform any major social justice text, for example, King's "Letter from a Birmingham Jail." Like King, Hermione believes that there are unjust laws, degrading not only to those they apply to but also to those who institute them; that these laws must be changed through civil disobedience; that positive peace, which is the presence of social justice, is preferable to negative peace, which is the absence of tension; and that all groups within society are interrelated. For Hermione, as for King, "[i]njustice anywhere is a threat to justice everywhere. ... Whatever affects one directly, affects all indirectly" (King, "Letter" par 4).

The sentiments expressed by Hermione lie at the heart of what modern people understand as social justice. A complex cultural construct, social justice emerged as a cluster of assumptions foundational for the *New Justice* paradigm and has since evolved to be its ideal for relations within any single society, which is based on a social contract made among equal—or at least equally rational—participants. This ideal has continued to elude us, but some progress has been made[1]. Issues of social justice have dominated Western moral philosophy from Kant through Rawls, and institutional change has eliminated, or at least criminalized, slavery—with some steps taken toward eliminating gender, class, and race inequality. Because the process is far from complete, the concern with social justice is central to all reform initiatives today. Social justice is the default understanding of justice in Western democracies, the most prevalent topic in its literature and arts, and the most comprehensive generic term used by scholars as an umbrella for all other types of justice. In this chapter, however, my understanding of social justice will be limited to relations within a single society. In this light, social justice concerns projected on relations between people and the environment are a domain of environmental justice. Social justice imperatives applied across different societies, cultures, or nations will be theorized in terms of global justice—my subject in the subsequent chapter.

My goal in the present chapter is to outline the evolution of the social justice script and then examine actualizations of its two dominant tracks. Because in any specific circumstances the ideal of a just society is a moving target, the primary cognitive advantage of the social justice script is to enable *perception* of specific inequalities as forms *of social injustice*, rather than provide us with tools to conclusively *achieve* social justice. The script may then offer models for remedying these injustices. In the rights track, characters change the social system from within or at least doggedly continue to resist and challenge it, even if large social change is not their immediate goal. In the freedom track, by contrast, characters are prompted to disengage from the oppressive social order. The choice of leaving rather than changing the system is then presented as a legitimate form of protest, which accounts for the prevalence of "rebel fugitive" plots in narratives based on this script.

SOCIAL AWARENESS, SOCIAL JUSTICE, AND THE SOCIAL JUSTICE SCRIPT

No other type of justice boasts as vigorous a tradition as social justice. The ideal of achieving justice for the entire society, across and among its various strata, was at the heart of the *New Justice* project, especially as voiced by Kant. In practice, though, social justice began to take shape only in the first part of the nineteenth century. It was a response at once to new inequalities created by the Industrial Revolution and to old inequalities inherent in the *Old Justice* social order, which supported traditional gender and race relations in European societies.

Although a product of the *New Justice* secularist thinking, social justice drew on an older religious tradition that offered different yet complementary rationale for its achievement. This Christian tradition encompasses social ideologies advocated by Protestant denominations such as Quakers and Methodists, both of which were among the earliest Christian social justice movements. The Quakers' pronounced opposition to slavery as well as the Methodist Church's impact on the formation of the working class in Britain and of the black churches in the US are just some of the ways in which these two challenged *Old Justice* hierarchies in the religious arena. This challenge was then expanded to the social sphere and found the most coherent expression in a body of Catholic social teaching that emerged in mid-nineteenth-century Italy. It was within this tradition that the term social justice was first coined in the 1840s in the writings of Jesuit scholar Luigi Taparelli D'Azeglio and Catholic priest Antonio Rosmini. The two Italians projected social justice as a Catholic response to the vision of society as a constant struggle among its various classes that was advocated by liberal capitalists and socialists alike. As Rosmini argued in *The Constitution of Social Justice* (1848),

> the time has come … when finally we [can] desist from attempting to organize society … as if social organization was a field of conquest, a domination of that class that was able to subjugate the others. Italians! … Be the first to let the world hear … this: The unique principle upon which a civil society must be organized is social justice, and not the brutal predominance of one class of society over another. (70)

Envisioning social justice as an alternative to brutal subjugation, Rosmini was the voice of *New Justice* tinted by religious sentiments. The cornerstone of social justice he proposed was taxation for everyone in the exact proportion of income and taxation of all properties without exception in their function. This postulate, obviously, was taken as a direct attack not only on aristocracy and the new industrialist classes but on the Church itself. Parts of Rosmini's work were thus officially condemned until quite recently; he himself was beatified in 2007. In the meantime, starting in 1891, elements of social justice such as the minimum wage and others were gradually embraced by the Church in six papal encyclicals. The most recent position, outlined in

Pope Benedict XVI's encyclical *Deus Caritas Est* (2006), is that the pursuit of justice in the political and social spheres is a central responsibility of the State. The responsibility of the Church, however, is "to help form consciences in political life and to stimulate greater insight into the authentic requirements of justice as well as greater readiness to act accordingly, even when this might involve conflict with situations of personal interest" (§28a). As evidenced in the encyclical's assertion that "the aim of a just social order is to guarantee to each person, according to the principle of subsidiarity, his share of the community's goods" (§26), the Catholic thought about social justice is based on Taparelli's idea of subsidiarity: different levels of society have the same rights but various duties and should cooperate rationally in the interest of common good. Over the course of the twentieth century social justice became the cornerstone of Catholic social teachings and spread throughout other Christian denominations as well. This religious leg of social justice explains why so many twentieth-century social justice movements were led or inspired by spiritual leaders—from Mohandas Gandhi, Dr. Martin Luther King, Jr., to Bishop Tutu, Dalai Lama, and Pope John Paul II.

The secular leg of social justice owes its emergence to early *New Justice* philosophers who theorized justice as contractarian, non-hierarchical, and absolutist: one universal quality deducible from the requirements of reasonableness. Because Enlightenment thinkers assumed that there is only one form of rationality, their own, they also assumed that there is only one justice, with the same principles informing distributive and rectificatory justice—as in Kant and Beccaria, respectively. So conceived justice for the Enlightened society was, of course, social justice, even though it was not qualified by any adjectives. It rested on the concepts of universal human rights, society as a social contract, and equality—later redefined by Rawls as fairness—as fundamental principles of any social interaction. This idea of justice became an important goal on the agenda of Modernity. Although in its pursuit of logical rigor and epistemological certainty the intellectual program of Modernity was a failure (Toulmin 172), it did help expand the circle of human rights and their application. From the abolition of slavery in the British Empire (1833) and first regulations about children's work in factories (also 1833), through the enshrinement of human rights as "as a common standard of achievement for all peoples and all nations" in the Universal Declaration of Human Rights (1948), and to the Rights revolutions of the 1960s through the 1990s—the *New Justice* equality imperative has been the fundamental driving force behind the ascendancy of social justice.

When specifically justice in this tradition acquired the "social" tag is less important than the fact that after Kant the overwhelming bulk of *New Justice* discussions focused on justice in its social/distributive aspects. Throughout the nineteenth century chartists, liberals, socialists, anarchists, nationalists, utilitarians, as well as proponents of human and women's rights all advanced their cases in response to what they saw as systemic inequalities inherent in the economic, political, class and other structures of early capitalist societies. Denouncing unfair distribution of costs and benefits for different social groups, these thinkers tended to base their arguments in the discourse of rights or social utility. In hindsight, they were demanding social justice even though they did not yet have a comprehensive term for it. Since Rosmini's and Taparelli's writings were largely unknown outside of Italy, the term does not appear in the nineteenth-century socialist classics—Friedrich Engels' *The Condition of the Working Class in England in 1844* (1845) or Karl Marx's first volume of *Capital* (1867). Yet, both Engels and Marx were seeking social justice for the working classes. The term is not featured, at least as far as I can assess, in the vast body of nationalist publications, even though much of the nationalists' quest was predicated on the demand of social justice for the disfranchised ethnic groups that lived in the then-dominant multi-ethnic empires. The term social justice was also absent in publications that championed human equality. Although *A Vindication of the Rights of Woman* (1792) identified the exclusion of women "from a participation of the natural rights of mankind" as "injustice and inconsistency" (Author's Introduction np) and *Uncle Tom's Cabin* (1852), in turn, denounced slavery as a "monstrous system of injustice that lies at the foundation of all our society" (309)—neither Wollstonecraft nor Stowe ever used the term social justice. The concept was in the air though. By the late 1860s major events on both sides of the Atlantic—the abolition of slavery in the US after the Union victory and the enfranchisement of the male working urban class in Britain's Second Reform Bill (1867)[2]—created paradigmatic models of social justice. It was only a matter of time before theoretical reflection caught up with it. The first to use the term social justice in the secular tradition was probably the utilitarian philosopher John Stuart Mill in *The Subjection of Women* (1869). Denouncing the many inequalities imposed upon women in Victorian Britain, Mill thundered that they were "contradictory to the first principles of social justice" (Chapter 4, par 5) but did not explain what these principles were and used the term only once[3].

On the other side of the Atlantic, as the era of robber barons was coming to a close, social justice became an important concept in the work of the dean of American political science, Westel Woodbury Willoughby. His *Social Justice* (1900), the first ever examination of the concept, identified social justice as indispensable to a stable social order and called for bringing the canons of distributive justice in line with "current conceptions of fairness and right" (10). Believing that "there is, or should be, an ethical justification for every social fact," Willoughby declared the scope of social justice to comprise "the proper distribution of economic goods; and the harmonizing of the principles of liberty and law, of freedom and coercion" (11). That, essentially, is also the position of John Rawls. Rawls' theory of justice as fairness, the most ambitious edifice in the *New Justice* tradition, is specifically a theory of *social* justice. This explains why, in his *Theory of Justice* (1971), Rawls unabashedly used "justice" and "social justice" interchangeably; the principles of justice that regulate a well-ordered society must, he insisted, "apply to the most important cases of social justice." Thus, a "conception of social justice ... is to be regarded as providing ... a standard whereby the distributive aspects of the basic structure of society are to be assessed" (9). By the time he published *Justice as Fairness* (2001), however, Rawls had jettisoned the term social justice altogether and replaced it with a more specific "justice as fairness." Nonetheless, his theory remains central to any reflection on social justice today.

What happened between 1971 and 2001 that made Rawls abandon the term social justice? I can only speculate, but two factors come to mind. First, by the mid-1980s what Urlich Beck has called "the first (national) modernity" (68)—a type of civil association, where social, political, and economic activities operate mainly within the nation-state—has been replaced by "the second modernity," where "territorial state guarantees of order ... lose their binding character" (102). The process triggered the rewriting of social and power relations, involving a "reformulation of the concept of society" (102), that made monolithic, universal *New Justice* visions of social justice either suspect or impossible. What social justice is possible in a world where global corporations have transformed the relationship between capital and labor, making money virtual and almost entirely independent of production and services (Barry 29)? How can social justice be sought after transnational players have created a global geography of social exclusion (Beck 103, Aristide) while severely limiting state governments' ability to do anything significant about it (Capeheart and Milovanovic 78)? Finally, how do we define social justice as a feasible ideal in the context of the dramatic rise of social inequality, polarization, and poverty, which, over the past forty years, has astronomically widened the gap between the rich and the poor in both the most and least developed countries (see Parenti and Goldsmith)? Part of Rawls' answer to these questions seems to have been a redefinition of social justice—which implies justice *within one society*—into somewhat broader "justice as fairness" as a universal principle applicable to justice *within* and perhaps even *across* societies[4].

The other reason why Rawls might have chosen to discontinue using the term is that in the last two decades of the twentieth century—i.e., with the rise of *Open Justice*—social justice had come to mean too many different things. The title of MacIntyre's 1988 *Whose Justice? Which Rationality?* was symptomatic of the *Open Justice* challenge to a single common standard that had been the foundation of social justice in the *New Justice* paradigm. As the big bang of justice expanded applications of social justice to emergent, often hybridic social arrangements, it became clear that social justice is not a monolithic, universal concept but a situated one that takes different forms in feminist, postcolonial, neo-historicist, Marxist, and other discourses concerned with social inequality and power relations. Rawls' withdrawal from using the term may have been a response to what he saw as a *de facto* expansion of social justice beyond its quantifiable *New Justice* scope. This broader, plural, and situational understanding of social justice has been delineated in *Open Justice* approaches. It accounts for why the study of social justice today moves beyond considerations of distributive justice—that is, allocation of burdens and benefits—and includes

> developing an understanding of ... retributive principles (appropriate responses to harm); how they relate to political economy and historical conditions; their local and global manifestations; the struggle for their institutionalization; how human well-being at the social and individual levels is enhanced by their institutionalization; and developing an evaluative criteria or processes by which we may measure their effects.
> (Capeheart and Milovanovic 2)

In light of this definition, *all* justice scripts are social justice scripts inasmuch as all types of justice can be seen as aspects of social justice. Although I do not have a quarrel with this broad understanding, my proposal to examine the social justice script necessitates distinguishing it from other scripts in terms of its principal markers. For purely analytical purposes, I thus take social justice to be primarily concerned with principles of distributive justice that specify fair terms of cooperation *within* any single society. Social justice as discussed in this chapter will mean an arrangement, described by Martha Nussbaum's Capabilities Approach, that secures to every member of society a minimum threshold of ten Central Capabilities—opportunities to act or be that are not reducible to a single metric, but each of which is indispensable for living human life with dignity (*Creating*, 33–34). Although the fair treatment of non-human animals and the natural environment as well as the fair treatment of people outside our own society are cognates of social justice. I examine them under separate rubrics: environmental justice and global justice, respectively.

The social justice script is a thought- and action-protocol for effecting positive change in one's own society through eliminating aspects of inequality or unfairness embedded in its structures or modes of functioning. This script is triggered when the protagonist experiences social injustice in the form of discrimination, abuse, or other mistreatment of self or others. For example, in the animated film *Robots* (2005), this happens when Rodney arrives in Robot City to find out that Bigweld Industries has been taken over by a greedy industrialist Phineas Ratchet. Ratchet plans to discontinue the manufacture of spare parts, thereby sentencing older robots to extinction. The protagonist is then cast in the role of a reformer or rebel, at odds with the status quo, and the narrative's interest lies in what he is going to do with that casting. This, in fact, occurs in *Robots*, where the inventor Rodney tries to become the leader of the revolution. Rodney's fight is for his own dream but even more for his friends, as they confront Ratchet's army of super-sweepers that are intended to destroy all older robots. In other words, the social justice script structures narratives where protagonists participate in a struggle for a more just society, not merely for the betterment of their own lot. As in Rodney's case, this script draws on a distinctly modern understanding of society as a web of relationships that ought to be mutually beneficial for all members and ought to be based on respect for citizens' rights or capabilities. Because a just society is always in the making, the social justice script is processual. Its goal is not to bring about a perfectly just society, even though this ideal may be invoked. Rather, the aim is to help characters and readers alike develop attitudes necessary for identifying and correcting instances of remediable social injustice. This is especially crucial in case of inequalities Nussbaum calls "corrosive disadvantages" (*Capabilities* 99)—capability failures that lead to failures in other areas—such as the paradigmatic injustice of slavery or, like in *Robots*, segregation through economic practice that condemns low-income robots to extinction. The social justice script proceeds through the protagonist's recognition of the political and economic underpinnings of social injustice, such as Rodney learning about Ratchet's scheme in *Robots*. The recognition then leads to action aimed to remedy that specific instance of social injustice, such as Rodney rallying other "Rusties" and convincing the legendary inventor Bigweld to join their cause. The script concludes when the protagonist either succeeds in transforming the unjust social order—as in *Robots*—or, if that transformation is impossible, disengages from the unjust society. Between these two extremes of "success" or "escape," the social justice script accommodates a range of paths and outcomes that blend these two elements in different degrees.

THE SOCIAL JUSTICE SCRIPT IN MODERN YA SPECULATIVE FICTION

The body of literature concerned with what today is recognized as social justice issues is immense. Class, birth, clan, and gender conflicts; prejudice, exclusion, inequality, and discrimination have been topical in fiction even before Fielding. Although many of those pre-modern authors condemned abuses of unequal power relations—like Defoe did for gender relations in *Moll Flanders* (1722)—rarely did they question social inequality as such. The conceptual framework to do so was to emerge only within the *New Justice* paradigm, in which "slavery had become the root metaphor of Western political philosophy, connoting everything that was evil about power relations" whereas freedom, its conceptual antithesis, was elevated to "the highest and universal political value" (Buck-Morss 21). Freedom and standing up against slavery were projected as the highest moral duties of man—only a white man, of course—in the revolutionary decades between the outbreak of the American War of Independence in 1775 and the fall of Napoleon in 1815. Even then, however, the budding call for social justice projected its achievement through an individual's change of heart—as in the novels by Jane Austin—rather than change in social institutions. One of the most vocal nineteenth-century literary calls for institutional social change came, of course, in Stowe's *Uncle Tom's Cabin* (1852). The affective power of this novel derived from the fact that Stowe was able to ignore insolvable complexities of the political-economic debate of her day and instead highlighted the evils of slavery in memorable vignettes. These, as Robin Bernstein has shown in *Racial Innocence* (2011), scripted a range of attitudes and a repertoire of behaviors not only for that historical moment, but for many decades after the abolition of slavery. Using the cognitive distance fiction offers, Stowe tapped literature's power to condemn a real-life case of social injustice. It bears noting that her "solution" to slavery was inconsequential. The leverage she achieved was through creating a set of attitudes and expectations[5].

While it may seem that realistic fiction is better positioned to address issues of social justice[6], speculative fiction has not ignored them either. In fact, the most ambitious vision of social justice in nineteenth-century American literature was not *Uncle Tom's Cabin* but a science fiction socialist utopia, Edward Bellamy's *Looking Backward* (1888). Bellamy's novel was probably the first to demonstrate the potential of speculative fiction for showcasing social justice issues. It was the third largest national bestseller of its time and even inspired a popular movement to propagate the book's ideas (Manton 326). However, because Bellamy dream-transported his young protagonist to the fair and equal America of 2000, his book outlined a utopian social justice without hints about how to achieve it. Another early dream of social justice, offered in the Superman comics of the early 1940s, did propose a solution but envisioned justice as administered along retributive lines by the society's "ultimate foreigner" (Bowers 29). Created by young authors "whose Jewish heritage deeply influenced [his] makeup" (28)[7], Superman was envisioned as "Champion of the Oppressed … sworn to … helping those in need" (41). For Jerry Siegel and Joe Shuster, this meant that Superman fought against what his creators experienced as anti-Semitism, nativism (116), and soon later Ku Klux Klan's hate-mongering. For instance, the 16-part Superman radio show "The Clan of the Fiery Cross" that aired just a month after the revival of the KKK in Atlanta in 1946 was so successful that it won praise across America for "rooting out hatred" and showing everyone "how important it is to respect each other's rights and to get along together" (141). While Superman was unquestionably a champion of social justice, the social justice he represented was implied to be unachievable within an ordinary society. It had to be forced on it by a Kryptonite outsider. This social justice was retributive, miraculous, and contingent on superhuman, if not colonial, intervention from the outside.

It was only in the wake of spectacular successes of the civil rights movement that first visions of viable social justice appeared in YA speculative fiction. Lloyd Alexander's *The High King* (1968), the closing volume of his Prydain Chronicles, concluded with a distinctly unmedieval, highly democratic vision of class equality as a foundation for a just society. Virginia Hamilton's *The House of Dies Drear* (1968) took up issues of racial equality and the legacy of slavery in North America. Ursula K. Le Guin's *The Left Hand of Darkness* (1969), one of the first works of feminist science fiction, presented a vision of gender equality in the androgynous society of the Gethenians. These and other novels ensured that social justice issues would be central to YA speculative fiction. In fact, the social justice focus of the time was so strong that even works not intended to raise social justice issues—such as *The Chronicles of Narnia* and *The Lord of the Rings*—became inspirational texts for large social justice movements, including the anti-Bomb protests and the environmental and then anti-globalization movements of the 1960s[8]. This protest tradition widened throughout the 1970s and the 1980s, when female and minority authors of speculative fiction such as Tamora Pierce, Ursula K. Le Guin, Marion Zimmer Bradley, Octavia E. Butler, and others[9] began redefining deep patriarchal patterns of the hero-tale by introducing the female perspective, questioning traditional gender roles, and subverting the gendered standards of what speculative fiction should be (Le Guin, *Revisioned* 6–7). The 1990s and the first decade of the twentieth century saw the expansion of postcolonial speculative fiction, with high-quality works by African, Caribbean, Asian, and indigenous Australian and North American authors (Attebery, *Stories* 169–185). These narratives have turned out to be so different from traditional Western speculative fiction that new generic names have been proposed to best capture their hybrid and situated uniqueness, such as "Caribbean Fabulist Fiction" (Hopkinson) or "Black Atlantic Speculative Fiction" (Thaler). These genres challenge the master narrative of *New Justice*, especially its dichotomic thinking in terms of clear divisions between the real and the unreal, the just and the unjust. They also subvert the *New Justice* insistence on universal standards—including standards of justice and rationality—which more often than not are white, patriarchal "universals." In the race and genre field this fight is far from over. Many works of modern speculative fiction—including the Harry Potter series—continue to champion racial hierarchies and heteronormative heroism even while on the surface they embrace ideals of equality and diversity[10]. On the whole, though, speculative fiction since the 1990s has absorbed the postcolonial perspective. It is increasingly preoccupied with issues of diversity, multiculturalism, hybridity, and cultural difference.

Championing these ideals has not been easy. As hopes for continuing social change were dashed by the coming apart of the Great Society in 1968, then buried under the weight of the Vietnam War, the oil crisis, and the economic free-fall of the 1970s, and then openly challenged by the conservative establishment through the 1980s, reasons for social optimism dwindled. Much speculative fiction since the mid-1970s has thus been defensive, dystopian, and reserved about the possibility of social change. Because of its thematic proclivities, fantasy has kept alive the belief in the transformative potential of human agency—albeit often at the price of being otherworldly. The bulk of YA science fiction, however, has become dystopian, extrapolating grim visions of future societies from the rise of corporate power, exclusionism, and the growing gap between the rich and the poor. Part of this dystopian impulse since the 1980s was the rise of technophobia examined by Noga Applebaum in *Representations of Technology in Science Fiction for Young People* (2010). "Young SF in particular," says Applebaum, has become "an anti-technology forum" (154): a narrative space that reflects adults' increasing anxiety regarding young people's use of technology but also projects technology in the service of wealthy elites or corporate states as the ultimate foundation of future oppression. For the system-dwarfed protagonists in such dystopias as Philip Reed's *Mortal Engines* (2001), M. T. Anderson's *Feed* (2003), or Susanne Collins' *The Hunger Games* (2008) no significant social change seems to be possible.

This is only partly true, however. In *The Hunger Games* and most other dystopias, even while protagonists seem to cooperate with the all-powerful system, they are rebels, and readers are expected to see them as such[11]. As noted by Nikolajeva, the genre of dystopia is a particularly "gratifying mode" for young readers not merely because it places protagonists in situations impossible or improbable in real life—like Katniss' gladiatorial combat in the arena. It is also gratifying by allowing an "exploration of the boundaries of a young person's body and mind" (155) that reflects the dynamics of power and repression—the key feature of the YA novel as described by Trites. For adolescent characters such as Katniss, identity-formation entails defining themselves in relation to Ideological State Apparatuses within which they function, such as the political power of the Capitol or the social power of the media. Whether or not the dystopian setting is fully convincing, the key cognitive challenge for protagonists and readers alike—projected against textual information about the system's seeming impregnability—is to figure out the best form of resistance to the oppressive status quo. These acts of resistance may be world-changing—Katniss' rebellion initially ensures only her and Peta's survival but eventually leads to a major revolution—or may merely transform the protagonists rather than the system. In most cases, dystopias employ the rebellion track of the retributive justice script, as in *The Hunger Games*. If they lead to social change—as in the Hunger Games trilogy—the retributive justice script becomes embedded in the social justice script. Most plots set in the dystopian, posthuman future offer "formulas for resistance through the youth subject, whether the character acts in isolation from her peers for herself or works for and with society as a collective whole" (Ventura 90).

These "formulas for resistance" are what constitutes the social justice script. The two behavioral models Ventura identifies—changing the system from within or fleeing the country/society until change will be possible—correspond, in turn, to the two dominant tracks of this script. Although elements of this script can be found in earlier works, I find reasons to believe that the modern social justice script emerged in the mid-1980s. It appears to be a blend of the *New Justice* recognition that the responsibility for social justice rests equally on all members of society, but especially on those from privileged groups, and of the *Open Justice* awareness that individuals' preferences are socially formed by possibilities available to them and social norms they have come to accept as their own. Building on these two, the social justice script became a permanent fixture of narratives where protagonists participate in a struggle for a more just society. This script is characterized by four signature phenomena. One, the story is set in a society built on some form of social injustice, usually a corrosive exclusion that condemns certain groups to subservience and denies them rights other groups enjoy. Two, the protagonist initially accepts the status quo as a norm but then, as a result, a seminal experience becomes aware of social injustice and determined to challenge it. Three, the protagonist takes action, becoming a social reformer, a rebel, or a fugitive. Four, at the story's conclusion the protagonist's society is transformed or the protagonist has found an alternative society free from that social injustice. Inasmuch as dominant forms of social injustice are racial, gender, and economic inequalities, most actualizations of this script will address one or more of these often interrelated issues.

In its cognitive substrate, the social justice script is the result of mapping concepts involving inequality or differences *among humans*—perceptions grounded in our embodied experience of interacting with other people in the social context—onto the conceptual domain of justice. Not all differences amount to inequality. For example, being short or tall does not qualify as inequality unless it is placed in a larger "integration network" (Fauconnier and Turner 72) that assigns discriminatory value to these qualities, as may be the case in the fashion modeling business. The complex cognitive operation of identifying conditions in which "difference" *means* "inequality" requires blending of these two domain-specific categories. Yet, it is only the first step toward higher-order conceptual compression involving cause and effect (76–82), whereby certain but not other kinds of inequality are mapped onto the domain of social injustice. The inequality of one person having a loving family whereas another having no family at all may be projected as an instance of social injustice only when the domain-specific features of "social injustice" are filled in with details of that person's life, which requires breaking up the unified state of "no family" into a causal chain made up of more elementary events that involve actors, intentions, and consequences. Only after all these operations are done and the social justice blend is in place does the script emerge. Other cases of inequality, such as the paradigmatic slavery, do not require such complex processing since they are already exemplar categories in the social injustice domain.

Established through cultural practice, the social justice script is a thought- and action-protocol of dealing with situations of social injustice. Since the goal projected by this script is "social change"—or, if change is impossible, disengagement from the oppressive system—the social justice script is no rocket science in terms of its structure. The creative part, for participants and readers alike, comes in running this script for specific cases. In fiction, the social justice script yields narratives whose plots are structured on the Trouble of social injustice, and whose processing involves cognitive challenges of identifying that injustice and recognizing the script or its non-standard variation. As I suggested earlier, young people are especially sensitive to issues of injustice. Because in their world, as Trites has shown, injustice usually means *social* injustice—and because script recognition happens by virtue of the world knowledge that particular script activates (Herman, *Story* 110)—young people's fiction is dominated by actualizations of the social justice script. Recognizing and completing the script embedded in a narrative not only enhances the young person's reading experience (Stephens, "Schemas" 15) but helps her "consolidate and reinforce" that script as a strategy for knowing and interacting with the world (Herman, *Story* 111). In other words, the goal of the social justice script is to make even those who are not themselves victims of discrimination come to develop a commitment to social justice and to put excluded or disempowered people into a position of agency and choice. Whether this choice is to leave a given social arrangement or to try to change it from within provides a useful distinction between the two dominant tracks of this script. Although these tracks can be identified in their model actualizations—as I demonstrate below—I want to note that in most of its actualizations the social justice script in modern YA speculative fiction is hybridized, alternating between the two tracks and often drawing on other justice scripts as well. I will attend to these nuanced models in the latter part of this chapter.

The freedom track of the social justice script informs narratives where social injustice is not immediately remediable and protagonists must leave their communities. Novels based on this track—for example, Monica Hughes' *Devil on My Back* (1984), Peter Dickinson's *Eva* (1988), Lois Lowry's *The Giver* (1993), Joanne DuPrau's *The City of Ember* (2003), or Timothee de Fombelle's *Toby Alone* (2009)—project breaking out from the oppressive society as a legitimate response to its inherent injustice. A similar pattern occurs in films where being different sets the protagonists apart from their communities, often resulting in their expulsion. This is the case in *The Incredibles* (2004) and *Frozen* (2013) but also in countless film franchises where anthropomorphized animals or toys are stand-ins for humans dealing with social justice issues, from *Toy Story* (1995), through *Ice Age* (2002), *Lilo and Stitch* (2002), to *Madagascar* (2005), and recently *The Lego Movie* (2014). The problem with exile as a reaction to social injustice, however, is that in order for it to be successful, the narrative must project a viable "outside" to the protagonist's oppressive society. This other world or society must then be positioned as an alternative that highlights the hows and whys of social injustice in the protagonist's native society.

Ursula K. Le Guin's *Powers* (2007) is a model exemplification of the freedom track. A closing volume of Annals of the Western Shore, the novel follows the quest for freedom and justice undertaken by a teenager slave Gavir. Gavir and his older sister Sallo grow up in the house of Arcamand, where they are groomed to become, respectively, the family teacher and a gift-girl for the eldest son of the family, Yaven. When Sallo is stolen away by the younger son Torm, raped, and murdered, Gavir's world collapses. Not realizing what he is looking for but knowing what he can no longer tolerate,

Gavir becomes a runaway slave on a quest that increasingly becomes one for freedom. He first comes to live with a wild hermit Cuga, then with a group called Forest Brothers, then in a larger community of runaway slaves, the Barnavites. The central question for him, as well as for the reader, is about the meaning of freedom and its relationship to justice. If Torm can kill Sallo without consequences, does his legal freedom to do so erase the act as murder? Likewise, is the collectivist freedom offered by the Barnavite society a reason to overlook Barna's sexual abuse of women? These are hard questions to which the participant narrator Gavir has no answers. Yet, his leaving each of these societies cues the reader that none of them represent the right balance of freedom and justice. Disappointed time and again, Gavir eventually reaches the city of Mesun, where he finds haven in the house of the poet Orrec Caspro.

Powers develops its argument about justice through showing what it is not. The key social injustice in the novel is slavery and its consequences, including internalization of enslavement. Accordingly, the social justice script structures the protagonist's journey from servility to freedom, which is not possible in his own society but may be found in another country. Since Gavir is a teenager, it is unrealistic to expect that he will overthrow the institution of slavery in his or neighboring countries where it is a norm. Gavir's only chance of protest is to run away. Because this journey is both physical and psychological, *Powers* represents the protagonist's growth by linking his embodiment and cognition—in what Trites calls "a script of psychological growth" that defines adolescent novels ("Growth" 77)—and then by inviting the reader to extrapolate about the importance of social justice from Gavir's experiences. Merely running away from something that is unpleasant or not fair, however, does not constitute the social justice script. For this script to come in place, the narrative must first establish the unfairness or inequality *as* social injustice. Then, within this framework, it must suggest ways of dealing with that unfairness and project a goal that would eliminate it.

In *Powers*, slavery is presented as a form of social injustice even before Gavir recognizes it as such. Readers are expected to bring their own background knowledge about slavery as a social evil—however general—and project it onto Gavir's early experiences in Chapters 1 through 6. This creates a gap between the reader's cognitive engagement with the story as narrated by Gavir and Gavir's own experience of the events as they happen to him. Whereas Gavir accepts slavery as a legitimate social arrangement and believes that "it's the way things are and must be" (Le Guin, *Powers* 496), the reader understands this to be a false belief. Since Gavir was brought up in a slave-owning society, his idea about its normativity is understood but not shared by the reader. The underlying discourse of the novel thus supports the concept foundational for the social justice blend, namely that slavery and its attendant mindset are not natural but result from socialization. These early chapters constitute the first signature marker of the social justice script and represent a period when Gavir remains oblivious to the injustice of slavery. Although this part of the novel recounts episodes that exemplify the evils of slavery, especially how unequal power relations between masters and slaves create space for abuse and disregard of slaves as persons, none of those events register on the protagonist's moral radar. When five-year-old slave child dies from Torm's blow, Gavir cannot understand why the young master is not even rebuked. Yet, he rests in the sure belief that "the Father and Mother and Ancestors of our House would not let anything go really wrong" (184). For the reader, these episodes foreground Gavir's eventual awakening to what slavery really means.

This wake-up call arrives in Chapter 7 when Sallo's murder awakens Gavir to the injustice of slavery. Losing her becomes Gavir's seminal experience, the second signature marker of this script. Although the situation is focalized through Gavir, our mind-reading ability allows us to understand his experience as a trauma of losing a sister. We also relate to Gavir's moral shock and outrage at the injustice when he realizes that nobody will see the murder for what it is. Gavir's reaction makes it clear that the responsibility for counteracting social injustice lies on everyone but especially on those who enjoy power. The "good" Father and Mother of the house are now seen as denying to others the dignity of a human person and are implicated in an unjust system. Gavir's belief in the righteousness of the Arca house is shattered. When he asks his teacher, a respected household slave, whether Torm will be punished for what he did, Everra "start[s] back, as if afraid of me" and manages to utter "For the death of a slave girl?"

This question, which is also the answer, throws Gavir into an abyss he describes metaphorically as "[s]ilence [that] enlarged around me, wider and deeper. I was in a pool, at the bottom of a pool, not of water but of silence and emptiness, and it went on to the end of the world" (196). The silence, as the reader is cued to understand, is the silence about the reality of slavery, which then takes Gavir two years, and a budding experience of freedom, to begin to understand. The reader is ahead of Gavir in this sense and can relate to the protagonist's explanation that he had been blinded with belief. "I had believed," Gavir recalls, "that the rule of the master and the obedience of the slave were a mutual and sacred trust. I had believed that justice could exist in a society founded on injustice" (293–4). That, he realizes, was a lie for "justice can exist only among people who found their relationships upon it" (294).

While this statement is inadequate to convey the complexity of Gavir's experience, it goes towards helping him define his goal: a quest for a society where relationships *are* built on justice. This is the third signature marker of the social justice script in *Powers*. Extending from Chapters 7 through 15, this part of the novel actualizes the freedom track sequence in which Gavir learns that freedom is more than merely the absence of slavery. Freedom is a positive value and a cluster of capabilities—"areas of freedom" (Nussbaum, *Creating* 31)—that everyone should enjoy. When seen through the lens used by Nussbaum, these chapters offer a narrative case for the Capabilities

Approach. First, slavery is shown as a "corrosive disadvantage," a type of capability failure that leads to failures in other areas, such as capabilities for bodily integrity, affiliation, or control over one's environment (99). Second, freedom from slavery is presented as a fertile capability, an "opportunit[y] that generate[s] other opportunities" (98), such as the ability to learn from other people, to freely move, choose one's goals, and grow as a human person. Third, not every kind of freedom is endorsed as concomitant with justice. Freedoms offered by Cuga, Forest Brothers, and the Marsh society shut the door on several capabilities Gavir considers so central that their removal would make his life unfulfilling: especially the ability to express himself through art, imagination, and play. Freedom offered by Barna, in turn, is corrupt inasmuch as it denies basic social entitlements to women and implicates Gavir in an ongoing social injustice. Although Barna declares that in his society nobody will live in slavery or in want (257), the commune he creates does not recognize rights for women whom Barna's takers kidnap to The Heart of the Forest. Gavir's eyes open to this fact when he witnesses Barna sexually abuse a young girl not unlike Gavir's sister. For the reader this and similar episodes evoke a cognitive-affective response that conveys important ethical knowledge: freedom that allows such abuse is indistinguishable from the abuse Gavir witnessed in Arcamand.

The social justice script in *Powers* concludes with the protagonist reaching the multicultural town of Mesun. As he makes home in a free society where all citizens irrespective of their race and gender are equal, Gavir has overcome, in his own life at least, the social injustice that made him leave Etra. Although he may face other forms of social injustice in the future, at the close of the novel Gavir has found freedom and justice.

If *Powers* can be taken as a model, albeit complex, actualization of the freedom track, a model actualization of the rights track—in which protagonists are able to challenge social injustice within their own communities—can be found in Margaret Paterson Haddix's *The Shadow Children* sequence. There are, of course, many single novels structured on this track, but a look at a series offers an additional benefit of illustrating how one script can provide an extended conceptual framework for several interrelated narrative sequences. Just as in the freedom track, stories that actualize the rights track represent social injustice as a denial of human or civil rights, but protagonists are cast into reformer or rebel roles, and a social transformation is projected as possible. In such novels protagonists are members of some ultra-privileged group, such as genetically enhanced Maxo from Nikki Singer's *Gem X* (2008), or members of some persecuted minority. These can include clones who are socially constructed as incomplete or unnatural humans or unwanted humans, such as the third-borns in Margaret Paterson Haddix's *The Shadow Children* sequence.

Comprising seven short volumes—*Among the Hidden* (1998), *Among the Impostors* (2001), *Among the Betrayed* (2002), *Among the Barons* (2003), *Among the Brave* (2004), *Among the Enemy* (2005), and *Among the Free* (2006)—the Shadow Children sequence is essentially one story. The narrative spans a year in the life of 12-year-old Luke Garner and concerns the fall of a totalitarian regime seen through his eyes in the first, second, fourth, and seventh book. The other three volumes are subplots complementary to Luke's narrative and are focalized through other "shadow children": Nina, Trey, and Matthias. Like Luke, these shadow children are the target of Population Police: in Luke's overpopulated and famine-prone society, no family can have more than two children. Taken as a whole, the series explores Luke's and other shadow children's struggle against exclusion. In depicting their journey from illegal "thirds" to legitimate citizens, the series offers an actualization of the rights track of the social justice script. Since in the series' future America there is no "outside" where the persecuted protagonists may flee, Luke and other shadow children have to make it or perish in their own society.

The key social injustice of the series, introduced through an account of Luke's early life in *Among the Hidden*, is discrimination against third-borns. Luke has lived in hiding his entire life. Confined to the attic of the house, he cannot go outside, play, go to school, have friends, do any work, or even eat with the family at the same table. He knows all this is to protect him, but he hates it. These and other aspects of Luke's life make the concept of exclusion palpably real for the reader, not only positioning exclusion as social injustice in Luke's world but also raising questions about its supposed unavoidability. Initially, Luke feels powerless and accepts the exclusion because he cannot see the alternative. This changes when he meets Jen Talbot, a third child of the Baron family that moves in to a newly erected mansion across from Luke's farm. Jen is a high-spirited rebel whose ambition is to organize all shadow children in the country and demand that the government "give us the same rights everybody else has" (83). This plan fails. Out of thousands of shadow children, only about forty show up in the capital, and they are all subsequently gunned down. Luke survives because he refused to come along—an act the reader may perceive as an instance of cowardice—but Jen's death becomes his seminal experience and the reader is expected to see it so. Leaving his parents to start a new life under a fake ID he had gotten from Jen's father, Luke declares he wants to "[f]igure out ways to help other third kids. ... Make a difference in the world" (149). With that determination, he becomes a quester for social justice.

The other three novels recount Luke's adventures as a rebel. In *Among the Impostors*, going by the Baron alias of Lee Grant, Luke attends Hendricks School for Boys and becomes part of the group there that helps other illegals. In *Among the Barons* he leaves the school, survives an assassination that kills his Baron "parents," and saves the youngest member of the Grant family. In *Among the Free*—set after the coup, in which the Population Police overthrew the government—Luke is drafted by the Population Police but refuses to shoot a defiant old woman, which sets the avalanche of civil disobedience that brings about the fall of the Population Police regime.

Although he is just a teenager and will not be part of a new government, by the end of this novel Luke and the reader alike are made aware that the best way to guard freedom is to guarantee the same rights to everyone. As the narrator puts it, Luke "didn't know if any of those things would really happen. But they were all possible now" (193–4).

The belief in the possibility of social change is central to the rights track and *The Shadow Children* sequence has all four of its markers. The protagonist grows up in an unjust society as a member of an excluded group and initially does not question his place in the system. His budding awareness of social injustice is then awakened as a result of Jen's death, a seminal experience that makes Luke determined to challenge the unjust status quo. He leaves home, adopts a new identity, and becomes involved in a movement aiming at overthrowing oppressive power structures represented by the totalitarian government and the Population Police. Eventually, although Luke is only one actor on the vast scene, his actions help achieve social transformation. While some episodes involve fighting, the regime falls in a non-violent way when citizens across the nation rise together in a mass movement of civil disobedience. Luke's America becomes a society based on social justice that respects everyone's equality and guarantees the threshold level of Capabilities to all citizens.

Although this may not be clear to the protagonist, one question the narrative raises is why Luke succeeds where Jen has failed. The reader is expected to compare his quest with Jen's—also in the context of Nina's, Trey's, and Matthias' stories—and understand that success or failure of social justice is contingent on how many people embrace its cause. Whereas Jen pushed for change only in the name of the excluded, the larger social movement that develops in the months following her death demands change for everyone and is embraced by the entire society. In this larger frame, exclusion and scapegoating of shadow children is just one aspect of government's oppressive policies that affect everyone.

This distinction in terms of societal support is a crucial factor for the social justice script. First, the scope of support for change reflects the difference between its two main tracks: in the freedom track Gavir's opposition to slavery, although shared by the reader, is not shared by people in Gavir's own world. His only option is to escape, but the unjust social order is not otherwise challenged. In the rights track, as in the Shadow Children sequence or the Hunger Games trilogy, social transformation does occur. However, it is shown as not merely the result of the protagonist's actions but arising out of a broad support for change the protagonist is able to elicit. Second, in narratives that do not project an "outside" world to escape to—and thus fail to create conditions for the unfolding of the freedom track—the rights track is bound to fail when the protagonists' goals as incompatible with societal goals. This happens to Jen, but perhaps a better example here is the failure of social justice in the Harry Potter series.

As demonstrated by numerous critics, the social order in the Hogwarts universe is one of racial hierarchy[12]. Although on the surface the story concerns an extended conflict against racism and exclusion, in which meritocratic multiculturalists clustered around Harry defeat totalitarian racists led by Voldemort, the anti-racist message is contradicted by the fact that Harry's magical society is profoundly exclusionist. The "good" wizards are blind to their own racism and speciesism. When "the hidden discourse of the novels supports and perpetuates racial and xenophobic prejudice" (Rana 46), the reader is effectively prevented from aligning with Hermione's position about house-elves. In Harry's world everyone except Hermione believes that, as Ron puts it, house-elves *enjoy* "being bossed around" (*Goblet* 112). Rowling does nothing to question this assumption. In fact, she consistently supports it by presenting other races' subservience as voluntary and natural rather than socially constructed and enforced. It is no surprise, then, that Hermione's isolated crusade—presented as well-meant but puerile, disrupting rather than improving the social order—fails to win any social support in her world and largely fails to win such support from the reader. If house-elves are complicit in their own subjugation, the reader is put in a position where respecting house-elves' rights implies respecting their right to remain slaves. This stance is embraced by everyone in Hermione's world, including Harry and Ron. The result is that the hierarchical social order of the wizard society is *not* acknowledged as based on social injustice. Although they defeat pureblood supremacists, Harry and his peers remain unaware of the more hidden aspects of social injustice in their own society. Rather than a social reformer, Harry becomes the defender of the status quo. At the conclusion of the series the wizard society, with its attendant social injustice, is upheld rather than changed. The outcome is the reinforcement of racial hierarchies and elevation of biological descent to the ultimate measure of one's place on the social scale. For all these reasons, and despite Rowling's at times apparent intentions to embrace the cause of social justice, the Harry Potter series creates an ideological framework that effectively dooms any attempts at executing the social justice script.

If *Powers* and the Shadow Children sequence are taken as model actualizations of the two dominant tracks of the social justice script, in between these two lie a number of hybrid cases that problematize representations of social justice and raise additional questions about its nature. I want to look at two types of such operations: participant valorization and track hybridization. Participant valorization includes strategies that prompt the reader to embrace or question the characters' moral judgments, including their perceptions of what constitutes social injustice. As I mentioned earlier, the social justice script can unfold only after a specific practice or arrangement has been established in the narrative *as* social injustice. The Harry Potter example suggests, however, that the author's intentions may not be enough. Rowling's metarepresentation of house-elves' slavery—"telling" about it through Hermione's reactions as well as Rowling's representation of it through "showing" us instances of house-elves' abuse at the hands of their masters—are insufficient to convince the reader. They are perceptions not validated by characters the reader is expected to identify with. Empathic identification means that we can understand characters' emotions without sharing their opinions (Nikolajeva 86). In situations where a number of characters differ on an issue, we make a choice based on textual clues. Thus, in the Harry Potter series, we are asked to choose between Hermione's morally right participant valorization and the valorization embraced by Harry, Ron, Dumbledore, and a number of other "good" characters, whose collective authority outweighs Hermione's. The majority's participant valorization is then reinforced by the fact that house-elves are shown to suffer only at the hands of characters that are cast by the narrative as evil but are not abused by anyone good. Even if slavery may be a paradigmatic instance of social injustice in the reader's knowledge repertoire, the series participant valorization cues the reader away from seeing house-elves' fate as slavery and toward perceiving it as free choice.

Participant valorization begs other questions too. What if, as in *Powers*, the majority is wrong and a given practice is socially unjust, at least by the reader's standards? Is it morally acceptable for the protagonist to force change on her society? This question is implied in Lois Lowry's postapocalyptic dystopia *The Giver*, where social injustice is not the consequence of inequality but of misconstrued equality. The 12-year-old Jonas lives in a community founded on the ideal of equality construed as "sameness." Supposedly serving to eliminate all sources of interpersonal and class conflicts, sameness is gradually revealed as a denial of basic human freedom, rights, emotions, memory, and creativity. Like Gavir, Jonas does not question the status quo until he undergoes a seminal experience—watching his father, by profession a nurturer of newborns, commit infanticide. Jonas realizes that the ethically neutral talk about "release" and "going Elsewhere" masks degrading practices that objectify human beings. He learns that the social order in which he lives is founded on lies. Trying to awaken the brainwashed community to its own injustice, Jonas and The Giver hatch a plan to force people to confront memories and start thinking for themselves. In order for this to succeed, however, Jonas must leave his community forever. He does so, taking along baby Gabriel slated for "release."

Although it is unclear whether at the end of the novel Jonas reaches a more humane society, the plot of the novel is an actualization of the freedom track, where the protagonist leaves behind the social injustice of his native community. Because the "sameness" world of the novel is the opposite of our lived experience of "difference," the narrative positions the reader as expected to employ life-to-text strategy—using her own life experience of color, personal memories, and being different from others—to understand Jonas' emotions and choices in the novel. Participant valorization occurring through Jonas and The Giver, reinforced by a life-to-text engagement with the story, invites the reader to see Jonas' escape as an act of "courage and compassion" (Latham 13) and socially responsible "political action against dystopian coercion" (Hintz 263). Little is made of the fact that Jonas' and The

Giver's scheme to force their community to become "normal"—as they and the reader might define it—ignores the community's democratic choice to live the way they do in order to avoid conflicts and wars in the future. Although Jonas' rebellion is something the adolescent reader can relate to (Steward 32), the novel employs the social justice script in a way that raises unanswerable questions about who has the right to define what social justice means.

Another textual operation that complicates actualizations of the social justice script pertains less to how the narrative defines social injustice and more to questions how it may be resolved. Track hybridization means that the story about the Trouble of social injustice moves through different levels of social injustice, on each level changing character roles, participant structures, solutions, and goals. Such a story interweaves smaller units—narratively organized action sequences that involve participants seeking to accomplish goal-directed plans—where injustice may mean different things. Only when all these action sequences, which may be subplots or stages of the main plot, are integrated by the reader into facets of the same narrative does the social justice script become visible.

A good case in point is Nancy Farmer's *The House of the Scorpion* (2002). Like most clone fiction—Kathryn Lasky's *Star Split* (1999), Ann Halam's *Taylor Five* (2002), Sonia Levitin's *The Goodness Gene* (2005), Patrick Cave's *Blow Away* (2005), or Chris Farnell's *Mark II* (2006)—Farmer's novel represents a type of story where fundamental entitlements are denied to characters representing groups socially constructed as non- or sub-human, and the plot focuses on the protagonists' resistance to this unequal treatment. Concerned with exclusion on the basis of conception, a posthuman version of racism, these novels raise questions about what rights should be granted to posthuman bodies created through technology. What is at stake in these novels is not the principles of social justice but their application.

In their complex negotiation of civil rights and social status of the posthuman, clone novels often move between the freedom and rights tracks of the social justice script. In *Scorpion*, the protagonist Matt's early childhood is spent in ignorance of his clone status. Until the age of six he lives in a small house among the poppy fields with an older woman Celia, whom he sees as his mother. Matt is then discovered by other children and brought to their house, where a tattoo on his foot reveals that he is a clone, "Property of the Alacrán Estate" (Farmer, *Scorpion* 23). Although he does not understand why, Matt is declared to be a beast and locked in a chicken coop. The servant who is supposed to take care of him abuses Matt by forcing him to go without clothes, sleep on and eat from the floor, live without bathroom or washing, and never speaks to him. This traumatic, six-month experience makes Matt a mute feral child. In that state, he is finally discovered by Celia. The woman rescues Matt by bringing his condition to the attention of El Patrón, a powerful drug lord who rules the country of Opium. Matt, as the readers learn, is the most recent in a line of many clones, whose transplanted organs have kept El Patrón alive for over 140 years. Although almost everyone in El Patrón's household sees Matt as a soulless abomination—despised for being a clone and hated for being a clone of his cruel original—Matt is precious to El Patrón. By his direct order Matt is elevated, overnight, from a prisoner to the prince of the household, who can enjoy almost as much power over others as El Patrón himself.

Matt's childhood is a mixture of extreme abuse and then extreme power. In the first part of the novel social injustice is child abuse, which the reader identifies despite other characters' claims that the clone is not a human. This section develops the rights track of the social justice script, but its resolution sets the stage for another set of questions about social justice. In the middle section, describing Matt's life between the age of 7 and 14, Matt experiences power but also social isolation. He and the reader are increasingly made aware of other aspects of social injustice that permeates the world of Opium. As one of the drug countries created on the border of Mexico and the US to stop illegal immigration, Opium intercepts anyone who attempts to cross over from Aztlán, formerly Mexico, to the US. Ruthless Farm Patrols, made up of criminal mercenaries, ambush illegal immigrants and install implants that turn immigrants into eejits: zombie-slaves that work the opium fields. In this part of the novel, social injustice is represented by the extremely hierarchical society of Opium, with the all-powerful El Patrón at the top and thousands of eejits at the bottom. Since Matt is a piece in this structure, his current privileges and future fate is also part of that injustice. Although he does not realize it, the readers and other characters know that Matt's heart will be harvested when he turns 14. Matt's awakening to this fact represents his seminal experience.

> So many hints! So many clues! ... Why had Tam Lin given him a chest full of supplies and maps? Why had María run from him when they found MacGregor's clone in the hospital? Because she knew! They all knew! Matt's education and accomplishments were a sham. It didn't matter how intelligent he was. In the end the only thing that mattered was how strong his *heart* was. (216)

After the threat to his life is made clear, Matt's primary goal becomes survival and this part of the novel actualizes the freedom track. Thanks to Celia, who had fed him poison over the years and made his heart intransplantable, Matt is temporarily saved. When El Patrón dies and his successor orders Matt to be put down, the protagonist is saved by his adopted father Tam Lin and manages to escape to Aztlán. At this point, the freedom track of the social justice script is concluded. However, the events in Aztlán and Opium cast Matt into the role of social reformer in another social justice script.

Having experienced the abuse suffered by Aztlán orphans whose parents had disappeared in Opium and having seen the inhumanity of eejiting and other evils that Opium represents, Matt understands the many faces of social injustice in his world. He is also well positioned to fight it. Since he is a genetic copy of his now dead original, he *is* El Patrón and the only person who can dismantle Opium from within. Backed by the governments of Aztlán and the US, Matt returns to Opium. The conclusion of *Scorpion* is thus an actualization of the rights track, or its promise. Just how the social order in Opium is going to be changed is the subject of the sequel, *The Lord of Opium* (2013), which also unfolds as an actualization of the rights track. Matt is helped by a small group of Opium people who survived El Patrón's dragon hoard-funeral and has to come up with practical solutions. One part of his effort to end Opium's injustice is to seek ways to reverse eejiting and free thousands of zombiefied people. Another is to destroy the drug empire and rebuild the country's economy, while remaining independent from the political control of the US or Aztlán. By the end of *Lord*, both goals are realized. The chips in eejits' brains are deactivated and eejits become free *paisanos*. Matt also discovers that during a century of total isolation, Opium has become an ecological treasure hoard that "holds the seeds of recovery for the entire planet" and can share biotechnology "to clean up polluted soil" all over the world (Farmer, *Lord* 398). Thanks to these resources, Matt strikes a deal with the UN that guarantees Opium's independence. This, in turn, may help undo a lot of injustice that happened in Opium but also all over the world.

When seen as an actualization of the social justice script, *Scorpion* blends three action sequences structured on the rights track, the freedom track, and the rights track again. Although they are connected through the character of Matt, each of these sequences casts him in a different role, each introduces readers to a different type of social injustice, and each projects a different range of possible solutions. For example, in the first sequence Matt experiences social injustice that affects him personally, but only in the second sequence does he become aware of how objectification and abuse impacts everyone else in Opium. Also, in the first and second sequence Matt accepts his sub-human status because almost everyone around him considers clones to be livestock[13]. This, as we come to realize, is a false belief. Nevertheless, we can relate to Matt's anguish that Farmer makes available to us through his thoughts and feelings. It takes Tam Lin's explanation that "[t]he idea of clones being inferior [or different from] people is a filthy lie" (Farmer, *Scorpion* 254) to make Matt see that. In the third sequence, finally, Matt acts not for himself only but out of moral duty and for society at large. He returns to Opium to dismantle the injustice El Patrón had caused. Farmer's use of three sequences, each structured on a different track of the social justice script, allows her not only to create an extended argument against objectifying clones but to challenge discrimination against any group: women, illegal immigrants, children, and the poor in general. The reader is expected to make a connection between the objectification of a single clone, the mass objectification of eejits, then the objectification of immigrants by the US government, the objectification of orphans by the government of Aztlán, and even the objectification of the natural environment that turns the Gulf of Mexico into a poisonous dump. Track hybridization enables Farmer to couch one form of discrimination within other forms to suggest that they are all related.

In theoretical reflection as well as in literary and filmic representations, social justice is a plural phenomenon that operates on many levels at once. In light of its definition offered in Nussbaum's Capabilities Approach—where social justice means securing to every member of society a minimum threshold of Central Capabilities—at least ten such levels can be identified. Unlike the actual world, however, fictional worlds are constructed from a limited set of events and characters. They highlight certain aspects but not others, and they do so by privileging the protagonist's perspective on what constitutes social injustice and what the best response to it may be. Each literary particularization of the characters' search for social justice is at the same time continuous with our ordinary cognitive processes of counterfactual thinking—our imagining hypothetical trajectories of actions and their consequences—and unique in its actualization of the social justice script it employs. Although actualizations of this script are many, in the end we are left with a few alternatives. Something is or is not projected by the text as social injustice. The protagonist can do something about it or not. In whatever track it is actualized, however, the social justice script affirms that no form of discrimination is natural or unavoidable and that the recognition of social injustice is the first step to an alternative way of behaving in the world. Even if there may never be a perfectly just society, the modern currency of the social justice script is a powerful recognition that we *all* have this dream.

NOTES

1. The ascendance of *New Justice*, what Stephen Toulmin has called the "experiential" trajectory of Modernity, has taught people in Europe and North America to identify and challenge a number of inequalities built into the social order. Since the era of the French and American revolutions, he says, "the emancipation of the classes which the New Cosmopolis has labeled as 'lower orders'—those human groups and interests were long disregarded without compunction—has been a consistent theme of political debate." This process, Toulmin adds, has been animated by "a growing perception that such inequalities cannot be justified by appeals to 'the Nature of Things' or 'the Will of God' or any other mere doctrine" (168)—a perception foundational to the *New Justice* challenge to the *Old Justice* paradigm.
2. Both events were closely interrelated in that the US Civil War made electoral reform in Britain inevitable. In 1775 Americans fought to secure themselves the rights and liberties of Englishmen; in 1867 English reformers fought to secure Englishmen the rights and liberties of Americans. For a more detailed discussion, see Phillips 489–510.
3. The force of his exposition rested on purely utilitarian arguments about the immediate greater good, the enrichment of society, and individual development that would follow the enfranchisement of women. Although Mill never again

referred to social justice, his other work—especially *On Liberty* (1859) and "On Social Freedom" (1907)—couched concerns specific to social justice firmly within the discourse of rights and freedoms individuals should enjoy as members of society.

4. Although "justice as fairness" blends elements of social and global justice, Rawls admits to having serious reservations about global justice. He shares Kant's view that a global government would either be tyrannical or ineffective and suggests that global justice, if possible at all, can only come about from an interaction of separate societies, each organized on the principles of justice as fairness (Rawls, *Fairness* 11–13).
5. If there is a link between Stowe's novel and the abolition of slavery in America, the same kind of cognitive-affective connection can be drawn between, for example, Theodore Dreiser's *Sister Carrie* (1900) and the enfranchisement of women, or between Harper Lee's *To Kill a Mockingbird* (1960) and the civil rights movement. Lee's novel, in fact, offers one of the earliest examples of the social justice script in its rights track: unlike Faulkner and other authors who wrote about racism as unavoidable, Lee saw nothing inevitable about it and offered a detail-rich story about how racism can be unlearnt.
6. By the mid-twentieth century, issues of social justice had become important topics explored in realist fiction and non-fiction—especially bestselling autobiographies, such as those of Booker T. Washington (1901) and Malcolm X (1956)—soon later even in poetry and drama. However, a glance at some excellent recent studies of social justice in and through literature, such as Dimock's *Residues of Justice*, Early's *Stirring Up Justice*, Bernstein's *Racial Innocence*, and Yasco Horsman's *Theaters of Justice,* reveals that the connection between social justice and speculative fiction has not been appreciated. As social justice activism is becoming part of academic education (see Holsinger), a number of studies have demonstrated that a commitment to social justice can be taught, even to people from privileged groups. See, for example, Mark Warren's *Fire in the Heart* and Diane Goodman's *Promoting Diversity and Social Justice.*
7. This fact was widely recognized. Even before the US entered the war against Germany, Nazi newspapers denounced Superman as a Jew and accused its creators of poisoning young American minds with Israelite propaganda (Bowers 100–1).
8. See Meredith Veldman's *Fantasy, the Bomb, and the Greening of Britain.*
9. For discussion of these early feminist works, see Thelma J. Shinn's *Worlds Within Women.*
10. See, for example, Pugh and Wallace's "Heteronormative Heroism and Queering the School Story in J. K. Rowling's Harry Potter Series."
11. Nikolajeva's discussion of *The Hunger Games* novel highlights ways in which the text makes Katniss, and thus the reader, contemplate the social order of her world with its social inequality among the districts and the capitol. See 216–224.
12. See articles on social justice in the Harry Potter series, for example Ostry's "Accepting Mudbloods: the Ambivalent Social Vision of J. K. Rowling's Fairy Tales," Patterson's "Kreacher's Lament: S.P.E.W. as a Parable on Discrimination, Indifference, and Social Justice," and Green's "Revealing Discrimination: Social Hierarchy and the Exclusion/Enslavement of the Other in the Harry Potter Novels."
13. Unlike other authors of "clone fiction," Farmer does not imply that cloning technology is evil. Instead, she points out that cloning may help create and sustain

attitudes that legitimize other technologies of objectification. The key example of such "other" technology in *Scorpion* is eejiting. For a discussion of *Scorpion* as part of fictional exploration of the posthuman age, see Ostry's "Is He Still Human?" and Crew's "Not So Brave a World."

WORKS CITED

Applebaum, Noga. *Representations of Technology in Science Fiction for Young People*. New York: Routledge, 2010.

Aristide, Jean-Bertrand. "Globalization: A View from Below." *Rethinking Globalization: Teaching for Justice in an Unjust World*. Eds. Bob Bigelow and Bob Peterson. Milwaukee, WI: Rethinking Schools Press, 2002: 9–13.

Attebery, Brian. *Stories about Stories: Fantasy and the Remaking of Myth*. New York: Oxford UP, 2014.

Barry, Brian. *Why Social Justice Matters*. Malden, MA: Polity Press, 2005.

Beck, Urlich. *What Is Globalization?* Malden, MA: Blackwell Publishing, 2000.

Bellamy, Edward. *Looking Backward from 2000 to 1887*. 1888. Project Gutenberg 2008. http://www.gutenberg.org/files/624/624-h/624-h.htm. July 12, 2014.

Benedict, XVI. *Encyclical Letter Deus Caritas Est*. 2006. http://www.vatican.va/holy_father/benedict_xvi/encyclicals/documents/hf_ben-xvi_enc_20051225_deus-caritas-est_en.html. June 22, 2014.

Bernstein, Robin. *Racial Innocence: Performing American Childhood from Slavery to Civil Rights*. New York: New York UP, 2011.

Bowers, Rick. *Superman Versus the Ku Klux Klan: The True Story of How the Iconic Superhero Battled the Men of Hate*. Washington, D.C., National Geographic, 2012.

Buck-Morss, Susan. *Hegel, Haiti, and Universal History*. Pittsburgh, PA: The U. of Pittsburgh P., 2009.

Capeheart, Loretta and Dragan Milovanovic. *Social Justice: Theories, Issues, and Movements*. Piscataway, NJ: Rutgers UP, 2007.

Collins, Susanne. *The Hunger Games*. New York: Scholastic, 2010.

Crew, Hilary S. "Not So Brave a World: The Representation of Human Cloning in Science Fiction for Young Adults." *The Lion and the Unicorn* 28 (2004): 203–221.

Dimock, Wei Chi. *Residues of Justice: Literature, Law, Philosophy*. Berkeley and Los Angeles: U of California P., 1996.

Early, Jessica. *Stirring Up Justice: Writing and Reading to Change the World*. Portsmouth, NH: Heinemann, 2006.

Farmer, Nancy. *The House of the Scorpion*. New York: Simon Pulse, 2004.

Farmer, Nancy. *The Lord of Opium*. New York: Atheneum Books, 2013.

Fauconnier, Gilles and Mark Turner. *The Way We Think: Conceptual Blending and the Mind's Hidden Complexities*. New York: Basic Books, 2002.

Green, Amy M. "Revealing Discrimination: Social Hierarchy and the Exclusion/Enslavement of the Other in the Harry Potter Novels." *The Looking Glass* 13.3 (2009). http://www.lib.latrobe.edu.au/ojs/index.php/tlg/article/view/162/161. August 10, 2014.

Goldsmith, Teddy. "Rethinking Development." *Rethinking Globalization: Teaching for Justice in an Unjust World*. Eds. Bob Bigelow and Bob Peterson. Milwaukee, WI: Rethinking Schools Press, 2002: 29–30.

Goodman, Diane. *Promoting Diversity and Social Justice: Educating People from Privileged Groups*. New York: Routledge, 2011.

Haddix, Margaret Peterson. *Among the Hidden*. New York: Simon and Shuster, 2000.

Haddix, Margaret Peterson. *Among the Free*. New York: Alladin Paperbacks, 2007.

Herman, David. *Story Logic: Problems and Possibilities of Narrative*. Lincoln, NE: U of Nebraska P., 2002.

Hintz, Carrie. "Monica Hughes, Lois Lowry, and Young Adult Dystopias." *The Lion and the Unicorn* 26 (2002): 254–64.

Holsinger, Kristi. *Teaching Justice: Solving Social Justice Problems through University Education*. Farnham, UK: Ashgate, 2012.

Hopkinson, Nalo. Ed. *Whispers from the Cotton-Tree Root: Caribbean Fabulist Fiction*. Montpelier, VT: Invisible Cities, 2000.

Horsman, Yasco. *Theaters of Justice: Judging, Staging, and Working through Arendt, Brecht and Delbo*. Stanford, CA: Stanford UP, 2012.

King, Martin Luther, Jr. "Letter from a Birmingham Jail." 1963. http://www.africa.upenn.edu/Articles_Gen/Letter_Birmingham.html. August 7, 2014.

King, Martin Luther, Jr. "I Have a Dream." August 28, 1963. http://www.americanrhetoric.com/speeches/mlkihaveadream.htm. Dec. 5, 2013.

Latham, Don. "Childhood under Siege: Lois Lowry's *Number the Stars* and *The Giver*." *The Lion and the Unicorn* 26 (2002): 1–15.

Le Guin, Ursula K. *Earthsea Revisioned*. Cambridge, UK: Labute Ltd., 1993.

Le Guin, Ursula K. *Powers*. Orlando, FL: Harcourt, 2007.

Manton, Kevin. "The British Nationalization of Labour Society and the Place of Edward Bellamy's *Looking Backward* in late Nineteenth-Century Socialism and Radicalism." *History of Political Thought* XXV.2 (2004): 325–347.

Mill, John Stuart. *The Subjection of Women*. 1869. http://ebooks.adelaide.edu.au/m/mill/john_stuart/m645s/contents.html. August 4, 2014.

Nikolajeva, Maria. *Reading for Learning: Cognitive Approaches to Children's Literature*. Amsterdam and Philadelphia: John Benjamin's Publishing Company, 2014.

Nussbaum, Martha. *Creating Capabilities: The Human Development Approach*. Cambridge, Mass.: Harvard UP, 2011.

Ostry, Elaine. "Accepting Mudbloods: the Ambivalent Social Vision of J.K. Rowling's Fairy Tales." *Reading Harry Potter: Critical Essays*. Ed. Giselle Lisa Anatol. Westport: Praeger, 2003: 89–101.

Ostry, Elaine. "Is He Still Human? Are You?: Young Adult Science Fiction in the Posthuman Age." *The Lion and the Unicorn* 28 (2004): 222–246.

Parenti, Michael. "Myths of Underdevelopment." *Rethinking Globalization: Teaching for Justice in an Unjust World*. Eds. Bob Bigelow and Bob Peterson. Milwaukee, WI: Rethinking Schools Press, 2002: 64–67.

Patterson, Steven W. "Kreacher's Lament: S.P.E.W. as a Parable on Discrimination, Indifference, and Social Justice." *Harry Potter and Philosophy: If Aristotle Ran Hogwarts*. Eds. David Baggett and Shawn E. Klein. Chicago: Open Court, 2004: 105–117.

Phillips, Kevin. *The Cousins' Wars: Religion, Politics, and the Triumph of Anglo-America*. New York: Basic Books, 1999.

Pugh, Tison and David L. Wallace. "Heteronormative Heroism and Queering the School Story in J. K. Rowling's Harry Potter Series." *Children's Literature Association Quarterly* 31:3 (2006): 260–281.

Rana, Marion. " 'The less you lot have ter do with these foreigners, the happier yeh'll be': Cultural and National Otherness in J. K. Rowling's *Harry Potter* Series." *International Research in Children's Literature* 4.1 (2011): 45–58.
Rawls, John. *Theory of Justice*. 1971. Cambridge, Mass: Harvard UP, 2005.
Rawls, John. *Justice as Fairness: A Restatement*. Cambridge, MA: Harvard UP, 2001.
Robots. Dir. Chris Wedge. Twentieth Century Fox, 2005.
Rosmini, Antonio. *The Constitution Under Social Justice*. Trans. Alberto Mingardi. Lanham, MD: Lexington Books, 2006.
Rowling, J. K. *Harry Potter and the Goblet of Fire*. London: Bloomsbury, 2000.
Shinn, Thelma J. *Worlds Within Women: Myth and Mythmaking in Fantastic Literature by Women*. New York: Greenwood Press, 1986.
Stephens, John. "Schemas and Scripts: Cognitive Instruments and the Representation of Cultural Diversity in Children's Literature." *Contemporary Children's Literature and Film: Engaging with Theory*. Eds. Kerry Mallan and Clare Bradford. Basingstoke, UK: Palgrave, 2011: 12–35.
Steward, Susan, Louise. "A Return to Normal: Lois Lowry's The Giver." *The Lion and the Unicorn* 31 (2007): 21–35.
Stowe, Harriet Beecher. *Uncle Tom's Cabin*. New York: Barnes and Noble Classics, 2004.
The Universal Declaration of Human Rights. 1948. http://www.un.org/en/documents/udhr/. August 13, 2014.
Thaler, Ingrid. *Black Atlantic Speculative Fictions: Octavia E. Butler, Jewelle Gomez, and Nalo Hopkinson*. New York: Routledge, 2010.
Toulmin, Stephen. *Cosmopolis: The Hidden Agenda of Modernity*. New York: The Free Press, 1990.
Trites, Roberta Seelinger. "Growth in Adolescent Literature: Metaphors, Scripts, and Cognitive Narratology." *International Research in Children's Literature* 5.1 (2012): 64–80.
Trites, Roberta Seelinger. *Disturbing the Universe: Power and Repression in Adolescent Literature*. Iowa City, IA: U. of Iowa P., 2000.
Veldman, Meredith. *Fantasy, the Bomb, and the Greening of Britain: Romantic Protest 1945–1980*. New York: Cambridge UP, 1994.
Ventura, Abbie. "Predicting a Better Situation? Three Young Adult Speculative Fictioin Texts and the Possibilities for Social Change." *Children's Literature Association Quarterly* 36.1 (2011): 89–103.
Warren, Mark R. *Fire in the Heart: How White Activists Embrace Racial Justice*. New York: Oxford UP, 2010.
Willoughby, Westel Woodbury. *Social Justice*. New York and London: Macmillan, 1900. http://archive.org/details/cu31924030246650. August 5, 2014.
Wollstonecraft, Mary. *A Vindication of the Rights of Woman*. 1792. http://oregonstate.edu/instruct/phl302/texts/wollstonecraft/woman-a.html. August 4, 2014.

8 Against Unseen Exploitation

Global Justice Scripts

The twenty-first century began on 9/11. The attacks targeted not just buildings emblematic of the current world order and globalization but the very ideal of dialogue among cultures. Whether or not this was the terrorists' intention, 9/11 did not stop the tide of globalization that is sweeping across contemporary societies. Instead, it raised awareness that the greatest quest of the century will involve envisioning a global civilization that can overcome the many negative aspects of globalization while enhancing its positive potential for spreading human rights, democracy, and the betterment of human life. Although the latter ideal has all too often been used by global corporate players whose operations—including unfair labor practices or destruction of the environment—are the root of the problem, we *are* faced with an ultimatum: either work toward global justice—an effort that will require reconceptualizing our understanding of humanity—or forego change and continue in a downward spiral of global injustice with its attendant ills that will only get worse. There is no single answer to this dilemma, but there is political urgency to it.

Global justice, environmental justice, and social justice are the offspring of the *New Justice* mindset. Environmental justice is the youngest of the three, conceived as an emerging blend in the mid-nineteenth-century, when the concept of rights was first mapped onto domains of the natural world and non-human animals. The seeds of social justice and global justice came earlier, their first formulations arriving on stage in the second part of the eighteenth century. However, whereas social justice was the long-awaited offspring whose coming of age became *New Justice*'s pride and ideal, its unexpected twin, global justice, was interred alive right after its birth. From Kant through Rawls, all *New Justice* theorists have dismissed global justice as impossible. It is hard to kill an idea, however, especially one that is a logical corollary of the interpretive framework and its core assumptions such as the ideal of universal human equality. The live burial of global justice was the consequence of the *New Justice* myopia that saw no discrepancy between the discourse of universal human rights and the practice of colonial slavery through which Western nations were establishing world dominance. In cognitive terms, the "universal" qualifier of human rights was mapped onto the domain of "all *rational* humans" that, in *New Justice* thinking, excluded "rationally challenged" beings such as children, women, and non-whites. All ills of modern globalization can be traced to this fateful blend and the history of colonialism it made possible. Although its nineteenth-century proponents would deny it, *New Justice* inherited and naturalized the *Old Justice* legacy of race, culture, and gender hierarchies. Only in the second part of the twentieth century, when these hierarchies of inequality were largely discarded, did the gravestone of global justice begin to crumble. Nevertheless, even in the first decade of the twenty-first century, *New Justice* theorists continued to pronounce global justice as unfeasible (Rawls, *Fairness* 13; Nagel 146).

It took *Open Justice*, with its claims that global justice can be pursued even in the absence of the global state and without trying to impose global uniformity or a single standard, to resurrect the idea of global justice from its *New Justice* tomb. When the term "global justice" became topical in the mid-1990s (Glasius 413), it was clearly theorized as an *Open Justice* concept. As used by communitarian philosophers, global justice advocates justice for a world of diversity and envisions securing the basic threshold of human rights to everyone across national borders as compatible with establishing fair terms of cooperation among different societies. Because the concept of global justice involves social *and* environmental justice applied to situations that cross national borders, in the globalized world of the twenty-first century these three types of justice are closely interrelated.

Predicated on the idea that obligations of justice extend to those outside of one's own society, the global justice script frames sequences of events in which the initially unequal relationship between two societies is altered so as to achieve a more fair balance. Although the means projected to achieve global justice may, at times, involve violence—especially in self-defense—the global justice script draws primarily on restorative solutions based on mutual recognition of needs and rights followed by dialogue, negotiation, and the rewriting of power relations between the two societies. The cognitive and affective appeal of this script derives from its activation of the readers' empathetic identification with the mistreated party that draws on our capacity for social cooperation and triggers the search for a viable solution. In YA speculative fiction, the global justice script comes in two dominant tracks. It structures stories about establishing just relationships between different human societies or cultures—usually after a legacy of oppression or exploitation—as in Le Guin's *Voices*, Gene Luen Yang's *Avatar, the Last Airbender: The Promise* (2013), Nahoko Uehashi's *Moribito: The Guardian of the Spirit* (1996/2008), Pratchett's *Nation*, or Farmer's *The Lord of Opium*. It also informs narratives that involve the encounter between different species that are thrown together and must overcome mutual fear and distrust, as in *Battle for Terra*, Jonathan Stroud's Bartimaeus Series (2003–10), the Ice Age movies, or *Monsters, Inc.* (2001). In narratives that actualize the global justice script there is no state or external authority that can "impose" fair terms of cooperation. Faced with human, alien, or animal others, the protagonist must resolve what justice claims may be extended to apply to those outside her own society. What is metaphorically worked through in these stories are the tensions and imperatives of global justice.

GLOBALIZATION, GLOBAL JUSTICE, AND THE GLOBAL JUSTICE SCRIPT

The concern with global justice is only in part a new idea. Its roots go to conceptions of international relations that emerged in the wake of the early-modern colonial expansion. Since globalization encompasses all large-scale processes that bring local humanity into increasingly global, embedded networks of exchange and interdependence, it is possible to see its earliest phase as continuous with the Columbian exchange that began in the fifteenth century[1]. An expression of the *Old Justice* "might-makes-right" mindset, this early mode of globalization was extremely brutal. It started with Portugal's attacks on Africa and then Spain's assault on the Americas that involved economic exploitation and enslaving of local populations. By the sixteenth century the Dutch were on board, replacing Spain and Portugal as world leader in the slave trade (Buck-Morss 23), but after a century they, in turn, had to acknowledge British hegemony (26). This period between the fifteenth and seventeenth centuries coincided with the rise of the European nation-state and a formation of early-modern nationalisms. As Hogan has shown, in cognitive terms the construct of nationalism involves "extensively elaborated sequences of events and actions undertaken in the pursuit of goals" that make nationalism understandable to us "in terms of stories" (Hogan, *Nationalism* 11). The story of nationalism, in turn, is structured on "three most common narrative prototypes" (12) that can also be theorized as three dominant tracks of the nationalism script. Although this script is outside of the scope of this study, in the case of each early-European nation that eventually became a colonial power this script was forged in the process of the state's consolidation of power. In other words, the emergence of nation-states made colonial conquests possible. The rise of European nation-states provided paradigmatic stories—actualizations of the nationalism script based on the cognitive and affective principles of "in-group/out-group division" (4)—with their motivational structures and models for colonization. Accordingly, Spain, which in 1492 had just won its long war against the Moors, came prepared for a new Reconquista and poured huge, well-armed forces into the New World. The Dutch, who until 1570 were a *de facto* Spanish colony and fought their own war of independence until 1609, had no resources for large-scale military operations so instead established a trading empire based on the same pattern of exterritorial economic exploitation they had experienced under the Spanish rule. The English, finally, used another model, which was the brutal pacification of Ireland completed between 1565 and 1576. Since the process involved subduing the local population and confiscating their land, the conquest of Ireland provided "not only practical experience in how to organize and finance a colonial venture but also a set of attitudes about cultural differences that were transferred to North America and applied to the Indians" (Boydston et al. 53).

Although the arrival of *New Justice* was revolutionary in many other ways[2], it took for granted the moral sufficiency of the nation-state established in the *Old Justice* paradigm as well as its nationalism script based on the sharp in-group/out-group division. When Locke and the Enlightenment philosophers asked questions about how the power of the state comes to be binding on its subjects, they ignored all colonial subjects, including slaves. Thus, globalization that continued to unfold under the banner of *New Justice* since about the 1760s inherited the *Old Justice* impulse of "might makes right," even though it now projected it as legitimate only outside of Europe. As noted by Susan Buck-Morss in her seminal *Hegel, Haiti, and Universal History* (2009), Kant, Rousseau, and other *New Justice* philosophers established "slavery" and its conceptual antithesis "freedom" as two root metaphors of Western political philosophy: the former "connoting everything that was evil about power relations" and the latter "as the highest and universal political virtue" (21). Yet, for all their devotion to logic, these *New Justice* theorists were curiously silent on the topic of non-metaphorical slavery. The paradox that "[t]he exploitation of millions of colonial slave laborers was accepted as part of the given world by the very thinkers who proclaimed freedom to be man's natural state and inalienable right" (22) is one of *New Justice*'s darkest legacies. Its discourse of freedom coupled with the practice of slavery informed the West's ascendance in global economy and characterized all of its colonial conquests in Africa, Asia, and America. Buck-Morss is correct in pointing out that it took the Haitian revolution (1791–1804), which translated into philosophical postulates in Hegel's Jena texts (1805–6), to help Europeans begin to recognize the barbarism of their own modernity and the fact that humanity includes non-white people as well.

Part of this recognition was the gradual abolition of slavery throughout the nineteenth century. Yet, by and large, the *New Justice* globalization until WWI was synonymous with colonial expansion and imperialist ideology. Already by the late eighteenth century much of the world outside of Europe had been forced into a subordinate position in the global economy just as most European nations had been subordinated into multinational empires. If world slavery was under the radar, imperial colonialism inside Europe, masquerading as global peace, was becoming an issue in philosophical reflection. The first attempts to describe the ideal of international peace—as in Charles-Irénée Castel de Saint-Pierre's *A Project for Settling an Everlasting Peace in Europe* (1712) and its critique by Rousseau in *Judgment on Perpetual Peace* (1756)—were soon expanded to encompass the world at large, as in Bentham's "A Plan for a Universal and Perpetual Peace" (1789, published in 1833) and Kant's "Perpetual Peace: A Philosophical Sketch" (1795). What linked these *New Justice* proposals was criticism of the use of force in international—or, more to the point, European—relations: even Kant and Bentham, who disagreed practically on everything, concurred on this point[3]. These postulates fell on deaf ears of European monarchs who continued their imperial expansion inside and outside of Europe, endorsing economic slavery long after the institution of slavery had been abolished. By 1897, when Queen Victoria celebrated her Diamond Jubilee, the total count of independent countries in the world had gone down to an all-time low of 57. Patterns of colonial domination established during that time lie at the root of inequalities of wealth and power in the modern world.

The imperialist phase of *New Justice* globalization received its first blow in World War I that brought about the fall of all empires in Europe and resulted in independence of several long-partitioned European nations. Although the war was "the first mass sensation of the global transition" (Adams and Carfagna 17), it did not challenge colonialism across the world. It took global economic depression, then the rise of totalitarianisms, and finally the carnage of World War II to bring this period of imperialist globalization to a close. The reinvented globalization that took its place was different. Its goals announced in the establishment of the Bretton Woods System, the United Nations Organization, and American military bases across the world, this globalization promised to be humanitarian. It was imagined to bring world peace and respect for universal human rights, besides ensuring global economic cooperation that would bring prosperity to all.

Although none of these goals have been fulfilled, this postcolonial globalization has not been a failure. The second part of the twentieth century did see massive decolonization, the rise of prosperity for many societies, a dramatic expansion in the number of sovereign states, and an equally dramatic expansion of electoral democracy. The number of independent states went up to 195 by 2015 ("Independent States in the World" Fact Sheet), and the near quadrupling of countries worldwide was paralleled by the spread of democracy. In 1900 most political systems in the world were monarchies and empires. Even countries with the most democratic systems, such as the UK and the US, were restricted democracies, with large segments of population disfranchised on the basis of gender and/or race. By 2000 democracy—as measured by the standard of universal suffrage and competitive multiparty elections—had become the statistical planetary norm, with 120 countries worldwide being liberal or electoral democracies. The social organization and practices associated with democracy have become, if not a global standard, then at least a global aspiration. They are central to all visions of a global sustainable civilization (see Rifkin, Capra, Pinker), global education, and global awareness (see Noddings; Adams and Carfagna).

For all these successes, globalization in the second part of the twentieth century has also acquired a menacing, neo-imperialist dimension. The watershed was probably the uncoupling of the dollar from the gold standard in 1972 that effectively ended the Bretton Woods system and freed global capitalism from its regulative moorings. Over the next three decades—boosted by the arrival of information technology—global capitalism became "a machine devouring our planet" (Aristide 9). One of its effects was to appropriate mechanisms of *New Justice* globalization—the idea of one standard and one economic system for all societies—in the service of economic neo-colonialism. While globalization involves much more than economic relations and is *not* synonymous with global capitalism, it is the operations of global capitalism that have provoked most hostile anti-globalization reactions (Adams and Carfagna 54–70).

The global justice movement is an expression of this protest. Commonly traced to 1999 Seattle demonstrations against the World Trade Organization, the GJM is "one of the largest and most diverse social change movements in existence today" (Highleyman, "Introduction" 1455). Bringing together individuals and groups from all over the world, this broad movement critiques structures, institutions, and policies of global capitalism as executed by international financial institutions and multinational corporations. Although global justice activists are a small group, foundational claims that inform the movement are widely shared and confirmed by countless studies. The first realization is that international organizations of global capitalism such as the IMF, World Bank, or WTO have not only failed to achieve their stated goal of alleviating worldwide poverty but have, in fact, assisted in worsening the gap between the rich and the poor (Adams and Carfagna 36–7)[4]. The second is that transnational corporations have claimed disproportionate power in shaping world affairs while effectively denying political and economic responsibility (Highleyman, "Introduction" 1455). The fact that transnational banks and corporations are playing by different rules at the public's expense was made clear in the 2008 meltdown and has been the primary motivation behind a number of protests since then, including the Brazilian public's protests against FIFA incredibly expensive 2014 World Cup whose cost was borne by Brazil.

What links the exploitative policies of institutions of global capitalism and the rising power of transnational players is their apparent collusion in bringing about a future of global oppression. As one among the mushrooming NGOs (79)—institutions devoid of physical force yet "whose moral opinions … are generally heard as stating 'the decent opinion of Humankind'" (Toulmin 197)—the global justice movement, or GJM, is a response to these developments. Inasmuch as global economic policies impact all areas of life, the movement's targets include *all* exploitative practices of international capital, from sweatshops, drug and plant patents, intellectual property rights, child labor, and food insecurity to developing-country debts, financial speculation, environmental degradation, and abuse of indigenous people's rights (see Bigelow and Peterson). Although GJM has been successful in its anti-sweatshop, living wage, and fair trade campaigns, its greatest challenge today to is to find "a better definition of what the movement is for—not just what it's against" (Highleyman, "Overview" 1469). Positively defining global justice, however, has been no easy task.

Although the idea of seeking fair relations with others across societies is easily within the ken of the neural hardware of most contemporary people, the practice of global justice has had to contend with two problems. In the framework of positive law theory, global justice is too narrow: it is limited only to what is agreed upon in a treaty signed among the states[5]. In the framework of natural law theory, by contrast, global justice is too broad: it may lead to an arbitrary imposition of one culture's values upon another[6]. Poised between these two extremes of political practice, global justice has also proven hard to define theoretically. Thus, in *Justice as Fairness*, Rawls refuses to discuss justice across societies at all, claiming that it must be "postponed until we have an account of political justice for a well-ordered democratic society" (13). Other *New Justice* philosophers, such as Thomas Nagel, have admitted that greater international authority would be desirable but have remained skeptical, assuming that any such project would first involve "the creation of patently unjust and illegitimate global structures of power" (146).

It was up to *Open Justice* philosophers to formulate the workable, open-ended definition of global justice. In the *Open Justice* paradigm, the demands of justice pre-exist institutional arrangements. If so, global democracy and global justice "can be pursued even without waiting for the global state" (Sen, *Idea* 410). This idea informs the most compelling communitarian theory of global justice to date. As advocated by British political theorist David Miller in *National Responsibility and Global Justice* (2007), global justice is compatible with state autonomy and does not seek to impose "global uniformity, in the sense of people everywhere enjoying the same bundle of rights, resources, and opportunities" (21). Rather, it stems from the recognition that there are obligations of justice that cross national boundaries. Global justice builds on the awareness that "a person's life chances ... depend much more on which society they belong to than on their individual choices, efforts, and talents" (7) and that we live in a world where our local functioning is contingent on global factors, of which we are largely unaware. Not global egalitarianism but "justice for a world of difference" (21), global justice requires two things: the universal protection of basic human rights and "fair terms of cooperation between societies, in particular terms of cooperation that allow weaker and less developed societies the opportunity to develop along paths of their own choosing" (267).

While this latter aspect is still to be achieved, global justice on its human-rights level has already become more than a mere theory. Even as scholars were discussing the possibility of global justice, some of its tenets were put into practice in the field of international justice. The year 1997 saw the first-ever conviction for the crime of genocide by the international war tribunal; 1998 witnessed the creation of the International Criminal Court—an unprecedented institution of global civil society. These two events introduced the term global justice into political, legal, and diplomatic discourses even before it became the clarion call of Seattle protests. Although the ICC is unlikely to end impunity across the globe, its actions have demonstrated that global justice is not impossible. They have had "a social and political impact on the situations on which [the ICC] focuses" (Glasius 414); they have provided important lessons about the need for flexibility and nuance in war crimes persecutions in the future; and they have been helping people everywhere "internalize norms that abhor and prosecute violations of international humanitarian law" (Moghalu 178)[7]. This last aspect of creating and reinforcing attitudes seems especially crucial, as it is attitudes that are in the lead of any change.

While the debate continues, three things seem clear. First, global justice involves protection of universal human rights across national borders and a search for fair cooperation among different societies/nations. Second, although it has roots in the *New Justice* ideas of global peace and human rights, global justice is an *Open Justice* concept that crystallized in the last decade of the twentieth century in response to the expansion of global capitalism. Instead of the *New Justice*, neo-liberalist ideal of "free" trade, which only benefits corporations, it demands the *Open Justice*-grounded ideal of fair trade, which benefits actual workers and their communities. Third, global justice shares a significant overlap with social, environmental, and restorative justice. In cognitive terms, global justice is a mirror network of social justice: a conceptual blend, in which the idea of rights as applied to one's subject position is mapped onto the domain of "others" who are equal to one as citizens of the planet. In affective terms, global justice is a step that extends our circle of empathy beyond the diversity encompassed under the header of nation-state society and onto international or global society. It includes the other, the alien, the unrecognized. If social justice is a protest against *visible* inequality within one society, global justice challenges *unseen* exploitation across societies. In all its forms global justice is a cultural adaptation for a multicultural world of many nations pulled into ever-more-complex interaction, convergence, and interdependence.

THE GLOBAL JUSTICE SCRIPT IN MODERN YA SPECULATIVE FICTION

The affective and cognitive function of the global justice script is to expose unequal relationships of power and unfair "cooperation" imposed by one society on another, and to either highlight the possibility of a just, non-exploitative arrangement or inspire those on the receiving end to challenge the ongoing injustice by refusing the benefits of exploitation. What kind of plot structure, Trouble, and character development are best to support one or both of these goals?

A good example is one of the most successful animated movies of the past decade, Pete Docter's *Monsters, Inc.* (2001). For all of its humor, the film exemplifies unseen exploitation across societies and is structured on the global justice script with an appropriate solution. From the outset the viewer is made aware that in a world where children are thought to be toxic, it takes more than daring to enter their lairs and risk exposure to deadly contamination. It takes civic responsibility. It takes understanding that the energy needs of Monstropolis cannot be compromised and that dedicated scarers are necessary to deliver good quality screams to sustain the economy. The energy sources are to be found in a different world, and so intrepid scarers must venture through space portals. There, they extract children's shrieks at risk of physical contact that would be, as they believe, fatal. This has been going on as long as anyone can remember. Monsters and children are bound in an exploitative relation, which seems even more unavoidable now, in times of energy crisis.

But suddenly everything changes. A human girl, Boo, infiltrates the plant and gets attached to the top scarer James P. "Sulley" Sullivan. When the initial shock wears off, Sulley finds out that she is not toxic at all. In fact, he develops a personal relationship with a being he so far regarded purely instrumentally. As Sulley and his friend Mike try to send Boo back to her world, they realize that children are persons not things and that scaring them is a form of abuse. They also learn that the CEO of Monsters, Inc., has built a torture machine and—in direct violation of Monstropolis regulations—is planning to torture children on site. Even though Monstropolis *needs* the power of high-pitched screams, Sulley and Mike realize that such a way of dealing with the falling production is unacceptable. They rescue Boo from the Scream Extractor and expose the CEO's scheme. Thanks to his experience with Boo, Sulley—now the chairman of Monsters, Inc.—comes up with a plan to solve the energy crisis. Having discovered that children's laughter is ten times more powerful than their screams, Sulley has his monster workers enter children's bedrooms to entertain them. The monsters' and children's worlds will remain interdependent, but the relation will now be one of mutual enhancement.

A major box-office success, *Monsters, Inc.,* was the first computer-animated movie for young audiences to build its Trouble around issues of unseen economic exploitation across interdependent societies. As in *Happy Feet* (2006), where a sentiment "they are just some birds at the bottom of the world" is rejected as endorsing outsourced unfairness, so too in *Monsters, Inc.,* unseen exploitation is condemned as injustice. The film communicates this message through cuing the audience to engage in a number of cognitive-interpretative steps. Drawing on the audience's factual knowledge that children *are* afraid of monsters—and that monsters supposedly come out of dark places such as closets—the film establishes its premise, making the monster world tangible and believable. It then invites the audience to see the same story from the monsters' side, offering answers to such questions as why monsters come out of the dark and why they scare children. In selecting events and characters, the film privileges the monsters' point of view for the story is told through Sulley and Mike as focalizers. Although Sulley and Mike win the audience's hearts, they do not realize—as the audience does—that their job involves hurting children. Soon the audience is faced with a cognitive dissonance: while the protagonists are cast as good "people"—especially in contrast to the slimy schemer Randall—they objectify children as an energy source. This, as the viewer understands, is a misconstrued belief, for children are beings with emotions and thoughts, not unlike the anthropomorphized monsters. It takes Boo's visit to make Sulley and Mike realize that they have feelings too. The audience is now emotionally invested in the protagonists' goal—taking the whole middle section of the movie, when the two friends fight with Randall and the company's CEO—which is to send Boo back to her world. That act alone, however, will not put an end to the large-scale abuse of other children unless a more comprehensive solution is found. That solution presents itself when Sulley discovers Boo's laughter, a nuclear power compared to the conventional energy of screams. This helps Sulley end the exploitation of *all* children through revisioning the company's strategy. By affectively engaging us in the Trouble resulting from Sulley's and Boo's relationship, the film makes global justice more knowable in the fictional world than in the real one. *Monsters, Inc.*, provides a storied example of unseen exploitation, projects it as an instance of global injustice, and articulates a solution, in which relations across societies are rewritten in terms of fair cooperation.

No matter how morally commendable it may seem, the vision of global justice featured in *Monsters, Inc.*, is a very recent construct. When depicted in fiction, interaction among different societies has traditionally been represented as conflict. From *The Iliad* to *War and Peace*, the greatest authors of the Western canon extolled the beauty and unavoidability of war between nations and civilizations. These accounts were informed by a clear us-versus-them division, in which the other was demonized or patronized. Compounded by cultural arrogance, in the nineteenth century the clash of nations, races, and cultures became "scientifically verified" in the framework of social Darwinism. Its cognates of racism and colonialism not only reinforced hostility to otherness but presented as virtues what, if acted upon in one's own society, would clearly be a vice. Thus, in 1876, a lawyer representing the city of San Francisco in hearings on the rights of Chinese immigrants argued in all earnestness that "the yellow races of China … are not to be permitted to steal from us what we robbed the American savage of" (qtd. in Pinker, *Angels* 659). In the social Darwinist mindset there was no place for global justice. Its proponents, including Nobel Peace Prize winner Theodore Roosevelt and British Prime Minister Winston Churchill, spoke with pride about taking part in wars of colonial conquest against what they deemed barbarous and uncivilized peoples. This compartmentalized morality, shared by their generation, blinded them to a perception that people of other races should be treated equally. Although the multiracial and multicultural world was already in the making, the appreciation of diversity was decades in the future.

Issues of global justice entered speculative literature through science fiction, shortly after Morris coined the retributive justice script in heroic fantasy. In 1898 H. G. Wells published a pioneering alien invasion novel *The War of the Worlds*, in which only a bacterial infection prevents the Martians' conquest of the Earth. Grafted on the retributivist paradigm, Wells' vision of a hostile inter-species encounter became paradigmatic for much science fiction, from Heinlein's *Starship Troopers* (1959) through Card's *Ender's Game* (1985) and Karen Traviss' The Wess'har Series (2003–8). But it did much more. First, it introduced the concept of aliens as "bugs"—repulsive anthropodian beings whose sentience precludes effective communication between them and humans. Second, it affirmed a causal link between global catastrophe and global peace. Like many pre-WWI authors, Wells believed that "the conflict-free new world … was possible through, and only through, war" and that before humanity can transition to the universal order "there had to be a nearly universal destruction to shake men free from bad thinking habits and the dead hand of existing institutions" (Turner, "Armed" 71, 73).

The vision of a peaceful, global world emerging from a cataclysmic conflict was not new. Rooted in the Christian idea of Apocalypse, it had been advocated by socialists, Marxists, and anarchists throughout the nineteenth century and was soon embraced by communists, fascists, and Nazis in the early twentieth century. For the proponents of this vision, the desire for universal peace was at the same time a desire for global war. What Wells did, however, was to spell out this vision in terms of a narrative plot pattern, the "paradigm [of] world war-miracle-world state" (76). This pattern starts with a description of the contemporary world on a brink of war, includes a middle section depicting the conflict, and concludes with a miracle, following which "humankind advances toward the world state" (74). Wells can thus be credited with putting together elements central to what later became the apocalyptic track of the global justice script: a worldwide conflict or civilizational collapse, two or more shattered societies in need of each other, and a reconstruction of the world based on their fair cooperation.

These elements are perhaps best exemplified in *The Shape of Things to Come* (1933), Wells' own "theory of world revolution" (446). In this novel, the achievement of global justice is projected through a world war, leading to universal collapse and disintegration of the existing order, leading to an oppressive world state, and eventually leading to a stateless, globally just society. However, what Wells outlines as global justice was nothing less than its *New Justice* specter that so scared Kant and Rawls. Locked within a set of assumptions inherent in the white, male, colonialist mindset—which explains his appreciation of fascism (Coupland 547)—Wells projected global justice in totalitarian terms, to be achieved by "an aggressive order of religiously devoted men and women who will try out and establish and impose a new pattern of living upon our race" (Wells 446). Besides resembling the restrictive arrangements of the dystopian societies of *The Giver* or *Feed*, this "new pattern of living" was a world, in which human individuality is curtailed to the extent that "the whole race is now … as much a colonial organism as any branching coral or polyp … [T]he individual differences of every one of these persons is like an exploring tentacle thrust out to … bring in new experiences for the common stock" (444). In short, Wells assumed that global justice *means* global uniformity and a total subjection of an individual to the needs of the state/collective. Appealing in the 1930s and embraced by all totalitarian regimes worldwide, that vision of global justice was no longer acceptable after 1945.

As the world was cautiously knitting itself together around the banners of democracy, universal human rights, and economic cooperation, the exploration of global justice in literature went in two directions, each critically responding to the vision proposed by Wells. One direction encompassed works in which the thematic preoccupation with the imminent collapse of civilization was uncoupled from the positive quest for global justice. In these works the postapocalyptic future was represented either as primitive, fragmented, and anarchic—as in Jack Vance's *Dying Earth* (1950), Russel Hoban's *Riddley Walker* (1980), or George Miller's three Mad Max movies (1979, 1981, and 1985)—or as organized into an oppressive empire, as in Isaac Asimov's *Foundation* (1951), Frank Herbert's *Dune* (1965), or George Lucas' *Star Wars* (1977). These and later works in the postapocalyptic tradition did not project a universal collapse as a prerequisite or a guarantee of global peace. In fact, they affirmed that global uniformity—whether or not legitimized by the requirements of global justice—is both unsustainable and unjust.

The other direction involved a search for narrative representations of global justice that would be free from two conceptual flaws inherent in Wells' vision: its inability to accommodate individualism, diversity, and multiculturalism; and its denial of the possibility of a slow, long-term improvement. One of the first works informed by the imperative to promote justice and human rights across societies, no matter how "invisible" these other societies may be from the dominant culture's perspective, was Dr. Seuss' *Horton Hears a Who!* (1954). Written in the aftermath of Seuss' visit to Japan—and often read as an allegory of the change in the character of American-Japanese relations—*Horton* is concerned with others' right to exist and with the evil of non-involvement when that right is threatened. The Whos' and Horton's worlds are different universes. The demise of the Whos' speck planet would seemingly not diminish the Jungle of Nool. Nevertheless, once Horton becomes aware of the Whos' existence, he recognizes the obligation of justice to save them from destruction. His advocacy of the invisible others costs Horton dearly: he is ridiculed, ostracized, and caged for championing the Whos' cause. Eventually, however, the Whos are heard and other animals in Horton's universe join him to protect their survival.

What makes *Horton* a milestone is that it spelled out three ideas essential to any modern conception of global justice. First, it suggested that some societies are "invisible" from where one stands and it takes breaking through one's own ethnocentrism to acknowledge their existence. Just as Horton's compatriots either negate the existence of the Whos or demonize them, so too the Whos, except for the Mayor, refuse to recognize that there may be a world outside of Who-ville. Second, it implied that no matter how seemingly insignificant other societies may be, they ought to be treated fairly. This imperative is reflected in Horton's argument, "[b]ecause, after all, // A person's a person no matter how small" (Seuss, *Horton* np). Third, it maintained that to achieve justice across societies a unilateral action such as Horton's is not sufficient. The Whos need to make themselves heard and in this justice-seeking process even the smallest voice has value. "They've proved they ARE persons, no matter how small. // And their whole world was saved by the Smallest of All!" (np). In all these ways Dr. Seuss was pointing at the idea of global justice that was no longer totalitarian. However, it was not fully modern either. A child of his time, Dr. Seuss projected global justice as somewhat patronizing. Since differences in power and size between the worlds of the Jungle and Whoville rule out partnership, justice is reduced to mercy. For all that, *Horton* was an important work in the rise of global justice awareness due to its message that obligations of justice are binding across societies, no matter how different they may be[8].

Whereas *Horton* affirmed that global justice can accommodate diversity, some works in the 1960s suggested that global justice may also be achieved through a peaceful evolution. Best exemplified by Ursula K. Le Guin's the Hainish Cycle, starting with *Rocannon's World* (1966) and by the Star Trek television series (since 1966), these works envisioned inter-planetary interaction of numerous races and societies of the future. Most of these societies are united into political-economic entities that resemble interstellar United Nations: it is not a coincidence, for example, that the flag of the United Federation of Planets in Star Trek is a copy of the UN flag in shape, design, and color, except that it features stars rather than Earth's continents. Le Guin's The League of All Worlds, later Ekumen, and the Federation are post-capitalist liberal democracies committed to inter-galactic cooperation based on equality, liberty, peace, and justice. Although both Le Guin's series and Star Trek involve fighting, these conflicts are predominantly defensive. The purpose of both organizations, as Genly Ai puts it while describing the Ekumen, is "[m]aterial profit. Increase of knowledge. The augmentation of the complexity and intensity of the field of intelligent life" (Le Guin, *Left* 335). Like the Federation, the Ekumen is not an empire but "a coordinator, a clearing house for trade and knowledge; without [which] communication between the worlds of men would be haphazard and trade very risky ..." (336). By envisioning large, human-rights based political entities such as the Ekumen and the Federation, authors in the 1960s helped disseminate the idea that global justice may emerge from higher organizational culture based on common principles, including gender and race equality[9].

Throughout the 1970s, as the concept of otherness was becoming one of degree, ideas of global justice spread. By the 1980s overcoming divisions and embracing difference in search of more just relationships among societies had become an important trope in all speculative fiction. In the 1960s the moral circle of cooperation was limited to humans and humanoid beings. In the 1970s, following the publication of Anne Rice's *Interview with the*

Vampire (1976) and Chelsea Quinn Yarbro's *Hôtel Transylvania* (1978), this circle expanded to include vampires. Soon later witches, werewolves, and other quasi-humanoid creatures traditionally labeled evil were redefined as merely different. Throughout the 1980s even non-human, bug-like intelligent species were reconsidered as possible partners rather than necessarily a threat. This trend began probably with Steven Spielberg's *E.T.* (1982), whose appeal was so great that it remained the highest grossing film of all time until 1993. In narrative fiction a more complex examination of inter-species relationship was first offered in Orson Scott Card's award-winning *Ender's Game* (1985) and *Speaker for the Dead* (1986)[10]. By the late 1990s, the moral circle of cooperation had expanded to encompass all beings encountered in speculative fiction. The "let's-get-along" vampires from Charlaine Harris' The Southern Vampire Mysteries (2001–continuing) and the friendly, though sometimes disgusting aliens from the Men in Black movies (1997, 2002, 2012) or *Monsters, Inc.* (2001) are all offspring of this trend, in which gender, race, or species difference does not release one from treating others fairly. As of 2015, the only "other" whose killing is projected as morally legitimate—sometimes even as an act of mercy—is the zombie.

Does this growing acceptance of otherness have anything to do with the rise of global justice awareness? I believe it does. If what is owed to those outside of one's own society is a question of global justice, then issues of global justice have been on the rise in speculative fiction at least since the 1960s. By the 1980s visions of global justice were enriched by insights from feminist, gender, and postcolonial perspectives. Throughout the 1990s they absorbed critical lessons drawn from the processes of globalization, increasingly locating challenges to global justice in the mechanisms of global capitalism and the commodification of individuals. By 1999, which saw the emergence of GJM, YA speculative fiction had already seen the formation of the global justice script.

Informing works that depict interaction between different societies, the global justice script is a cognitive programming that prompts agents to seek and expect fair terms of cooperation across societies. It calls audiences to recognize and respect the same type of rights for themselves and for those in other societies, whether it is a parallel society of American vampires from The Southern Vampire Mysteries or the alien society of Hoots from Carol Emshwiller's *The Mount* (2002). Although the root of global injustice in the modern world is economic inequality enforced by global capitalism, in most works of YA speculative fiction—as in Philip Reeve's *Mortal Engines* (2001), Anderson's *Feed* (2002), and Pete Hautman's *Rash* (2006)—economic exploitation is explored within a single society. In these cases, it is approached through one or another track of the social justice script. The global justice script, in turn, informs stories where the conflict arises in relations across clearly distinct societies or species: where the we-versus-them divide is taken for granted and maintained throughout the story. This is the case in narratives about an encounter between human societies separated by a major cultural/technological divide or societies from parallel worlds, distant planets or planes of existence, including the world of the dead. The global justice script also frames stories about interaction of humans and non-human aliens. Stories about such encounters offer a unique possibility to explore the duty of fairness binding on us in relation with the non-human other. They also draw attention to the relationship between ways of marginalizing the non-human other and the ways in which the discourse of otherness is used to marginalize humans as well. Like its social justice counterpart, the global justice script is geared primarily toward identifying and challenging instances of injustice. Its goal is not to blend the two societies into a new organism but to specify fair terms of cooperation without imposing cultural uniformity.

As the above suggests, the line between the social and global justice script is blurred, ultimately depending on whether the story positions—and the reader interprets—a given conflict as framing relationships within or across societies. Whereas *Monsters, Inc.*, is rather clear in its construction of the human and monster societies as being different, this distinction is a matter of interpretation in works such as Le Guin's *Voices*, where the occupied Ansulians and their Ald occupiers live in the same city. Are they one society or two? Since the construction of otherness operates on the same cognitive premises for distinctions within and outside of society (Hogan, *Nationalism* 24–44), this question is not important for the reader. In their actualizations, social, global, and even environmental justice scripts overlap to a degree that they can seamlessly coexist in the same work. Pratchett's *The Amazing Maurice* employs the environmental justice script as a framing structure inasmuch as it concludes with a redefined human-animal relationship. However, if the humans and the Clan are seen as representing two different "societies," the novel may also be read as an actualization of the global justice script. Likewise, if the focus is on the method of achieving justice, actualizations of the global justice script can embed the restorative justice script, as in Le Guin's *Voices*, or the retributive justice script—usually in its defensive war track—as in Farmer's *The Lord of Opium*.

The global justice script is an important narrative pattern in contemporary speculative fiction. In all actualizations, this script is characterized by four signature markers. First, the story concerns two or more societies of unequal power in such interaction that one side is exploited or treated unfairly. Second, the protagonist becomes aware of the current arrangement as unjust and challenges it by envisioning alternatives to the status quo. Third, the protagonist becomes an advocate for justice across—and a link between—the two societies, working to convince both sides that their relation can be fairer and mutually beneficial. Fourth, both societies are transformed and their relationship is redefined along global justice lines. This transformation always occurs through dialogue rather than force. Since the global justice script requires not merely tolerance of but a positive appreciation of difference, diversity is projected as enriching and the stress is on seeking fair terms cooperation despite differences.

The so-defined global justice script operates in two main versions, the rights track and the apocalyptic track. The key difference between them is in the manner the global justice impulse is set in motion. The rights track departs from the protagonists' ignorance of the ongoing injustice and proceeds through its acknowledgment to the search for the solution. In this track, whose elements can be found in *Horton*, the possibility of justice rests primarily on the protagonist's realization of the plight of the other society; often, as in *Monsters, Inc.*, this realization leads to organizational or structural adjustments in the functioning of global enterprises, transforming their bases from exploitation to mutual enhancement. In the apocalyptic track, by contrast, the protagonists from two or more societies are literally thrown into close contact by unprecedented and cataclysmic developments. In this track, different societies or their members learn fair cooperation under a common threat and/or a shared goal of survival in the new world. Whether they are different human societies as in Pratchett's *Nation*, different animal species as in the Ice Age movies, or humans and aliens as in *Battle for Terra*, characters in these narratives learn to appreciate that the obligations of justice transcend the we-versus-them divide.

An interesting actualization of the rights track—structurally similar to Le Guin's *Voices*—is Carol Emshwiller's *The Mount*, in which humans are pet animals for the powerful alien species. Set in some unidentified future, centuries after the alien Hoots colonized the Earth, *The Mount* tells a story of a human mount for a Hoot rider. The Hoots are two-foot-tall beings whose senses are far more acute than those of the humans and whose underdeveloped legs are compensated by extremely strong hands. Their technology and their sensory advantages help Hoots overcome the human civilization, pushing small bands of human incorrigibles into the wilderness. Most humans die as a result of a mysterious virus. The rest are adopted by Hoots as mounts, selectively bred for racing, transportation, communication, and other purposes. The protagonist is an 11-year-old mount of the most prestigious Seattle breed who first goes by the Hoot name Smiley and later by his person name Charley. Imprinted from a young age to train with his rider, Charley develops a rapport with Little Master, who happens to be the Hoot heir, About-To-Be-The-Ruler-Of-Us-All. When wild humans raze the Hoot town, Charley saves Little Master. The two are then taken by Charley's father, leader of the Wilds, to the mountain hideout where they experience discomforts of "freedom" but at the same time learn greater self-reliance. Sent back to a Hoot capital, Charley and Little Master become central to the transformation that ensues. While one human faction presses for exterminating Hoots and the surrounded Hoots offer humans degrees of freedom, Charley and Little Master offer a solution of fair cooperation that will benefit each side without compromising its cultural distinctness.

The Mount offers a paradigmatic example of the rights track of the global justice script, but presents a number of cognitive challenges for the reader. The novel opens with a chapter narrated by an unnamed Hoot, and the reader is given an insight into the Hoot perspective. It requires a substantial cognitive effort on the reader's part to figure out that Sams and Sues the Hoot talks about are male and female humans. It is also important that the reader mind-reads the Hoot's words and thoughts to understand Hoot's dependence on humans as mounts, especially their admiration of human legs. This, for instance, is foregrounded not only in the Hoot's "rider" language but also in contrasts between what the Hoot says and what he thinks. For example, extolling their friendship and togetherness, the Hoot concludes with the question, "where would I be without you?" The next line, representing the Hoot's thoughts, reads: "Here by myself, helpless on my wobbly legs, but I don't say that" (8). Perhaps the crucial function of this chapter, besides introducing the relationship between the two races, is to help the reader understand that Hoots had convinced themselves, and most humans, that there is nothing unfair about their relationship. As the Hoot rider explains to his mount, "we're not against you, we're for. In fact we're built for you and you for us ... born for each other even though we come from different worlds" (1). According to Hoots, humans are "recipient[s] of our kindness, our wealth and knowledge, our intelligence, our good growth of greens. Without us, you'd not exist" (2). This rhetoric of cooperative partnership is, however, undermined by hidden domination. Humans are not allowed to learn reading or counting, they are not allowed to speak or express opinions, they are not allowed to mate with persons of their own choice—all this and more while being told they are "*all* free" (8). Human reactions to the discrepancy of rhetoric and reality are brushed away as misperceptions. "You will never *fully* understand," the Hoot coos, "but you must trust us, that we *always* have our reasons" (3).

Beginning with its second chapter, the novel's focalization moves to its human protagonist, who becomes the first person narrator. This enables our empathetic identification with Charley's emotions and motivations, but we do not share his interpretation of events. Standing in the way is Charley's ignorance of his own humanness. He has internalized enslavement and is proud to be a proper mount. He speaks about himself as a Sam—a male mount in the Hoot terminology (16)—and does not question his subservience. Like Gavir in *Powers*, Charley recalls experiences that the reader easily identifies as abuse—for example, how his mother was "poled" by Hoots with electric cattle prods for refusing to give him up—but does not see them as such. When the Wilds raid the Hoot settlement, Charley sees them as savages, a deranged group of his own species. He believes in Hoot benevolence and wonders "[h]ow can my kind turn against them!" (29). After saving Little Master, Charley is taken to the mountains to live with free humans. That experience—Chapters 3 through 10—helps them overcome cultural conditioning and realize that the arrangement between Hoots and humans is, in fact, unjust. At this stage, Charley learns autonomy in functioning and thinking without Little Master; the Hoot, in turn, learns autonomy by practicing walking and acknowledging that humans are not a lower form of life.

Caught in the inter-species hostility, Charley and Little Master are repeatedly confused about which side they are on, eventually agreeing that "[w]e're each others' only ones" (113). The odd pair become advocates of justice for both societies: a third way between Hoot dominance and human resistance. Charley's father Heron realizes that the way Charley and Little Master are together as equals "*is a whole new thing ... a solution I haven't thought about*" (192, italics in the original). As he and a small group of the Wilds start working toward convincing other humans of the vision of the two species coexisting in peace, Charley and Little Master are sent to the Hoot capital to do the same for the Hoot side.

This development encompasses Chapters 11 through 14. Although Little Master is formally recognized as the Present-Ruler-Of-Us-All, the Hoot establishment refuses to give up. Instead, they attempt to lure humans with "freedom" in the form of managing their own stalls and racing arenas. Charley is happy to accept this, but Little Master, speaking for the human side, rejects these concessions. The negotiation that concludes with an agreement spans Chapters 15 and 16. Initially the Hoots want to stay in control. They allow humans to mate with whoever they please and then give them freedom to go or stay but add that those who stay as mounts will receive gifts. This offer still positions sides as unequal: one as giving and the other as receiving. Charley then reverses the hierarchy. If the Hoots are here to stay, he declares, "it's *us* who should be saying what we'll give to *them*. ... It's *us* who will be kind. We'll have the tidbits. It's us who'll imprint *their* babies" (227). The shock this idea sends through humans and Hoots alike prepares them to appreciate the justice of equal relationship then announced by Little Master: "We'll *all* have shoes. We'll *all* have diamonds. We'll imprint each other. This mount and I have already done it" (227). Hoots and humans agree to live together as equals. Hoots will learn to walk, but humans will help them build moving stools. Humans will learn to be free, but Hoots will share their technology. The novel thus ends "with the transformation of the social order, looking toward a new cooperative relationship between Hoots and humans in which true friendship rather than species hierarchy structures the exchange" (Vint 180).

As an actualization of the global justice script, the plot of *The Mount* proceeds through the four stages of the rights track: from the protagonists' blindness to inequality, through its acknowledgment, and on to successful redress in which both societies are transformed. Because the two protagonist-narrators are a human and a Hoot, the reader is offered insights into each side's thinking, value system, and motivations. The reader is also cued to empathetically identify more with Charley, in part because of his humanness and in part because he represents to subjugated party. Yet, it is Charley's genuine friendship with the Little Master that commands our interest and becomes the foundation for the redefined, voluntary relation between the two species. It is Little Master, not Charley, who first recognizes that the Hoots have been mistreating humans and that "half a kindness" (205) will not do. The meaning of justice across societies is also reinforced in other ways. First, *The Mount* explores the dialectic of freedom and dependence, suggesting that neither excludes the other as long as they are embraced voluntarily and with the recognition of the other side as one's equal. Second, it suggests that constructing others as radically unlike oneself is always the alibi for unfairness, as when Hoots patronize humans as "just primates" (99), while humans demonize Hoots as inhuman aliens. Third, the novel demonstrates that othering operates in the same way across and within societies. In *The Mount*, divisions between Hoots and humans are replicated by those between Wilds and Tames, and then breeds and non-pedigreed nothings. For example, when pure-bred Charley falls in love with mongrel "nothing" Lily and befriends hobbled Blue Bob—a purebred made worthless by leg injury—he unlearns judging other humans by their breed or physical condition in the same way that his friendship with Little Master dissolves the absolute distinction between humans and Hoots. If the renunciation of othering creates space for the appreciation of difference, the removal of assumptions about "natural" inequality eliminates the grounds for exploitation and opens the door to reciprocity. Whether or not humans and Hoots will achieve global justice, Charley and Little Master have certainly enacted a script that can make it possible.

The same kind of motivational pattern informs other actualizations of the global justice script. In Gene Luen Yang's *Avatar: The Last Airbender—The Promise*, the plot begins at a point where *The Mount*'s plot ends: that is after the sides have come to an agreement. In the Avatar TV series, a prequel to the graphic novel, the Fire Nation's attack on the Earth Nation, the Water Nation, and the Air Nation is stopped. The new Fire Lord Zuko apologizes for the aggression and withdraws his troops. In the graphic novel, in order to restore the four nations to harmony, the protagonists take steps to remove Fire Nation's colonies that were built in the Earth Kingdom. The Trouble, however, is that in over a century of existence, these colonies have developed hybrid identities, where their citizens identify *both* with Fire and Earth Nations. The whole "harmony restoration movement" (12) turns out to be based on a misconstrued notion of global justice *as separation*. This separation turns out to be increasingly more problematic as the story develops. The reader is gradually led from an initial assumption that "[h]armony requires four separate nations to balance each other out" (66) to an appreciation that peace may include all four nations within a larger whole. On one level separation is questioned as an ideal by the fact that the protagonists are an international group, including airbender Avatar Aang, his fiancée, waterbender Katara, firebender Prince Zuko, and earthbender Toph. On another level, the reader is asked to extrapolate the consequences of separation for the protagonist's lives: "If the nations have to be separate," Katara asks Aang, "what will that mean for us?" (197). All the while, the narrative cues the readers about other benefits of cooperation among various nations and elements.

For example, the new practice of metalbending emerges from a combination of earthbending and firebending (135) and the city of Yu Dao gives rise to Fire/Earth multiculturalism in the world of formerly segregated cultures. A segregated world is shown to entail either-or dilemmas that lose their edge in a multicultural world that allows hybridity. Prince Zuko faces a dilemma of whether he should protect Fire Nation's citizens, even if it means continuing the war against the Earth Nation—a war he just brought to a peaceful end. Aang faces a dilemma of whether he should save world peace, even if it means killing his friend Zuko—a promise he made to Zuko earlier and something that appears to be his duty as an avatar (200). These seeming contradictions in a separated world can be reconciled in what has become a world of the four cultures. The war to remove what used to be a colony is a war "against a whole new kind of world" (215), where cultures can coexist together without being separate. This new world demands new justice, global justice, where harmony means togetherness and accommodates hybridity. Yu Dao is left in peace, and the war is over. In the concluding panels the new global justice standard of international cooperation is confirmed when the "last airbender," Aang, accepts a group of people from other nations to teach them his culture and ensure its hybridic survival in the future.

As demonstrated by *The Mount*, and *Avatar: The Last Airbender—The Promise* in the rights track the fairer relationship is largely reasoned out as a better alternative to exploitation or war; in the apocalyptic track, however, the search for global justice is carried out less willingly and under a common threat. Novels and movies that actualize this track feature protagonists representing various species, such as the animal characters in the Ice Age movies, various human societies, such as Kargs and Archipelagians in Ursula K. Le Guin's Earthsea Sequence (1964–2001), or humans versus non-human others, such as human colonists on the planet of Austar and indigenous dragons in Jane Yolen's the Pit Dragon Chronicles (1982–2009). In Le Guin's and Yolen's series, where the struggle for equality and fairness is simultaneously played out among human groups as well as between humans and dragons, issues of global justice overlap with those of social and environmental justice.

Such, too, is the case in Jonathan Stroud's the Bartimaeus Series (2003–10), in which the quest for social justice informing the first two books turns out to be unresolvable unless the protagonists deal with issues of global justice, as happens in book three. The series recounts the education and the political career of the magician Nathaniel, known under his magical pseudonym John Mandrake, but it is focalized in part through djinni Bartimaeus and in part through an omniscient third-person narrator. Since the first-person narration is more emotionally engaging, Bartimaeus' perspective is privileged, even though he is not human. This combination allows the reader to appreciate that Nathaniel's world is based on unseen exploitation. The magicians' enslavement of immaterial spirits, which they call "demons," easily translates into enslavement of ordinary humans. In the series' universe the British Empire is ruled by magicians, "the most conniving, jealous, duplicitous group of people on earth, even including lawyers and academics" (Stroud, *Amulet* 85). Always power-hungry, magicians thrive on global and social injustice. They create political empires that oppress smaller nations. They also create a caste society, with commoners as a semi-enslaved working class. As magicians plot against their own superiors, they also have to contend with the growing resistance from commoners.

In *The Amulet of Samarkand* (2003) and *The Golem's Eye* (2004) the magicians' exploitation of spirits is merely a background for human power struggles. The consequences of this exploitation, however, are the central theme of *Ptolemy's Gate* (2006)[11]. The plot tells of an attempted takeover of the Earth by spirits, whose insurgence is made possible by the magicians' coup against their own government. This time, conspirators led by Quentin Makepeace choose to call powerful spirits into their own bodies, hoping to become superhuman, invincible, and deathless. This hope is not fulfilled. The minds of the seven magicians are devoured by terrifying spirit entities who take over their bodies. The most powerful spirit, called Nouda, announces that in return for centuries of pain at human hands, the spirits will now take revenge on humankind. "We have entered the world on our own terms. ... Soon we shall roam in hundreds about the world and feed, feed, feed upon its people" (Stroud, *Ptolemy* 342, 343). Faced with the imminent Apocalypse, the protagonists unite to stop the slaughter. Nathaniel gets hold of Gladsone's Staff, the most powerful magical object on earth. Kitty travels to the Other Place—the spirits' own realm—to request Bartimaeus' return. Lastly, in an unprecedented gesture of trust, Bartimaeus and Nathaniel share the boy's body and together eliminate most of the spirit-possessed magicians. Nouda, however, has grown too powerful; to destroy him, Nathaniel has to unleash the Staff's full energy by breaking it while being devoured. Seconds before this sacrifice, he dismisses Bartimaeus, thus saving his life.

As an actualization of the apocalyptic track, *Ptolemy's Gate* deals with the consequences of what seems to be an unavoidable master-slave relationship between humans and spirits. In the series, the human and spirit worlds have always been locked in an exploitative relation, as magicians use spells to bring spirit entities to earth and force them to do their bidding. While summoning is hurtful to beings who are immaterial essences, magicians have always ignored spirits' complaints that they suffer essence fringing during encasement in materiality. They call spirits "demons"—a racist term, according to Bartimaeus—and treat them instrumentally. Spirits, in turn, obey magicians only because they are hurt if they do not cooperate (Stroud, *Amulet* 160). Even then their enslavement on earth is "nothing *but* a series of penalties!" (257).

Is there an alternative to exploitation? Yes. Described in detail in *The Ring of Salomon* (2009), a prequel to *Ptolemy's Gate*, the alternative was first envisioned by Ptolemy, Bartimaeus' only master-friend in his five thousand years of service to humans. Ptolemy believed that the enmity between their kinds could be replaced by enrichment. He was the first human to visit the Other Place and understand a spirit being in its home world. In return for Bartimaeus' hospitality, Ptolemy had summoned the djinni to Earth and gave Bartimaeus freedom to explore it, combined with the freedom to return home whenever Bartimaeus wished. This relationship was cut short by Ptolemy's death, and the whole idea was forgotten for millennia. It is envisioned again when the commoner Kitty, seeking an alliance with spirits against magicians, calls on Bartimaeus in *Ptolemy's Gate*. "[W]e're *all* victims here, both djinni and commoners" she says. "The magicians subjugate us the same whether we're human or spirit. So, we can team up and defeat them" (208). At that time Bartimaeus rejects the proposal. The spirits have suffered too much pain at human hands to "*ever* ... view a human as an ally" (209).

The possibility of equal and fair relation emerges only when both sides are forced to consider it under an apocalyptic threat: the Trouble of *Ptolemy's Gate*. For humans the risk is no less than the annihilation of their species. For spirits, however, revenge on humans is a death sentence too. As Bartimaeus realizes, any spirit who comes into the human flesh will be protected from material elements only so long as the host's body holds. If the body dies or otherwise fails to contain the spirit's essence, the spirit will be condemned to painfully wither on Earth. Unable to enter another body—unless called into it—it will be unable to return home, unless dismissed by the magician who had called it but is now dead. Nouda's revenge, if executed, will thus mean the end for humanity but also for any spirit who enters the material world.

It is in such circumstances, when both humans and spirits face annihilation, that the protagonists enact the apocalyptic track of the global justice script. In *Ptolemy's Gate* the goal of global justice takes the form of a promise, voiced by Kitty, that if Bartimaeus helps commoners defeat magicians, they would eliminate the source of spirit enslavement (212). Humans and spirits would be free to live in their separate worlds. If they chose to interact, it would be as equals, visiting and learning from each other. The steps taken toward this goal are fourfold. The first is the recognition of the threat by the human side. As vengeful spirits ravage London, magicians acknowledge that the spirits' reaction is a consequence of centuries of abuse at magicians' hands. The second step is the establishment of a bond of trust between the two kinds that enables cooperation. As Kitty travels in spirit to talk to Bartimaeus in the Other Place, her gesture renews Ptolemy's bond with the djinni. Although made between individuals, the bond is an open-ended declaration made in the name of two kinds, that "djinni and humans might one day act together, without malice, without treachery, without slaughter" (425). The third step toward a new relationship between humans and spirits is the cooperation between Bartimaeus and Nathaniel, which eliminates the immediate threat and helps commoners take over the government in London. Speaking about Faquarl, Nouda and their entourage, Bartimaeus admits they have become corrupted by long slavery, their personalities "*brutalized, dull, vindictive* ... [and] *defin*[ing] *themselves now by their hatred* ... [in] *a final capitulation to the values of your world*" (402). Likewise, Nathaniel and Kitty admit that magicians too had become corrupted by enslaving the spirits. An additional incentive for transforming an abusive relationship into a mutually enriching one is Nathaniel's and Bartimaeus' experience of free and equal cooperation in one body, which both of them find exhilarating. The fourth step, after Nathaniel's death and Bartimaeus' return to the Other Place, is the political transformation of the British Empire. Commoners and magicians sit in the government together and negotiate terms of fair cooperation, including ways to implement Ptolemy's vision that would change the way magicians work and redefine the relationship between humans and spirits. Although the change is not described, as the novel ends the reader is given to understand that Kitty and Bartimaeus will meet again. The bridge they threw over time and space is just the beginning of amazing possibilities that lie ahead.

These four signature markers of the global justice script are played out through a dense plot that offers several cognitive challenges. First, as a number of minds become transparent to the reader—Nathaniel's, Kitty's, Bartimaeus', as well as these of other spirits and magicians—immersive identification with any of the characters is subverted, and the diversity and discrepancy of many subjective perceptions becomes clear. While the reader may empathize with Nathaniel, Kitty, and Bartimaeus, she may not be able to identify with any of them. For all their good intentions, Nathaniel is implicated in magicians' oppressive policies, Kitty knows very little about spirit-human relations, and Bartimaeus—although first-person narrator—is not human, so his consciousness remains alien. Second, the reader is cued to make connections between various forms of oppression that characterize relations between the human and spirit worlds as well as relations between magicians and commoners. This cluster of abuse leads to a situation in which the survival of the human world, and that of many spirits too, is in jeopardy. Third, as suggested by Bartimaeus' recollections of Ptolemy's reconciliatory gesture—now repeated by Kitty—cooperation between humans and spirits is not just possible but linked to their enrichment. Nathaniel's death in the name of this future cooperation is not unlike Stanton's sacrifice in *Battle for Terra*: both ensure the survival of the two species and create a framework for global togetherness that respects difference.

This stress on mutual enhancement of both societies—once they are brought to fair cooperation—is perhaps the strongest link between the rights and apocalyptic tracks. Because fair cooperation requires trust, an important feature in all actualizations of the global justice script is close bonds between characters representing different societies: Charley and Little Master; Aang, Zuko, Katara, and Toph; Kitty, Nathaniel, and Bartimaeus. In each case it is those unprecedented, new model relationships that make the two societies appreciate global justice. In all stories that actualize the global justice script at issue is not merely whether justice claims apply to those outside of our own society, but the more far-reaching question of the validity of the grounds upon which any society claims these same rights and justice for themselves. The social justice script and the global justice script both have to do with the art of cooperating with others. As such, they are detail-rich test runs of strategies studied under the rubric of the iterated Prisoner's Dilemma—strategies of cooperation self-interested agents employ when interacting with other, equally self-interested agents in open-ended sequences of actions with a limited set of partners. Like its social justice counterpart, the global justice script is not a humanistic plea for a "let's-get-along" ethics but a cognitive protocol-in-the-making driven by the search for the Nash Equilibrium of social relations in the global world.

The emergence of the global justice script is not a guarantee that our world will achieve global justice. It does, however, offer us a flexible motivational and behavioral model for navigating a noise-laden world of many cultures and races. Global justice is a recent blend of social justice praxis mapped on the domain of global awareness. Although still in the making, global justice is becoming crucial for tackling issues that extend across political borders in a world where communication and transportation technologies facilitate complex and sustained interaction among heterogeneous societies. According to Rifkin, "we are in the long end game of including 'the other,' 'the alien,' 'the unrecognized' … and the simple fact that our empathetic extension is now exploring previously unexplorable domains is a triumph of the human evolutionary journey" (26). Part of this empathetic extension is the global justice script that helps popularize the rhetoric of fairness and the equality imperative. This script creates attitudes affirming that the claims of justice in a multicultural, global world require attention to inequality and exploitation that may not be apparent. It validates interest in the global, transnational dimension of justice. And it stimulates reflection about one's own situatedness, for we are all "others" in someone else's eyes. The radical character of the cognitive programing encoded by the global justice script cannot be stressed enough. In a global world, the key challenge for the future is not a return to isolated localism but making our local lives part of global relations built on fair cooperation and mutual responsibility. The first step toward this goal is to acknowledge the current global injustice, stir emotions and intellect to challenge it, and imagine principles on which it can be redressed. And that, in short, is the cultural, affective-cognitive function of the global justice script.

NOTES

1. The term "globalization" was first used in 1962 (Adams and Carfagna 23), whereas "the Columbian Exchange" was first theorized a decade later in Alfred W. Crosby's *The Columbian Exchange: Biological and Cultural Consequences of 1492* (1973).
2. Until 1789 and the French Revolution, the word "revolution" in European languages had a completely different semantic range and meant a return to the former, better, or natural state, "a restoration of the *status quo ante*" (Toulmin 176). It was the furthest away from implying a transformation of the political system or social order from the ground up the way revolution was defined by the events in France.
3. Writing in the wake of the American Revolution, Bentham argued that world peace requires abandoning all colonies and the colonial mindset, at least colonies where it is white people who are the colonized. Writing in the wake of the dissolution of Poland—partitioned by a treaty among Prussia, Austria, and Russia—Kant specified the first two basic requirements for perpetual peace to be that "No treaty of peace shall be held valid in which there is tacitly reserved matter for a future war" and that "No independent states, large or small, shall come under the dominion of another state by inheritance, exchange, purchase, or donation" (Kant, "Perpetual" np).
4. In 1900, for example, the wealth ratio between the ten richest and poorest countries on earth was 9:1; by 1960 it grew to 30:1, and by 2005 it had reached 131:1 (Carter 179).
5. In this currently dominant view, the legal rights of individuals are granted by the state and the legal rights of states in relation to other states are specified by international treaties. Thus, within legal positivism, violations of human or civil rights are legitimate as long as they are legalized by a state, no matter how oppressive; violations of rights of other states are legitimate as long as they were agreed on in a treaty, no matter how forced.
6. In natural law theory the legal rights of individuals are derived from natural rights inherent in persons as members of society and it is assumed that states have no right to bargain away these rights in domestic or international law. The problem remains, though, what these natural rights are and who decides when they have been violated.
7. Also outside of international criminal law, the flows of trade, technology, and capital in the era of globalization have made more urgent the search for standards of comparing conflicting and interacting national laws—a global justice reform that, according to Hiram E. Chodosh in *Global Justice Reform* (2005), is as pressing as it is still in the making.
8. The film version of *Horton Hears a Who!* (2008) updated the original's message by highlighting complacency and lack of imagination as standing in the way of recognizing the obligations of global justice. The kangaroo's claims "we have standards, you know," are mirrored by the Whoville's council's mantra that "nothing ever goes wrong in Whoville, never has, never will." Both reflect complacent positions that deny a posibility of new challenges and condemn imagination as troublemaking. By the end of the film, the inhabitants of the jungle acknowledge their responsibility to Whos, and Whos acknowledge they must make themselves heard. The film ends by zooming out into the Earth as a

speck in space, thus suggesting that what seems large or small is only a question of perspective and reinforcing its global justice message about equal basic rights for all societies.

9. One of those principles, and an important contribution of Hainish novels and Star Trek to shaping the foundations of global justice, was that in both series the possibility of global justice was explicitly tied to removing inequality and exploitation in gender and race relations. The scriptwriters for Star Trek were not as progressive as Le Guin, but they consistently included multicultural crews, and the episode "Plato's Stepchildren" (1968) featured one of the first inter-racial kisses on American television. Le Guin raised the bar even higher. In *Planet of Exile* (1966) the Terran colonists and the Teveran natives overcome centuries of separatism, among others, by opening to inter-racial marriages. In *The Left Hand of Darkness* a black male envoy of united humanity invites androgynous Gethenians to join the Ekumen universum. Although Gethenians initially perceive permanently gendered outsiders as "perverts" (Le Guin, *Left* 337)—and the outsider Genly Ai is endlessly confused by Gethenians' androgyny—both sides discover that the lines of race, gender, and even different physiology can be crossed and successful communication can be achieved.
10. In *Ender's Game* the Formics' attempted invasion of the Earth results from misunderstanding. They had assumed humans were a non-sentient life form because they did not have a hive mind. By the time the Formics realized their mistake, the humans were already exterminating them. *Ender's Game* thus concludes with a Formic Queen communicating telepathically with Ender, asking him to take her egg to a planet where her race can reproduce again. In *Speaker for the Dead* and *Xenocide* (1991) the protagonists not only release the Queen but also establish contact, cooperate with, and protect the Pequeninos—an alien species who metamorphose from animal to plant.
11. The prequel, *The Ring of Salomon* (2009), is woven around the theme of cooperation between a human and a spirit, not unlike that between Charley and Little Master in *The Mound*. As such, *The Ring* offers an actualization of the rights track of the global justice script.

WORKS CITED

Adams, Michael J. and Angelo Carfagna. *Coming of Age in a Globalized World: The Next Generation*. Bloomfield, CT: Kumarian Press, 2006.

Anderson, Benedict. *Imagined Communities: Reflections on the Origin and Spread of Nationalism*. 1983. 2nd (revised) ed. London: Verso 2006.

Aristide, Jean-Bertrand. "Globalization: A View from Below." *Rethinking Globalization: Teaching for Justice in an Unjust World*. Eds. Bob Bigelow and Bob Peterson. Milwaukee, WI: Rethinking Schools Press, 2002: 9–13.

Battle for Terra. Dir. Aristomenis Tsirbas. SnootToons and MeniThings Productions, 2007.

Boydston, Jeanne, Nick Cullather, Jan Lewis, and Michael McGerr. *Making a Nation: The United States and Its People*. Upper Saddle River, NJ: Prentice Hall, 2002.

Buck-Morss, Susan. *Hegel, Haiti, and Universal History*. Pittsburgh, PA: The U. of Pittsburgh P., 2009.

Capra, Fritjof. *The Hidden Connections: A Science for Sustainable Living*. New York: Anchor Books, 2004.

Carter, Jimmy. *Our Endangered Values: America's Moral Crisis*. New York: Simon & Schuster, 2005.

Chodosh, Hiram E. *Global Justice Reform: A Comparative Methodology*. New York: New York UP, 2005.

Coupland, Philip. "H. G. Wells's 'Liberal Fascism'." *Journal of Contemporary History* 35.4 (2000): 541–558.

Crosby, Alfred W. *The Columbian Exchange: Biological and Cultural Consequences of 1492. 30th Anniversary Edition*. Westport, CT: Praeger Publishers 2003.

Dr. Seuss. *Horton Hears a Who!* New York: Random House, 1982.

Emshwiller, Carol. *The Mount*. Brooklyn, NY: Small Beer Press, 2002.

Farmer, Nancy. *The Lord of Opium*. New York: Atheneum Books, 2013.

Glasius, Marlie. "Global Justice Meets Local Civil Society: The International Criminal Court's Investigation in the Central African Republic." *Alternatives* 33.4 (2008) 413–33.

Happy Feet. Dir. George Miller. Kennedy Miller Productions and Animal Logic Films, 2006.

Highleyman, Liz. "Global Justice Movement: Overview." *Encyclopedia of American Social Movements*. Vol. 4. Ed. Immanuel Ness. Armonk, NY: M. E. Sharpe Inc., 2004: 1458–70.

Highleyman, Liz. "Global Justice Movement: Introduction." *Encyclopedia of American Social Movements*. Vol. 4. Immanuel Ness. Ed. Armonk, NY: M. E. Sharpe Inc., 2004: 1455–57.

Hogan, Patrick Colm. *Understanding Nationalism: On Narrative Cognitive Science and Identity*. Columbus, OH: The Ohio State UP, 2009.

Horton Hears a Who! Dir. Jimmy Hayward and Steve Martino. Twentieth Century Fox Animations and Blue Sky Studios, 2008.

"Independent States in the World" Fact Sheet. Bureau of Intelligence and Research. The US Department of State. Washington DC, Dec. 9, 2013. http://www.state.gov/s/inr/rls/4250.htm. August 15, 2014.

Kant, Immanuel. "Perpetual Peace: A Philosophical Sketch." 1795. http://www.constitution.org/kant/perpeace.htm. August 19, 2014.

Le Guin, Ursula K. *The Left Hand of Darkness*. Le Guin. *Five Complete Novels*. New York: Avenel Books, 1985: 311–491.

Miller, David. *National Responsibility and Global Justice*. New York: Oxford UP, 2007.

Moghalu, Kingsley Chiedu. *Global Justice: The Politics of War Crimes Trials*. Westport, CT: Praeger Security International, 2006.

Monsters, Inc. Dir. Pete Docter. Walt Disney Pictures and Pixar Animation Studios, 2001.

Nagel, Thomas. "The Problem of Global Justice." *Philosophy & Public Affairs* 33.2 (2005): 113–47.

Noddings, Nel. Ed. *Educating Citizens for Global Awareness*. New York: Teachers College Press, 2005.

Pinker, Steven. *The Better Angels of Our Nature: Why Violence Has Declined*. New York: Viking, 2011.

Pratchett, Terry. *Nation*. New York: Harper Collins, 2008.

Rawls, John. *Justice as Fairness: A Restatement*. Cambridge, MA: Harvard UP, 2001.

Rifkin, Jeremy. *The Empathic Civilization: The Race to Global Consciousness in a World in Crisis*. New York: Penguin, 2009.
Sen, Amartya. *The Idea of Justice*. Cambridge, Mass.: Harvard UP, 2009.
Stroud, Jonathan. *The Amulet of Samarkand*. New York: Hyperion Books, 2003.
Stroud, Jonathan. *Ptolemy's Gate*. New York: Hyperion Books, 2006.
Toulmin, Stephen. *Cosmopolis: The Hidden Agenda of Modernity*. New York: The Free Press, 1990.
Turner, Arthur Campbell. "Armed Conflict in the Science Fiction of H. G. Wells." *Fights of Fancy: Armed Conflict in Science Fiction and Fantasy*. George Slusser and Eric S. Rabkin. Eds. Athens, GA: The U of Georgia P., 1993: 70–78.
Wells, Herbert George. *The Shape of Things to Come*. London and New York: Penguin Classics, 2005.
Yang, Gene Luen and Gurihiru. *Avatar: The Last Airbender—The Promise*. Milwaukie, OR: Dark Horse Books, 2013.

Index

Printed in the United States
by Baker & Taylor Publisher Services